ACTINOBACTERIA

Application in Bioremediation and Production of Industrial Enzymes

ACTINOBACTERIA

Application in Bioremediation and Production of Industrial Enzymes

Editors

María Julia Amoroso
Universidad Nacional de Tucumán/PROIMI
(Planta Piloto de Procesos Industriales y
Microbiológicos-CONICET-Tucumán-Argentina)
Tucumán
Argentina

Claudia Susana Benimeli
Universidad del Norte Santo Tomás de
Aquino/PROIMI (Planta Piloto de
Procesos Industriales y Microbiológicos-
CONICET-Tucumán-Argentina)
Tucumán
Argentina

Sergio Antonio Cuozzo
Universidad Nacional de Tucumán/PROIMI
(Planta Piloto de Procesos Industriales y
Microbiológicos-CONICET-Tucumán-Argentina)
Tucumán
Argentina

CRC Press is an imprint of the
Taylor & Francis Group, an **informa** business
A SCIENCE PUBLISHERS BOOK

CRC Press
Taylor & Francis Group
6000 Broken Sound Parkway NW, Suite 300
Boca Raton, FL 33487-2742

Printed in the United States of America on acid-free paper

International Standard Book Number: 978-1-4665-7873-9 (Hardback)

Visit the Taylor & Francis Web site at
http://www.taylorandfrancis.com

CRC Press Web site at
http://www.crcpress.com

Science Publishers Web site at
http://www.scipub.net

Preface

This book provides useful information on actinobacteria phylum, a bacterial group with a great metabolic versatility, and which has a wide spectrum of useful applications: in industry, in medicine and in environmental protection. Actinobacteria are abundant in soil, and are responsible for much of the breakdown of resistant carbohydrates such as chitin and cellulose. Many actinobacteria are well known as degraders of toxic materials and are used in bioremediation. They are particularly well adapted to survival in harsh environments. Some are able to grow at elevated temperatures and are essential to the composting process. Several actinobacteria are important human, animal and plant pathogens. Morever, they are major producers of medically important antibiotics, especially members of the genus *Streptomyces*, the most abundant group.

This book describes isolated actinobacteria from different environments, and how these can be used to bioremediate heavy metals and pesticides in contaminated sites. It also describes how free-living actinobacteria acquire the capability to produce nodules in plants and how this factor could be important for accelerating the degradation of pesticides in soils or slurries. Some chapters show how actinobacteria can be used to produce industrial enzymes and metabolites under different physicochemical conditions for use in the food industry. This book will interest professionals involved with waste management, environmental protection and pollution abatement.

Tucumán
Argentina

María Julia Amoroso
Claudia Susana Benimeli
Sergio Antonio Cuozzo

November 2012

Contents

CHAPTER 1

Biology of Actinomycetes in the Rhizosphere of Nitrogen-Fixing Plants

Mariana Solans* and Gernot Vobis

Introduction

The general characteristics of the actinomycetes have been known for a long time. However, during the last few years, it was possible to learn more about their manifold functions in nature. Two genera, *Streptomyces* and *Frankia*, were detected in the second half of the 19th century and *Micromonospora* and *Actinoplanes* during the first half of the 20th century. In 1875, Cohn illustrated for the first time the typical sporulating hyphae of the genus *Streptomyces* (Miyadoh et al. 1997). The knowledge of this important soil inhabiting actinomycete was enlarged fundamentally by the taxonomic description of the species *Streptomyces scabies*, the pathogene of the potato scab (Thaxter 1891). The most typical saprophytic characters of this genus, the life cycle, the physiological requirements and the metabolic activities were studied intensively during the following decades, in which the discovery of antibiotics produced by streptomycetes strains was of

Departamento de Botánica, Centro Regional Universitario Bariloche, Universidad Nacional del Comahue. INIBIOMA—CONICET. Quintral 1250, 8400 S.C. de Bariloche, Río Negro, Argentina.
*Corresponding author: marianasolans2005@hotmail.com; msolans@comahue-conicet.gob.ar

prime importance (Waksman 1950). The genus *Frankia* was described by Brunchorst (1886), as this "fungal-like" microorganism which can cause the production of nodules on young roots of certain trees like *Alnus* and *Hippophae*. The hyphal development and formation of small vesicular structures inside the infected cells of the host plant could be presented clearly. Nobbe and Hiltner (1904) were able to prove the symbiotic character and the capacity to fix atmospheric N_2 *in situ*, but it took many years to obtain pure cultures of *Frankia* strains (Quispel 1990). The genus name *Micromonospora* was introduced by Ørskov (1923), emphasizing the type of sporulation with single spores on short sporophores for representatives of those actinomycetes. *Micromonospora* strains can be isolated commonly from soil; however their predominant incidence seems to be in aquatic ecosystems. They show a high capacity to degrade biopolymers (Vobis 1992). With the methods usually applied for isolating aquatic fungi, Couch (1950) detected the genus *Actinoplanes* from soil samples. *Actinoplanes* is closely related to *Micromonospora* regarding the colonial characteristics; however it is very different in the type of sporulation. It produces sporangia releasing flagellated spores when immersed in water. The whole life cycle of this saprophytic actinomycete can be described as "aero-aquatic" (Vobis et al. 2012).

The ecological importance of soil microorganisms and their mutual influence on the development of plants was first recognized fully by Hiltner (1904). This pioneer of "rhizosphere investigation" dedicated his extensive studies to still higher actual themes, including not only the rhizobial-, actinorhizal- and mycorrhizal symbiotic systems, but also the physiological and phytopathological aspects of the saprophytic components of the soil microflora (Hartmann 2005).

The aim of the present chapter is to show the helper effect on the growth of a few nitrogen-fixing plants by saprophytically living strains of *Streptomyces*, *Micromonospora* and *Actinoplanes*. The promotion operates indirectly via actinorhizal or rhizobial symbioses. The plants are of ecological interest in their capacity to improve the nutrient soil quality and stabilize soil systems under destroyed conditions.

Actinomycetes Isolated from the Rhizosphere of *Ochetophila trinervis*

Ochetophila trinervis (Gillies ex Hook. & Arn.) Poepp. ex Miers, for many years known as *Discaria trinervis* (Kellermann et al. 2005) (Family *Rhamnaceae*), is a native actinorhizal plant from South America (Tortosa 1983). In north-west Patagonia, it grows along watercourses (Reyes et al. 2011). *O. trinervis* plants are nodulated by the nitrogen-fixing actinomycete

Frankia and this interaction is an example of an actinorhizal symbiosis (Chaia 1998, Wall 2000) with intercellular root invasion and an infection pathway (Valverde and Wall 1999a) that implies no root hair deformation process. Although this symbiosis is well studied, little is known about the interaction with other actinomycetes of the rhizosphere. For this reason, the aim was to isolate rhizospheric actinomycetes from *O. trinervis* plants growing in natural ecosystem.

The diversity of isolated actinomycetes is largely dependent on isolation methods. For this, the following techniques were applied: 1) soil-dilution-plate, using 0.01 g of rhizospheric soil (Vobis 1992); 2) distribution of small particles, using 0.5 g of root pieces and 0.5 g of soil (Nonomura 1989); 3) stamping technique with 1 g of roots and nodules (Hunter et al. 1984); 4) natural baits, using 0.1 g of soil (Schäfer 1973); and 5) chemotactic method, with 2 g of soil (Palleroni 1980). The two last methods were used for organisms with zoospores. A total of 122 strains of saprophytic actinomycetes were isolated from both rhizosphere and rhizoplane of the actinorhizal shrub *O. trinervis*.

By employing morphological criteria, the strains could be arranged into the six genera *Streptomyces* (54 strains), *Actinoplanes* (27), *Micromonospora* (20), *Actinomadura* (7), *Pilimelia* (4), *Streptosporangium* (1) and nocardioform organisms (9). Most of the strains (62.3%) were isolated from the rhizosphere (Solans and Vobis 2003).

A symbiotic *Frankia* strain was isolated by Chaia from *O. trinervis* nodules. It has been shown to be infective (Nod+) and effective (Fix+) in its proper host plant species (Chaia 1998).

Selected Strains for Plant Growth Studies

The isolates were first classified by morphological criteria, considering microscopial aspects of spores development, presence or absence of mobility, development of substrate and aerial mycelium, pigmentation and consistency of colonies (Shirling and Gottlieb 1966, Lechevalier 1989, Vobis 1992, Miyadoh et al. 1997). Then, three saprophytic strains out of 122, with the highest enzymatic activity were selected for plant growth promoting studies. Their systematic generic positions were additionally confirmed by molecular analysis based on 16S rRNA gene sequence (Solans 2008): *Streptomyces* sp. (BCRU-MM40), *Actinoplanes* sp. (BCRU-ME3) and *Micromonospora* sp. (BCRU-MM18); GenBank accesion numbers: FJ771041, FJ771040, FJ771042. The symbiotic N_2-fixing *Frankia* strain BCU110501 was used for these experiments. All strains are stored at the culture collection of the Herbarium BCRU, Department of Botany, Centro Regional Universitario Bariloche, Universidad Nacional del Comahue, 8400 S.C. de

Bariloche, Argentina (Website: http://www.sciweb.nybg.org/science2/IndexHerbariorum.asp).

Some morphological characteristics of the actinomycete strains in study are shown in Fig. 1.

Diffusible melanoid were observed in *Streptomyces* and *Actinoplanes* after incubation at 28°C for 14 days, but not in *Micromonospora*. After culturing *Frankia* BCU110501 for three weeks, it was possible to observe the presence of irregularly formed reddish globular crystalloids and elongated pink crystals. Cultures in liquid medium after one month or more, produced granular red irregular pigments. These observations agree with those found by Chaia

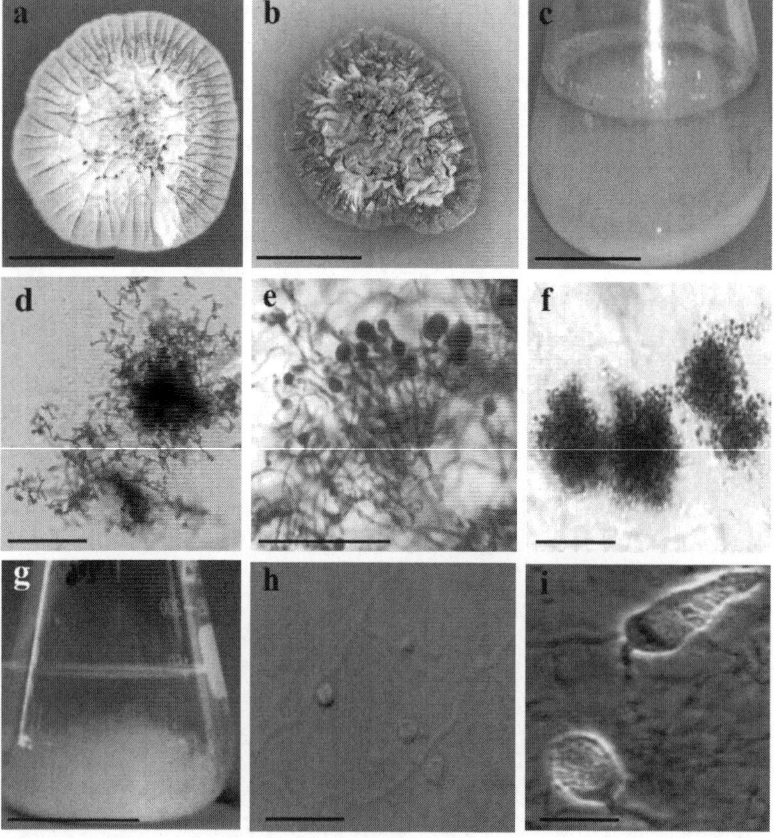

Figure 1. Morphological aspects of the selected actinomycete strains. a) and b) Colonial growth with substrate and aerial mycelium of *Streptomyces* and *Actinoplanes* on agar medium. c) Growth of *Micromonospora* in liquid medium. d) Development of spores in chains by *Streptomyces*. e) Production of globose sporangia by *Actinoplanes*. f) Single spores in *Micromonospora* strain. g) Growth of *Frankia* strain in liquid medium. h) Development of vesicles on hyphae by *Frankia* sp. i) Subglobose and elongated sporangia of *Frankia*. Scales: a, b) 15 mm. c, g) 40 mm. d) 45 µm. e) 100 µm. f) 15 µm. h) 10 µm. i) 25 µm.

(1998). The production of red, yellow, orange, pink, brown, greenish and black pigments is usual for *Frankia* strains (Lechevalier et al. 1982, Dobritsa 1998). Under experimental conditions, two red-orange crystalline quinonoid pigments were obtained from a *Frankia* strain isolated from the nodules of a *Casuarina equisetifolia* plant (Gerber and Lechevalier 1984).

Physiological activities

In general it is known, that the saprophytic actinomycetes form an important part of the microbial community in the soil environment, responsible for degradation and cycling of natural biopolymers, such as cellulose, lignin and chitin (Semédo et al. 2001). On the other hand, they are also the source of a wide range of other types of bioactive compounds for biotechnological applications (Okami and Hotta 1988, Bull et al. 1992).

To know more about the isolated strains and their possible plant growth-promoting rhizobacteria (PGPR) function or helper effect on N-fixing symbioses, following physiological tests were realized: growth rate at different temperatures (6–45°C) and pH (4–8.5), degradation of natural substrates, production of phytohormones, antifungal effects, and solubilization of inorganic P.

Culture growth conditions

The actinomycetes are characterized as predominantly aerobic bacteria; however some microaerophylic species have been described (Goodfellow and Williams 1983). Like many other soil microorganisms, the actinomycetes have an optimal mesophylic growth from 25–30°C. Generally, they operate on the ground at a pH between 5 and 9, with an optimal near neutral (Vobis and Chaia 1998).

Streptomyces MM40 presented an optimal growth between 17–35°C at pH 6–7 (8.5). This strain can also grow at 45°C, but developing only substrate mycelia in a thermotolerant manner. Strains *Actinoplanes* ME3 and *Micromonospora* MM18 presented an optimal growth between 17–28°C at pH 6–8.5. Also, the latter strain grew at 35°C, developing only non-sporulating mycelia. For *Frankia* BCU110501, the optimal growth was at 28°C and pH 7, presenting cottony appearance in liquid medium after 3 weeks (Fig. 1g).

Degradation of natural substrates

Streptomyces strains are known to produce hydrolytic enzymes such as cellulases, hemicellulases, ligninases, chitinases, amylases and glucanases (Antai and Crawford 1981, Yuan and Crawford 1995). In the present degradation studies, various natural substrates were used: filter paper as cellulose source, seeds of *Phoenix dactylifera* as hemicellulose, leave tissues of *Trisetum* sp. for pectinase test, dead wood of *Nothofagus* sp. as source of lignin, and mouse hairs as keratine (Solans and Vobis 2003). Each one of the 122 isolates were placed on artificial soil extract agar medium (Vobis 1992) in contact with the different substrates incubated at 28°C during 5–6 weeks (Fig. 2).

Pectin was degraded by 65.6% of the strains, cellulose (59%), hemicellulose (28.7%), keratine (3.3%) and 15.6% had strong affinities to use lignin as a preferable sustrate (Solans and Vobis 2003). The three saprophytic strains MM40, ME3 and MM18 were selected, because they showed the highest enzymatic activity of decomposing starch, cellulose, hemicellulose and pectin (Fig. 2). The strains MM40 and ME3 had additionally positive affinity to lignin, colonizing densely the wooden tissue (Table 3).

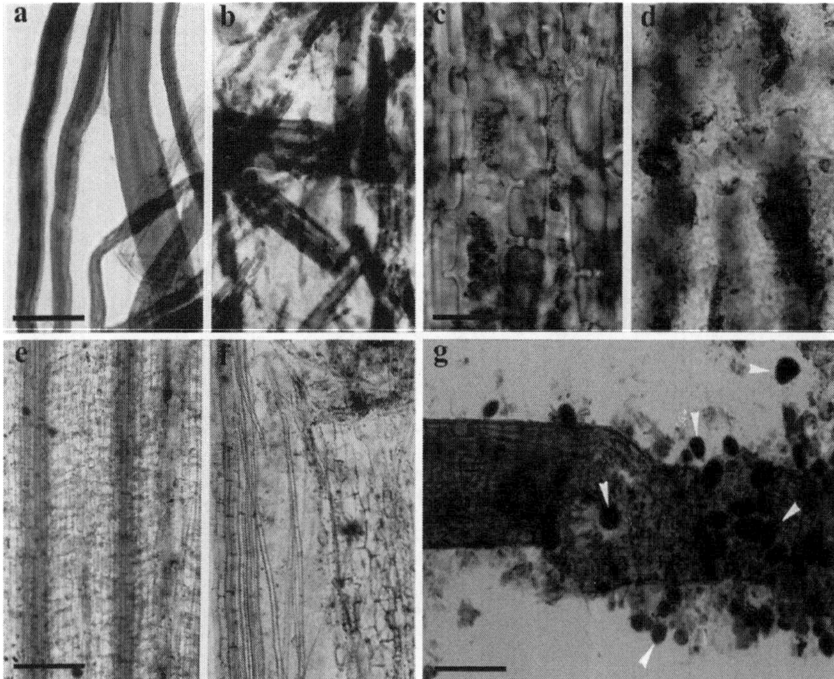

Figure 2. Effects of degradation on various biological substrates and structures by actinomycete strains isolated from rhizosphere of *O. trinervis*. a) Intact cellulose fibres. b) Fibres degraded by *Streptomyces* sp. c) Walls of endosperm cells of *Phoenix dactylifera*. d) Walls decomposed by *Streptomyces* sp. e) Intact tissues of grass leaf. f) Pectinous middle lamella desintegraded by *Streptomyces* sp. g) Mouse hair colonized by *Pilimelia* sp. forming sporangia (arrows) and structure of hair destroyed. Scales: a, b, e, f) 10 µm. c, d) 5 µm. g) 30 µm.

Production of phytohormones by actinomycetes

Diverse genera of actinobacteria, including *Streptomyces* and *Micromonospora* produce phytohormones (Ghodhbane-Gtari et al. 2010). It has been shown that in the interaction between plants and free-living PGPR, bacterial phytohormones are involved (Liste 1993, Höflich et al. 1994, Glick 1995). The most important ones are IAA and cytokinins.

The three saprophytic strains, together with the symbiotic *Frankia* strain, were evaluated regarding the production of phytohormones. The supernatant of the liquid medium of the strains cultured was used in the exponential growth phase. Indole 3-acetic acid (IAA) and gibberellic acid (GA_3) were analyzed by gas chromatography-mass spectrometry (GC-MS), while zeatine (Z) production was determined by gas chromatography-flame ionization detector and high performance liquid chromatography (HPLC fluorescent and UV), according to the methods used by Cassán et al. (2009). The levels of the three phytohormones produced by the saprophytic actinomycetes were higher than those produced by the symbiotic *Frankia* strain. Zeatine biosynthesis was thousand times higher than IAA and GA_3 (Table 1). The *Micromonospora* strain produced the highest levels of these phytohormones (Solans et al. 2011).

The phytohormones produced by free-living microorganisms are not only involved directly as growth factors into the development of plants, but also indirectly in regulation of nodulation (Kuhad et al. 2004, Barea et al. 2005). *Frankia* secretes several auxins including phenyl acetic acid, which could influence cell division and nodule development in actinorhizal symbiosis (Berry et al. 1989, Hammad et al. 2003). In rhizobial symbiosis, auxin accumulates in lateral-root initials and in the early stages of nodule differentation (Mathesius et al. 2000). Current evidence suggests that nod factors lead to an increase both in auxin and cytokinins, in the progenitor cells of the nodule in legumes. These two phytohormones may then act synergistically to initiate cell divisions to form nodule primordia (Mudler et al. 2005). Moreover, a pathway for nodulation that is independent of nod factors has recently been described and the signal involved seems to be phytohormone related (Giraud et al. 2007).

Table 1. Phytohormones production by the selected actinomycete strains.

Strains	IAA (ng ml^{-1})	GA_3 (ng ml^{-1})	Z (ng ml^{-1})
Streptomyces sp. MM40	0.75	0.96	240,000
Actinoplanes sp. ME3	0.27	1.53	310,000
Micromonospora sp. MM18	9.03	3.73	270,000
Frankia sp. BCU110501	0.92	1.76	15,000

IAA, indole 3-acetic acid; GA_3, gibberellic acid; Z, zeatine.

Antifungal compounds

The actinomycetes, particularly *Streptomyces* species, are well-known saprophytic bacteria quantitatively and qualitatively important in the rhizosphere, where they may influence plant growth and protect plant roots against invasion by root pathogenic fungi (Lechevalier 1989, Crawford et al. 1993, Yuan and Crawford 1995). *Streptomyces* species have been shown to protect several different plants in various degrees from soil-borne fungal pathogens (Yuan and Crawford 1995, Tokala et al. 2002).

The antifungal effects of the three saprophytic actinomycete strains as potential biocontrol agent were studied. They showed some antifungal effects, represented qualitatively by inhibition of the fungal growth, causing a halo in the contact area of the fungal culture. Strain *Streptomyces* MM40 presented the most antagonistic effect against the tested fungal strains: *Alternaria* sp. (VP60), *Cladosporium* sp. (VP61) and *Pestalotiopsis* sp. (VP59) (Table 2, Fig. 3a, d, g). *Micromonospora* MM18 presented antifungal effects for *Cladosporium* and *Pestalotiopsis*, while *Actinoplanes* ME3 showed a milder effect (Table 2, Fig. 3c, f, i and b, e, h).

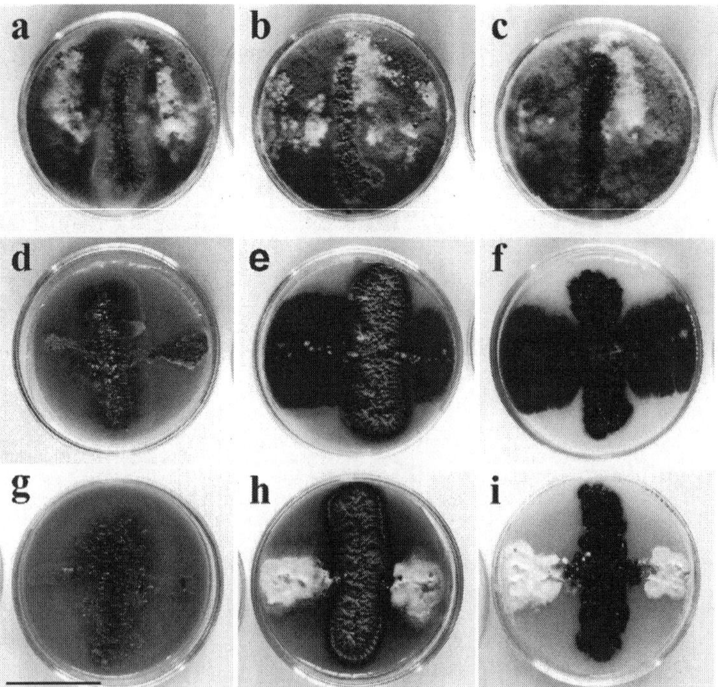

Figure 3. Antifungal effect of saprophytic actinomycetes on *Alternaria* (above), *Cladosporium* (center) and *Pestalotiopsis* (below). a, d, g) *Streptomyces* MM40. b, e, h) *Actinoplanes* ME3. c, f, i) *Micromonospora* MM18. Scale: 2.5 cm.

Table 2. Estimation of antifungal effects of the saprophytic actinomycetes on *Alternaria*, *Cladosporium*, and *Pestalotiopsis*.

Saprophytic strains	Fungal strains		
	Alternaria sp.	*Cladosporium* sp.	*Pestalotiopsis* sp.
MM40	++	++	++
ME3	-	+	+
MM18	-	++	++

++, with distinct halo inhibition of the fungal growth; +, without halo, but inhibition of growth; -, no effect.

Solubilization of inorganic phosphorus

The solubilization of phosphorus in the rhizosphere is the most common mode of action implicated in PGPR that increase nutrient availability to host plants (Vessey 2003). Phosphate-solubilizing microorganisms, such as bacteria, fungi and actinomycetes are common in the rhizosphere (Frioni 2006). However, the ability to solubilize P by no means indicates that a rhizospheric bacterium will constitute a PGPR. For example, Cattelan et al. (1999) found only two of the five rhizospheric isolates positive for P solubilization actually had a positive effect on soybean seedling growth. The three saprophytic strains show negative activity on P solubilization, using the method of Katznelson and Bose (1959).

A summary of the physiological characteristics of actinomycetes studied, is given in Table 3.

Table 3. Comparison of physiological characteristics of actinomycete strains.

Actinomycetes strains	Growth conditions	Degradation of biopolymers S C H P L K	Production of phytohormones (ng / ml)	Antifungal effects on A C P	Solubilization of inorganic P
Streptomyces sp. MM40	17 - 45° C pH 8.5	+ + + + + -	IAA: 0.75 GA3: 0.96 Z: 240000	+ + +	-
Actinoplanes sp. ME3	17 - 28° C pH 6 - 8.5	+ + + + + -	IAA: 0.27 GA3: 1.53 Z: 310000	- + +	-
Micromonospora sp. MM18	17 - 35° C pH 6 - 8.5	+ + + + - -	IAA: 9.03 GA3: 3.73 Z: 270000	- + +	-
Frankia sp. BCU110501	28° C pH 7	nd	IAA: 0.92 GA3: 1.76 Z: 15000	nd	nd

S, starch; C, cellulose; H, hemicellulose; P, pectine; L, lignin; K, keratine.
IAA, indole 3-acetic acid; GA_3, gibberellic acid; Z, zeatine.
Fungal strains: A, *Alternaria* sp.; C, *Cladosporium* sp.; P, *Pestalotiopsis* sp.
+, positive reaction; -, negative reaction; nd, not determined.

Comparison Between Rhizobial and Actinorhizal Symbioses

The biological reduction of atmospheric dinitrogen (nitrogen fixation) is the fundamental process that provides the essential nitrogen to the biosphere, as ammonium in the utilizable form to plants. Some prokaryotes have a remarkable capacity to fix atmospheric nitrogen. This capacity is determined by a highly conserved enzyme complex called nitrogenase, which is inactived in the presence of oxygen (Angus and Hirsch 2010). Nitrogenase probably arose in the Archean age and, throughout evolution, has been maintained in several genera that are collectively known as diazotrophic micro-organisms (Dixon and Kahn 2004). Diazotrophs are found in a variety of phylogenetic groups such as green sulphur bacteria, firmibacteria, actinomycetes, cyanobacteria and all subdivisions of the proteobacteria (Dixon and Kahn 2004, Angus and Hisrch 2010). The symbiotic association between diazotrophic bacteria commonly known as rhizobia with plants of the family *Leguminosae* is one of the more important symbiotic nitrogen fixation systems, together with *Frankia* and woody plants species, collectively called actinorhizal plants. The latter belong to the Eurosid I clade where legumes and *Parasponia* are also placed (Soltis et al. 1995), suggesting a common evolutionary origin of root nodule symbioses. In these symbioses, the host plants form root nodules, highly-specialized organs, wherein the bacteria carry out nitrogen fixation while being supplied by the plant with photosynthetically-derived carbon.

In the case of rhizobial symbioses, eleven genera of Gram-negative unicellular soil bacteria, *Rhizobium, Allorhizobium, Azorhizobium, Bradyrhizobium, Mesorhizobium, Sinorhizobium, Methylobacterium, Blastobacter, Devosia, Burkholderia and Ralstonia,* induce nodules on the roots of legumes and non-legume, *Parasponia* sp. (Family *Ulmaceae*) (Moulin et al. 2001, Young et al. 2001, Pawlowski and Sprent 2008). About 700 known plant genera with 17000 species form the rhizobial symbioses (Werner 1987). In the case of actinorhizal symbioses, Gram-positive actinomycetous of the only genus *Frankia* induce nodules on the roots of dicotyledonous perennal plants, mostly trees or woody shrubs, comprising 200 species, belonging to 25 genera in eight families: *Betulaceae, Casuarinaceae, Myricaceae, Elaegnaceae, Rhamnaceae, Rosaceae, Coriariaceae* and *Datisticaceae* (Chaia et al. 2010). *Frankia* is characterized by the following morphological structures: the hyphae (or filaments), sporangia with spores, and vesicles, the site of nitrogen fixation (Chaia et al. 2010).

In both, legume and actinorhizal symbioses, infection can take place either intracellularly via root hairs or intercellularly via penetration of root epidermis and bacterial colonization of the root cortex. The pathway by which the bacteria enter the plant depends on the host-plant species. In most legumes examined, rhizobia enter the plant intracellularly via root hairs. However, different intercellular infection mechanisms have been characterized for some tropical legumes, most of them woody (Pawlowski and Sprent 2008).

Two types of legume nodules are known, determinate and indeterminate. Indeterminate nodules are elongated and have a persistent meristem that continually gives rise to new nodule cells that are subsequently infected by rhizobia residing in the nodule (Gage 2004). These nodules are present in the forage legume *Medicago sativa* (alfalfa), *Medicago truncatula*, *Pisum sativa* (pea), *Vicia* species (vetches) and *Trifolium* species (clovers) and have been used historically as a model for studying the formation of indeterminate nodules (Newcomb 1981). The determinate nodules lack a persistent meristm, and are usually round legumes that form determinate nodules which are typically tropical in origin and include *Glycine max* (soybean), *Vicia faba* (bean), and *Lotus japonicus* (forage). In both types, legume nodule primordia are formed in the root cortex and develop into stem-like organs with a peripheral vascular system and infected cells in the central issue (Sprent 1995, Gage 2004).

In contrast, actinorhizal nodule primordia are formed in the root pericycle, like lateral root primordia. Mature actinorhizal nodules are coralloid organs, composed of multiple lobes, each of which represents a modified lateral root without root cap, but covered with peridermal tissue. Among actinorhizal symbioses, three of the eight host families (*Betulaceae, Casuarinaceae* and *Myricaceae)* are nodulated by *Frankia* via intracellular infection pathway. In five of the families (*Elaegnaceae, Rhamnaceae, Rosaceae, Datiscaceae* and *Coriariaceae*), early nodule initiation occurs via intercellular colonization (Wall and Berry 2008). During this type of infection, *Frankia* hyphae enter the root by penetration between epidermal cells and colonize the root cortex inter-cellulary (Miller and Baker 1986, Berry and Sunell 1990). This is the case in *O. trinervis* plants. In contrast to rhizobia, *Frankia* can fix N_2 in the free-living state, and hence could be expected to be less dependent in their distribution on their macrosymbionts than are rhizobia (Pawlowski and Sprent 2008).

Helper Effect of Saprophytic Actinomycetes on N-fixing Symbioses

In the last few years, symbiosis with rhizobia and mycorrhizal fungi have been studied extensively (Barea et al. 2004, Kuhad et al. 2004), but little has been investigated about their relationship with other functional groups, notwithstanding that many other interactions exist in the rhizosphere and are ecologically important.

The rhizosphere is the volume of soil under the influence of plant roots that contains an increased microbial biomass and activity compared with bulk soil and where very important and intensive microbe-plant interactions take place (Lynch and Whipps 1990). Some bacteria occurring there, termed rhizobacteria, exhibit different functions, exerting beneficial effects on plant development and are therefore referred to as Plant Growth Promoting Rhizobacteria (PGPR), because their applications are often associated with increased rates of plant growth (Kloepper and Schroth 1978). PGPR can affect plant growth by producing and releasing secondary metabolites, which either decrease or prevent the deleterious effects of phytopatogenic organisms in the rhizosphere, and/or by facilitating the availability and uptake of certain nutrients from the root environment (Glick 1995). It is common to observe the effects of biofertilizing-PGPR by synergism or promotion of the beneficial effects of a third-party rhizospheric microorganism. In these cases, the bacteria aid the other host-symbiotic relationship and is often referred to as helper bacteria (Vessey 2003, Banerjee et al. 2006). The majority of studies investigating PGPR as aids of other host-symbiont relationship involve either legume-rhizobia symbioses or plant-fungi symbioses (Garbaye 1994, Vessey 2003). Similarly, we use the term helper effect of actinomycetes (or just helper actinomycetes), referring to these bacteria as aids of actinorhizal or rhizobial symbioses.

The symbiotic establishment of mycorrhizal fungi on plant roots is affected in various ways by other rhizospheric microorganisms and more especially by bacteria. Some of these bacteria, which consistently promote mycorrhizal development, lead to the concept of mycorrhization helper bacteria (MHB) (Garbaye 1994). To date, many bacterial strains have been reported to be able to promote either arbuscular or ectomycorrhizal symbioses (Garbaye 1994, Barea et al. 2002, Barea et al. 2004, Duponnois 2006). The MHB strains which were identified belong to diverse bacterial groups and genera, such as *Agrobacterium, Azozpirillum, Azotobacter, Pseudomonas, Burkholderia, Bradyrhizobium, Rhizobium, Bacillus* and *Streptomyces*, among other bacterial genera (Frey-Klett et al. 2007).

In relation to another host-symbiotic relationship, it was demonstrated that other microorganisms besides *Frankia* or rhizobia are involved in the dynamics of the process of nodule development. In legumes plants, such as pea (*Pisum sativum*), the microbe rhizosphere interaction, and involving *Streptomyces* sp. promoted the nodulation by rhizobio (Tokala et al. 2002). Other example of promotion of nodulation by actinomycetes was observed by Gregor et al. (2003) in soybean (*Gycine max*) co-inoculated with *Bradyrhizobium japonicum*.

It is known that the actinomycetes are common rizoplane- and rhizosphere-colonizing bacteria (Frioni 2006, Solans and Vobis 2003), which have a high capacity to synthesize an array of biodegradative enzymes, antibiotics, phytohormones, and antifungal metabolites (Goodfellow and Cross 1974, Takana and Omura 1990, Tokala et al. 2002, Gregor et al. 2003, Solans et al. 2011). Also, they are the most widely distributed group as saprophytic soil inhabitants (Takisawa et al. 1993) and can promote plant growth by producing promoters such as indole-3-acetic acid (IAA) to help growth of roots, or produce siderophores to improve nutrient uptake (Merckx et al. 1987, Nimnoi et al. 2010).

In the case of actinorhizal symbioses, the helper bacteria effect was studied in *Alnus* plants. For example, the nodulation in axenic seedling cultures of *Alnus rubra* (family *Betulaceae*) was increased when rhizospheric bacterium, *Pseudomonas cepacia,* were coinoculated with the infective *Frankia* strain (Knowlton et al. 1980). This helper bacterium caused substantial host root-hair deformation, even in the absence of *Frankia,* suggesting that their main effect on nodulation was through preconditioning of host cellular process and was independent of direct interaction with *Frankia* (Berry and Torrey 1983). *P. cepacia* promoted the nodulation of *Alnus rubra* by *Frankia,* and it is proposed that this bacteria aid in the infection process at the host root hair surface, by causing root hair curling, allowing an intimate contact between *Frankia* and the hair wall (Knowlton et al. 1980, Knowlton and Dawson 1983). Soil microorganisms, therefore, seem to play a critical ecological role in enhancing the actinorhizal nodulation process in nature, however, the organisms and process underlying this enhacement remains to be fully elucidated.

In our studies, the helper effect could be demonstrated in experimental assays under controled conditions, using *O. trinervis* plants growing in tubes, pots and pouches system and inoculated with *Frankia* and coinoculated with saprophytic actinomycetes strains (Solans 2008). It could be observed, that saprophytic strains *Streptomyces* MM40, *Actinoplanes* ME3 and *Micromonospora* MM18 act as helper bacteria on both, actinorhizal and rhizobial N_2-fixing symbioses (Solans 2007, Solans et al. 2009), these strains clearly produce phytohormones (Table 1), but the real responsible metabolites are still unknown.

The nodulation kinetic of the actinorhizal symbiosis in *O. trinervis* was compared directly with rhizobial symbiosis in *M. sativa* in pouches cultures, applying the saprophytic helper actinomycetes. In both symbioses, the highest effect could be observed, when *Actinoplanes* or *Micromonospora* were coinoculated (Fig. 4). After 7 weeks of inoculation, not only the number, but also the dry mass of nodules increased significantly in comparison to single inoculation with *Frankia* (Fig. 4a) or *Sinorhizobium* (Fig. 4b).

In *O. trinervis*, the nodulation was higher during the first three weeks, when only *Frankia* was inoculated, and later surpassed by coinoculations

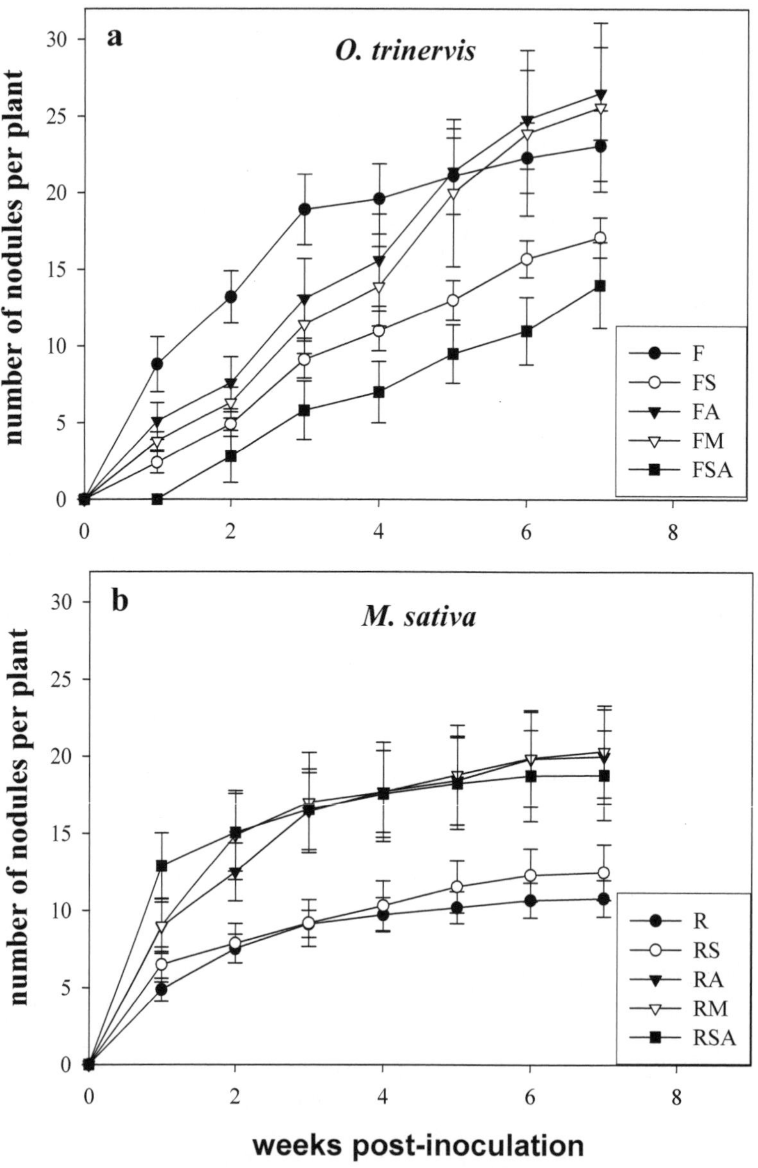

Figure 4. Comparison of nodulation kinetics of *O. trinervis* and *M. sativa* inoculated with helper actinomycetes. F, *Frankia* (●); FS, *Frankia* + *Streptomyces* (○); FA, *Frankia* + *Actinoplanes* (▼); FM, *Frankia* + *Micromonospora* (∇); FSA, *Frankia* + *Streptomyces* + *Actinoplanes* (■); R, *Sinorhizobium meliloti* (●); RS, *S. meliloti* + *Streptomyces* (○); RA: *S. meliloti* +*Actinoplanes* (▼); RM: *S. meliloti* + *Micromonospora* (∇); RSA, *S. meliloti* + *Streptomyces* + *Actinoplanes* (■). Values represent means ± SE, n = 8.

(Fig. 4a). On the contrary, in *M.sativa*, the single inoculation with *S. meliloti* remained from the beginning in a lower nodulation rate than with coinoculations (Fig. 4b).

In others culture experiments like those in glass tubes and pots, the nodulation and growth of plants was promoted by coinoculations, too (Solans 2007, Solans et al. 2009, Solans et al. 2011).

Apparently, the promotion of nodulation by helper actinomycetes is caused by a common effect in both symbioses systems, although *O. trinervis* is infected via intercellular invasion (Valverde and Wall 1999a,b) and *M. sativa* via root hair (Rhijn and Vanderleyden 1995). It seems that the helper effect operates in the same manner at the autoregulation level of the plant nodulation process (Solans et al. 2009). Recent studies on nodulation kinetics analysis in *Frankia-O. trinervis* symbiosis revealed different factors involved in the nodulation process, suggesting the existence of more than one signal of bacterial origin involved in the process (Gabbarini and Wall 2008).

Another example of rhizobial symbiosis is shown in Table 4. *Lotus glaber*, another forage legume, forms nodules in symbiosis with *Mesorhizobium* bacterium. The most remarkable observation was the significative increase of the shoot and root dry weight, when cultured with *Streptomyces* and *Actinoplanes* as coinoculum. The coinoculated plants developed flowers, whereas the single inoculated plants with *Mesorhizobium* not.

The helper effect of actinomycetes on actinorhizal and rhizobial symbioses under laboratory conditions are a first step to show the broad spectrum action of these actinobacteria in plant-microbe interactions. Further studies under controlled and field conditions, will be needed to know more about their role in the rhizosphere.

Table 4. Effects of actinomycetes on growth and nodulation of *Lotus* plants grown in pots after 11 weeks.

Treatments	Shoot length (cm)	Root length (cm)	No. Nodules	Shoot dry weight (mg)	Roots dry weight (mg)
R	41.8 ± 2.3	21.7 ± 3.5	31.6 ± 4.8	376.6 ± 56.4	117.7 ± 11.5
RS	46.0 ± 2.3	12.1 ± 1.2	44.3 ± 5.6	691.0 ± 40.2	222.7 ± 14.9
RA	42.2 ± 2.1	14.3 ± 2.1	29.3 ± 6.0	712.1 ± 60.4	233.4 ± 32.8
RM	43.8 ± 3.2	13.0 ± 0.4	36.7 ± 3.5	381.2 ± 43.9	97.5 ± 14.9
RSAM	49.7 ± 1.9	14.0 ± 1.9	35.8 ± 4.9	600.8 ± 42.1	253.7 ± 32.0

R, *Mesorhizo*bium; RS, *Mesorhizo*bium + *Streptomyces;* RA, *Mesorhizo*bium + *Actinoplanes;* RM, *Mesorhizo*bium + *Micromonospora;* RSAM, *Mesorhizo*bium + *Streptomyces* + *Actinoplanes* + *Micromonospora*. Values represent means ± SE, n = 7.

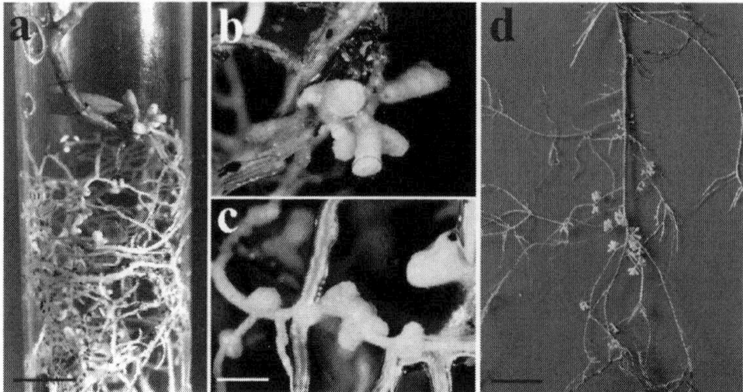

Figure 5. Nodulated roots of *O. trinervis*, coinoculated with *Frankia+Actinoplanes*. a) Glass tube culture after removing the sand-vermiculite substrate. b, c) Root nodules grown in glass tubes. d) Nodules developed in pouche culture. Scale: a) 10 mm, b, d) 2.5 mm, d) 15 mm.

Aspects of ecological importance and applications of rhizospheric actinomycetes

Availability of nutrients in the rhizosphere is controlled by the combined effects of soil properties, plant characteristics, and the interactions of plant roots with microorganisms and the surrounding soil (Bowen and Rovira 1992). The rhizosphere is characterized by a high concentration of easily degradable substrates in root exudades, such as sugar, organic acids, amino acids and others (Lynch and Whipps 1990), which leads to a proliferation of microorganisms and a higher density compared to the bulk soil (Foster 1986, Marschner and Rengel 2007). Microorganisms like bacteria, fungi, actinomycetes, and microalgae play a key role in organic matter decomposition, nutrient cycling and other chemical transformations in soil, particularly in the rhizosphere (Andrade 2004, Murphy et al. 2007).

Actinomycetes are the most widely distributed group as saprophytic soil inhabitants (Takisawa et al. 1993). They are known to be common rizoplane and rhizosphere-colonizing bacteria (Frioni 2006, Solans and Vobis 2003), which have a high capacity to produce several types of extracellular enzymes to degrade complex macromolecules (McCarthy 1989), playing an important role in the decomposition of recalcitrant biopolymers of plant waste (McCarthy and Williams 1992). Also, between biopolymers that degrade are starch, cellulose, hemicellulose, pectin, lignin, lignocellulose, keratin (Solans and Vobis 2003), humus, lignin and chitin (Semèdo et al. 2001). At the end of the degradation processes, the products are again available in the soil (Vobis and Chaia 1998). Actinomycetes utilize a wide range of carbon sources, degrade complex polymers such as lignin and possess advantageous characteristics of fungi, i.e., mycelial growth, production of spores resistent to drought and production of enzymes (McCarthy and Williams 1992, April et al. 2000).

Historically, actinomycetes have been the origin of the largest number of a new antibiotic drug candidates and lead molecules with applications in many other therapeutic areas (Genilloud et al. 2011). Members of the genus *Streptomyces* produce useful compounds, notably antibiotics, enzymes, enzyme inhibitors and pharmacologically active agents (Bérdy 2005, Khamna et al. 2009, Zhao et al. 2011). Besides a wide metabolic versatility, they may represent an underexplored reservoir of novel species of potential interest in the discovery of new compounds useful for agricultural technology and pharmaceuticals industry (Qin et al. 2011). For these reasons, the actinomycetes are good candidates for application in soil bioremediation and present an important potential for biotransformation and biodegradation of several contaminants, such as pesticides (Fuentes et al. 2010, Álvarez et al. 2012). Many genera have the ability to degrade different organic pollutants, such as polycyclic aromatic hydrocarbons (Pizzul et al. 2006). There are studies showing that actinomycetes and Streptomycetes in particular, have the ability to oxidize, partially dechlorinated herbicides such as atrazine or metolachlor (Speedie et al. 1987, Liu et al. 1990, Pogell 1992). These microorganisms are well suited for inoculation of soil, because of their metabolic diversity, its growth rate and colonization of semi-selected substrates relatively fast (Hsu and Lockwood 1975) and its ability to be genetically engineered (Pogell et al. 1991).

The other genus of actinomycetes also important in the soil, involved in the mineral recycled and plant nutrition, is *Frankia*. This bacterium can fix N_2 in the free-living state and in symbioses (Pawlowski and Sprent 2008). Fixed nitrogen (fixed-N) availability is commonly limiting to primary productivity and other processes of ecosystems throughout the world. Nitrogen fixation by actinorhizal plants is a major source of fixed-N in diverse and widespread terrestrial ecosystems, including forests, bogs, swamps, coastal dunes, landslides, glacial deposits, riparian zones, shrublands, prairies and deserts (Dawson 1986). Actinorhizal plants play important roles in wild-land ecosystem function and are used in land reclamation, range management, forestry, agroforestry and horticulture. It is probable that actinorhizal plants contribute a high proportion of the total amount of N_2 fixed globally, primarily in wild or extensively managed lands. Dixon and Wheeler (1986) estimated that the contribution of actinorhizal plants to terrestrial global nitrogen fixation could be as much as 25% of the total. While legumes are often considered to be major nitrogen-fixing systems, as they may drive up to 90% of their nitrogen from N_2 (e.g., faba bean, lupin, soybean, groundnut), estimated rates of actinorhizal nitrogen fixation are comparable to those of legumes (Franche et al. 2009).

A common ecological niche for actinorhizal plants is where soil nitrogen availability is low. Low levels of available combined-N can critically limit the establishment and growth of plants possessing symbiotic associations with N_2-fixing bacteria are often favored. Available combined-N is severely limiting to plant growth in soils low in organic matter, the major pool of soil N. The early stage of plant succession following a disturbance, which creates a N-limited situation, is a common temporal niche for actinorhizal plants, specially in degraded and eroded soils in Patagonian area (Chaia et al. 2010). The actinorhizal plants are thus often early successional species that play a key role in ecosystem disturbed by natural events (e.g., fires, landslides), by conditioning soil and rendering it capable of supporting other species of plants (Roy et al. 2007). For this, actinorhizal plants may have potential for wider use in soil reclamation (Chaia et al. 2010).

Besides actinorhizal plants, legumes have traditionally been used in soil regeneration, owing to their capacity to increase soil nitrogen. Recently, legumes have attracted attention for their role in remediation of metal-contamined soils. Legumes accumulate heavy metals mainly in roots and show a low level of metal translocation to the shoot. The main application of these plants is thus in metal phytostabilization (Pajuelo et al. 2011). Inoculation of legume plants with appropriate inocula containing rhizobia and heavy metal-resistant plant growth-promoting rhizobacteria (PGPR), and/or mycorrhizal has been found an interesting option to improve plant performance under stressed conditions (Pajuelo et al. 2011).

Beneficial plant-associated bacteria play a key role in supporting and/or incresing plant health and growth. PGPR can be applied in agricultural production or for the phytoremediation of pollutants (Wenzel 2009, Glick 2010, Compant et al. 2010). The use of PGPR in phytoremediation technologies, which can aid plant growth on contaminated sites (Burd et al. 2000, Gerhardt et al. 2009) and enhance detoxification of soil is relatively new. The properties of plants like high biomass production, low-level contaminant uptake, plant nutrition and health, used for phytoremediation, can be improved by PGPR but it is important to choose PGPR that can survive and colonize, when used in phytoremediation practices (Compant et al. 2010).

PGPR are commonly used to improve crop yields. In addition to their proven usefulness in agriculture, they possess potential in solving environmental problems. For example, PGPR may prevent soil erosion in arid zones by improving growth of desert plants in reforestation programs. This, in turn, reduces dust pollution. Thus, PGPR are involved in phytoremediation techniques to decontaminate soils and water (Glick 2003, 2010).

Technologies based on the use of microorganisms with broad capabilities biodegradable use in the prevention and remediation of pollution and also as a source of many novel compounds of important economic activities, have been developed. Although there are several studies about the role of specific strains of PGPR and rhizobia in plant growth promotion, N_2 fixation, biofertilizer activities, biological control and bioremediation, further studies are needed to understand the importance of actinomycetes in the rhizosphere and their potential use in the environment.

Until now, the helper actinomycetes could be applied in cultures of actinorhizal and legume plants to promote the symbiosis, growth rate and biomass production (Solans 2007, Solans et al. 2009). But it is imagined that they are also useful in the process of phytoremediation like PGPR, known for the effect of cleaning the environment (Glick 2003, 2010, Gerhardt et al. 2009).

The helper effect of the actinomycetes on N-fixing plants open a wide range of applications in land reclamation and landscape ecology, in the case of this pioneer actinorhizal shrubs and trees, since degraded soil systems can be fertilized and stabilized with higher effectivity than with the symbiotic actinomycete *Frankia* alone. Also, the application in agriculture, for example, the growth rate and production of N-fixing legume plants may be improved by the presence of these helper actinomycetes and used as a potential tool for bioremediation of polluted soils.

Acknowledgements

Financial support for this study was provided through grant by Universidad Nacional del Comahue 04/B140. We thank Dra. Eugenia Chaia for providing the *Frankia* BCU110501 strain, constructive suggestions, and corrections of the manuscript. M. Solans is member of CONICET.

References Cited

Álvarez, A., M.L. Yañez, C.S. Benimeli and M.J. Amoroso. 2012. Maize plants (*Zea mays*) root exudates enhance lindane removal by native *Streptomyces* strains. Int. Biodet. Biodegradation. 66: 14–18.

Andrade, G. Role of functional groups of microorganisms on the rhizosphere microcosm dynamics. pp. 51–69. *In:* A. Varma, L. Abbot, D. Werner, R. Hampp (eds.). 2004. Plant Surface Microbiology. Springer-Verlag. Berlin. Germany.

Angus, A.A. and A.M. Hirsch. 2010. Insights into the history of the legume-betaproteobacterial symbiosis. Mol. Ecol. 19: 28–30.

Antai, S.P. and D.L. Crawford. 1981. Degradation of soft wood, hardwood and grass lignocellulose by two *Streptomyces* strains. Appl. Environ. Microbiol. 42: 378–380.

April, F.M., J.M. Foght and R.S. Currah. 2000. Hydrocarbon-degrading filamentous fungi isolated from flare pit soils in northern and western Canada. Can. J. Microbiol. 46: 38–49.

Banerjee, M.R., L. Yesmin and J.K. Vessey. Plant-growth-promoting rhizobacteria as biofertilizers and biopesticides. pp. 137–181. *In*: M.K. Raid (ed.). 2006. Handbook of microbial biofertilizers Food Products. Academic Press, New York, USA.

Barea, J.M., R. Azcón and C. Azcón-Aguilar. 2002. Mycorrhizosphere interactions to improve plant fitness and soil quality. Antonie van Leeu. 81: 343–351.

Barea, J.M., R. Azcón and C. Azcón-Aguilar. Mycorrhizal fungi and plant growth promoting rhizobacteria. pp. 351–371. *In*: A. Varma, L. Abbott, D. Werner, R. Hampp (eds.). 2004. Plant Surface Microbiology. Springer-Verlag, Berlin, Germany.

Barea, J.M., M.J. Pozo, R. Azcón and C. Azcón-Aguilar. 2005. Microbial co-operation in the rhizosphere. J. Exp. Botany. 56: 1761–1778.

Bérdy, J. 2005. Bioactive microbial metabolites. J. Antibiot. 58: 1–26.

Berry, A.M. and J.G.Torrey. 1983. Root hair deformation in the infection process of *Alnus rubra*. Can. J. Bot. 61: 2863–2876.

Berry, A.M. and L. Sunell. The infection process and nodule development. pp. 61–81. *In*: C.S. Schwitzer, J.D. Tjepkema (eds.). 1990. The biology of *Frankia* and actinorhizal plants. Academic Press, San Diego, USA.

Berry, A.M, R.K.S. Kahn and M.C. Booth. 1989. Identification of indole compounds secreted by *Frankia* HFPArl3 in defined culture medium. Plant Soil. 118: 205–209.

Bowen, G.D. and A.D. Rovira. The rhizosphere: the hidden half of the hidden half. pp. 641–669. *In*: Y. Waisel, A. Eshel, U. Kafkafi (eds.). 1992. Roots: The hidden half. Marcel Dekker Inc., New York, USA.

Brunchorst, J. 1886. Über einige Wurzelanschwellungen, besonders diejenigen von *Alnus* und den *Elaeagnaceen*. Unters. Bot. Inst. Tübingen 2: 151–176.

Bull, A.T., M. Goodfellow and J.H. Slater. 1992. Biodiversity as a source of innovation in biotechnology. Annu. Rev. Microbiol. 46: 219–252.

Burd, G.I., D.G. Dixon and B.R. Glick. 2000. Plant growth-promoting bacteria that decrease heavy metal toxicity in plants. Can. J. Microbiol. 46: 237–245.

Cassán, F., D. Perrig, V. Sgroy, O. Masciarelli, C. Penna and V. Luna. 2009. *Azospirillum brasilense* Az39 and *Bradyrhizobium japonicum* E109, inoculated singly or in combination, promote seed germination and early seedling growth in corn (*Zea may* L.) and soybean (*Glycine max* L.). European. J. Soil Biol. 45: 28–35.

Cattelan, A.J., P.G. Hartel and J.J. Fuhrmann. 1999. Screening for plant growth-promoting rhizobacteria to promote early soybean growth. Soil Sci. Soc. Am. J. 63: 1670–1680.

Chaia, E.E. 1998. Isolation of an effective strain of *Frankia* from nodules of *Discaria trinervis* (Rhamnaceae). Plant Soil. 205: 99–102.

Chaia, E.E., L.G. Wall and K. Huss-Danell. 2010. Life in soil by the actinorhizal root nodule endophyte *Frankia*. A review. Symbiosis. 51: 201–226.

Compant, S., C. Clément and A. Sessitch. 2010. Plant growth-promoting bacteria in the rhizo- and endosphere of plants: their role, colonization, mechanisms involved and prospects for utilization. Soil Biol. Biochem. 42: 669–678.

Couch, J.N. 1950. *Actinoplanes*, a new genus of the *Actinomycetales*. J. Elisha Mitchell Sci. Soc. 66: 87–92.

Crawford, D.L., J.M. Lynch, J.M. Whipps and M.A. Ousley. 1993. Isolation and characterization of actinomycete antagonist of a fungal root pathogen. Appl. Environ. Microbiol. 59: 3899–3905.

Dawson, J.O. 1986. Actinorhizal plants: Their use in forestry and agriculture. Outlook in agriculture 15: 503–535.

Dixon, R. and D. Kahn. 2004. Genetic regulation of biological nitrogen fixation. Nat. Rev. Microbiol. 2: 621–631.

Dixon, R. and C.T. Wheeler. 1986. Nitrogen fixation in plants. New York, Chapman and Hall.

Dobritsa, S.V. 1998. Grouping of *Frankia* strains on the basis of susceptibility to antibiotics, pigment production and host specificity. Int. J. Syst. Bacteriol. 48: 1265–1275.

Duponnois, R. Bacteria helping mycorrhiza development. pp. 297–310. *In*: K.G. Mukerji, C. Manoharachary, J. Sing (eds.). 2006. Microbial activity in the rhizosphere. Springer, Berlin, Germany.

Foster, R.C. 1986. The ultrastructure of the rhizoplane and rhizosphere. Ann. Rev. Phypathol. 24: 211–234.

Franche, C., K. Lindström and C. Elmerich. 2009. Nitrogen-fixing bacteria associated with leguminous and non-leguminous plants. Plant Soil. 321: 35–59.

Frey-Klett, P., J. Garbaye and M. Tarkka. 2007. The mycorrhiza helper bacteria revisited. New Phytol. 176: 22–36.

Frioni, L. 2006. Microbiología: básica, ambiental y agrícola. Universidad de la República, Facultad de Agronomía, Uruguay. Montevideo, Uruguay.

Fuentes, M.S., C.S. Benemeli, S.A. Cuozzo and M.J. Amoroso. 2010. Isolation of pesticide-degrading actinomycetes from a contaminated site: bacterial growth, renoval and dechlorination of organochlorine pesticidas. Int. Biodet. Biodegradation. 64: 434–441.

Gabbarini, L. and L.G. Wall. 2008. Analysis of nodulation kinetics in *Frankia–Discaria trinervis* symbiosis reveals different factors involved in the nodulation process. Physiol. Plant. 133: 776–785.

Gage, D.J. 2004. Infection and invasion of roots by symbiotic, nitrogen-fixing rhizobia during nodulation of temperate legumes. Microbiol. Molecular Biol. Reviews. 68: 280–300.

Garbaye, J. 1994. Tansley Review No 76. Helper bacteria: a new dimension to the mycorrhizal symbiosis. New Phytol. 128: 197–210.

Genilloud, O., I. González, O. Salazar, J. Martín, J.R. Tormo and F. Vicente. 2011. Current approaches to exploit actinomycetes as a source of novel natural products. J. Ind. Microbiol. Biotechnol. 38: 375–389.

Gerber, N.N and M.P. Lechevalier. 1984. Novel benzol α-naphthacene quinones from an actinomycete, *Frankia* G-2 (ORS 020604). Can. J. Chem. 62: 2818–2821.

Gerhardt, K.E., X-D. Huang, B.R. Glick and B.M. Greenberg. 2009. Phytoremediation and rhizoremediation of organic soil contaminants: potential and challenges. Plant. Sci. 176: 20–30.

Giraud, E., L. Moulin, D. Vallenet, V. Barbe, E. Cytryn, J. Avarre, M. Jaubet, D. Simon et al. 2007. Legumes symbioses: absence of nod genes in photosynthetic bradyrhizobia. Science. 316: 1307–1312.

Ghodhbane-Gtari, F., I. Essoussi, M. Chattaoui, B. Chouaia, A. Jaouani, D. Daffonchio, A. Boudabous and M. Gtari. 2010. Isolation and characterization of non-*Frankia* actinobacteria from root nodules of *Alnus glutinosa, Casuarina glauca* and *Elaeagnus angustifolia*. Symbiosis. 50: 51–57.

Glick, B.R. 1995. The enhancement of plant growth by free-living bacteria. Can. J. Microbiol. 41: 109–117.

Glick, B.R. 2003. Phytoremediation: synergistic use of plants and bacteria to clean up the environment. Biotechnol. Advances. 21: 383–393.

Glick, B.R. 2010. Using soil bacteria to facilite phytoremediation. Biotechnol. Adv. 28: 367–374.

Goodfellow, M. and T. Cross. Actinomycetes. pp. 269–289. *In*: C.H. Dickinson, G.J.F. Pugh. (eds.). 1974. Biology of plant litter decomposition. Academic Press, London.

Goodfellow, M. and S.T. Williams. 1983. Ecology of actinomycetes. Ann. Rev. Microbiol. 37: 189–216.

Gregor, A.K., B. Klubek and E.C. Varsa. 2003. Identification and use of actinomycetes for enhanced nodulation of soybean co-inoculated with *Bradyrhizobium japonicum*. Can. J. Microbiol. 49: 483–491.

Hammad, Y., R. Nalin, J. Marechal, K. Fiasson, R. Pepin et al. 2003. A possible role for phenil acetic acid (PAA) on *Alnus glutinosa* nodulation by *Frankia*. Plant Soil. 254: 193–205.

Hartmann, A. 2005. Lorenz Hiltner, Pionier der Bodenbakteriologie und Rhizosphärenforschung. Biospektrum. 11: 191–192.

Hiltner, L. 1904. Über neuere Erfahrungen und Probleme auf dem Gebiete der Bodenbakteriologie unter besonderer Berücksichtigung der Grundüngung und Brache. Arbeiten der Deutschen Landwirtschafts-Gesellschaft. 98: 59–78.

Höflich, G., W. Wiehe and G. Kühn. 1994. Plant growth stimulation by inoculation with symbiotic and associative rhizosphere microorganisms. Experiencia. 50: 897–905.

Hsu, S.C. and J.L. Lockwood. 1975. Powered chitin agar as a selective medium for enumeration of actinomycetes in water and soil. Appl. Environ. Microbiol. 29: 422–426.

Hunter, J.C., M. Fonda, L. Sotos, B. Toso and A. Belt. 1984. Ecological approaches to isolation. Dev. Ind. Microbiol. 25: 247–266.

Katznelson, H. and B. Bose. 1959. Metabolic activity and phosphate-dissolving capability of bacterial isolates from wheat roots, rhizosphere, and non-rhizosphere soil. Can. J. Microbiol. 5: 79–85.

Kellerman, J., D. Medan, L. Aagesen and H.H. Hilger. 2005. Rehabilitation of the South American genus *Ochetophila* Poepp. Ex. Endl. (Rhamnaceae: Colletieae). New Zeoland J. Bot. 43: 865–869.

Khamna, S., A. Yokota and S. Lumyong. 2009. Actinomycetes isolated from medicinal plant rhizosphere soils: diversity and screening of antifungal compounds, indole-3-acetic acid and siderophore production. World J. Microbiol. Biotechnol. 25: 649–655.

Kloepper, J.W. and M.N. Schroth. Plant Growth-promoting Rhizobacteria on radishes. *In*: F.B. Meting. Jr. (ed.). 1978. Soil Microbial Ecology. Marcel Dekker.

Knowlton, S. and J.O. Dawson. 1983. Effect of *Pseudomonas cepacia* and cultural factors on the nodulation of *Alnus rubra* roots by *Frankia*. Can. J. Bot. 61: 2877–2882.

Knowlton, S., A. Berry and J.G. Torrey. 1980. Evidence that associated soil bacteria may influence root hair infection of actinorhizal plants by *Frankia*. Can. J. Microbiol. 26: 971–977.

Kuhad, R.C., D.M. Kothamasi, K.K. Tripathi and A. Singh. Diversity and functions of soil microflora in development of plant. pp. 71–98. *In*: A. Varma, L. Abbott, D. Werner, R. Hampp (eds.). 2004. Plant Surface Microbiology. Springer-Verlag, Berlin.

Lechevalier, H.A. The Actinomycetes III. A practical guide to generic identification of Actinomycetes. pp. 2344–2347. *In*: S.T. Williams (ed.). 1989. Bergey`s manual of systematic bacteriology, Vol 4. Williams & Wilkins. Baltimore.

Lechevalier, M.P., F. Horriere and L. Lechevalier. 1982. The biology of *Frankia* and related organisms. Dev. Ind. Microbiol. 23: 51– 60.

Liste, H.-H. 1993. Stimulation of symbiosis and growth of Lucerne by combined inoculation with *Rhizobium meliloti* and *Pseudomonas fluorescens*. Zentralbl. Mikrobiol. 148: 163–176.

Liu, S.Y., M.H. Lu and J.M. Bollag. 1990. Transformation of metachlor in soil inoculated with *Streptomyces* sp. Biodegradation. 1: 9–17.

Lynch, J.M. and J.M. Whipps. 1990. Substrate flow in the rhizosphere. Plant Soil. 129: 1–10.

McCarthy, A.J. 1989. Lignocellulose-degrading actinomycetes. FEMS Microbiol. Rev. 46: 145–163.

MacCarthy, A.J. and S.T. Williams. 1992. Actinomycetes as agent of biodegradation in the environment—a review. Gene. 115: 189–192.

Marschner, P. and Z. Rengel. Contributions of rhizosphere interactions to soil biological fertility. pp. 81–98. *In*: L.K. Abbott, D.V. Murphy (eds.). 2007. Soil Biological Fertility. A key to Sustainable Land Use in Agriculture. Springer, The Netherlands.

Mathesius, U., C. Charon, B.G. Rolfe, A. Kondorosi and M.D. Crespi. 2000. Temporal and spatial order of events during the induction of cortical cell division in white clover by *Rhizobium leguminosarum* bv. trifolii inoculation or localized cytokinin addition. Mol. Plant Microbe Interact. 13: 617–628.

Merckx, R., A. Dijkra, A.D. Hartong and J.A.V. Veen. 1987. Production of root-derivated material and associated microbial growth in soil at different nutrient levels. Biol. Fertil. Soils. 5: 126–132.

Miller, I.M. and D.D. Baker. 1986. Nodulation of actinorhizal plants by *Frankia* strains capable of both root hair infection and intercellular penetration. Protoplasma. 131: 82–91.

Miyadoh, S., M. Hamada, K. Hotta, T. Kudo, T. Seino, G. Vobis and A. Yokota (eds.). 1997. Atlas of Actinomycetes. The Society for Actinomycetes Japan, Asakuva, Tokyo.

Moulin, L., A. Munive, B. Dreyfus and C. Boivin-Masson. 2001. Nodulation of legumes by members of the ß-subclase of Proteobacteria. Nature. 411: 948–950.

Mudler, L., B. How, A. Bersoult and J.V. Cullimore. 2005. Integration of signalling pathways in the establishment of the legume-rhizobia symbiosis. Physiol. Plantarum. 123: 207–218.

Murphy, D.V., E.A. Stockdale, P.C. Brookes and K.W.T. Goulding. Impact of microorganisms on chemical transformations in soil. pp. 37–59. *In*: L.K. Abbot, D.V. Murphy (eds.). 2007. Soil Biological Fertility. A Key to Sustainable Land Use in Agriculture. Springer.

Newcomb, W. 1981. Nodule morphogenesis and differentiation. Int. Rev. Cytol. 13: 247–297.

Nimnoi, P., N. Pongsilp and S. Lumyoung. 2010. Endophytic actinomycetes isolated from *Aquilaria crassna* Pierre ex Lec and screening of plant growth promoters production. World J. Microbiol. Biotechnol. 26: 193–203.

Nobbe, F. and L. Hiltner. 1904. Über das Stickstoffsammlungsvermögen der Erlen und *Elaeagnaceen*. Naturwiss. Z. Forst-Landwirtsch. 2: 366–369.

Nonomura, H. Genus *Streptosporangium* Couch 1955. 148^{AL}. pp. 2545–2551. *In*: St. Williams, M.E. Sharpe, J.C. Holt (eds.). 1989. Bergey's manual of systematic bacteriology. Vol. 4. Williams & Wilkins. Baltimore.

Okami, Y. and K. Hotta. Search and discovery of new antibiotics. pp. 37–67. *In*: M. Goodfellow, S.T. Williams and M. Mordaski (eds.). 1988. Actinomycetes in Biotechnology. Academic Press, London. UK.

Ørskov, J. 1923. Investigations into the morphology of the ray fungi. Levin & Munksgaard, Copenhagen.

Pajuelo, E., I.D. Rodríguez-Llorente, A. Lafuente and M.A. Caviedes. Legume-Rhizobium simbiosis as a tool for bioremediation of heavy metal polluted soils. 2011. *In*: M.S. Khan et al. (eds.). 2011. Biomanagement of Metal-Contaminated Soils, Envieronmental Pollution 20. Chapter 4. doi 10.1007/978-94-007-1914-9-4. Springer.

Palleroni, N.J. 1980. A chemotactic method for the isolation of Actinoplanaceae. Arch. Microbiol. 128: 53–55.

Pawlowski, K. and J.I. Sprent. Comparison between actinorhizal and legume symbiosis. pp. 261–288. *In*: K. Pawlowski, W.E. Newton (eds.). 2008. Nitrogen-fixing Actinorhizal Symbioses. Springer.

Pizzul, L., M.C. Castillo and J. Stenström. 2006. Characterization of selected actinomycetes degrading polyaromatic hydrocarbons in liquid culture and spiked soil. World J. Microbiol. Biotechnol. 22: 745–752.

Pogell, B.M., H-L. Zhang and Y-M. Feng. 1991. Expression of veratryl alcohol oxidase activity and cloned fungal lignin peroxidase in Streptomyces lividans. Int Symposium on Biology of Actinomycetes. Madison, Wi.

Pogell, B.M. 1992. N-gealkylation and complete degradation of atrazine by microorganisms. Conf. On Genetics and Molecular Biology of Industrial Microorganisms. B 16: 16.

Qin S., K. Xing, J.H. Jiang and l.H. Lu. 2011. Biodiversity, bioactive natural products and biotechnological potential of plant-associated endophytic actinobacteria. Appl. Microbiol. Biotecnol. 89: 457–473.

Quispel, A. Discoveries, discussions and trends in research on actinorhizal root nodule symbioses before 1978. pp. 15–33. *In*: C.R. Schwinter, J.D. Tjepkema (eds.). 1990. The biology of Frankia and actinorhizal plants. Academic Press, New York, USA.

Reyes, F.M., M.E. Gobbi and E.E. Chaia. 2011. Reproductive ecology of Ochetophila trinervis in Northwest Patagonia. Functional Plant Biol. 38: 720–727.

Rhijn, P.V. and J. Vanderleyden. 1995. The Rhizobium-plant symbiosis. Microbiol. Rev. 59: 124–142.

Roy, S., D.P. Khasa and C.W. Greer. 2007. Combining alders, frankiae, and mycorrhizae for the revegetation and remediationof contamined ecosystems. Can. J. Bot. 85: 237–251.

Semédo, L.T.A.S., A.A. Linhares, R.C. Gomes, G.P. Manfio, C.S. Alviano, L.F. Linhares and R.R.R. Coelho. 2001. Isolation and characterization of actinomycetes from Brazilian tropical soils. Microbiol. Res. 155: 291–299.

Schäfer, D. 1973. Beiträge zur Klassifizierung und Taxonomie der Actinoplanaceen. Dissertation, Marburg.

Shirling, E.B. and D. Gottlieb. 1966. Methods for characterization of *Streptomyces* species. Int. J. Syst. Bacteriol. 16: 313–340.

Solans, M. and G. Vobis. 2003. Actinomycetes saprofíticos asociados a la rizósfera y rizoplano de *Discaria trinervis*. Ecol. Austral. 13: 97–107.

Solans, M. 2007. *Discaria trinervis-Frankia* symbiosis promotion by saprophytic actinomycetes. J. Basic Microbiol. 47: 243–250.

Solans, M. 2008. Influencia de rizoactinomicetes nativos sobre el desarrollo de la planta actinorrícica *Ochetophila trinervis*. PhD Thesis. Universidad Nacional del Comahue. Bariloche, Argentina.

Solans, M., G. Vobis and L.G. Wall. 2009. Saprophytic actinomycetes promote nodulation in *Medicago sativa-Sinorhizobium meliloti* symbiosis in the presence of high N. J. Plant Growth Regul. 28: 106–114.

Solans, M., G. Vobis, F. Cassán, V. Luna and L.G. Wall. 2011. Production of phytohormones by root-associated saprophytic actinomycetes isolated from the actinorhizal plant *Ochetophila trinervis*. World J. Microbiol. Biotechnol. 27: 2195–2202.

Soltis, D.E., P.S. Soltis, D.R. Morgan, S.M. Swensen, B.C. Mullin, J.M. Dowd and P.G. Martin. 1995. Chloroplast gene sequence data suggest a single origin of the predisposition for symbiotic nitrogen fixation in angiosperms. Proc. Natl. Acad. Sci. USA. 92: 2647–2651.

Speedie, M.K., B.M. Pogell, M.J. Mc Donald, R. Kline and Y.I. Huang. 1987. Potential usefulness of *Streptomyces* for the detoxification of recalcitrant organochlorines and other pollutants. Actinomycet. 20: 315–335.

Sprent, J.I. 1995. Legume trees and shrubs in the tropics: N_2 fixation in perspective. Soil Biol. Biochem. 27: 401–407.

Takana, Y. and S. Omura. 1990. Metabolism and products of actinomycetes—An Introduction. Actinomycetol. 4: 13–14.

Takisawa, M., R.R. Colwell and R.T. Hill. 1993. Isolation and diversity of actinomycetes in the *Chesapeake Bay*. Appl. Environ. Microbiol. 59: 997–1002.

Thaxter, R. 1891. Potato scab. 15th Annual Rept. Connect. Agric. Exptl. Stat. 153–160.

Tokala, R.K., J.L. Strap, J.L., C.M. Jung, D.L. Crawford, M.H. Salove, L.A. Deobald, J.F. Bailey and M.J. Morra. 2002. Novel plant-microbe rhizosphere interaction involving *Streptomyces lydicus* WYEC108 and the pea plant (*Pisum sativum*). Appl. Environ. Microbiol. 68: 2162–2171.

Tortosa, R.D. 1983. El género *Discaria* (Rhamnaceae). Bol. Soc. Arg. Bot. 22: 301–335.

Valverde, C. and L.G. Wall. 1999a. Regulation of nodulation in *Discaria trinervis* (Rhamnaceae) -*Frankia* symbiosis. Can. J. Bot. 77: 1302–1310.

Valverde, C. and L.G. Wall. 1999b. Time course of nodule development in *Discaria trinervis* (Rhamnaceae)-*Frankia* symbiosis. New Phytol. 141: 345–354.

Vessey, J.K. 2003. Plant growth promoting rhizobacteria as biofertilizers. Plant Soil 255: 571–586.

Vobis, G. The genus *Actinoplanes* and related genera. pp. 1029–1060. *In*: A. Ballows, H.G. Trüper, M. Dworkin, W. Harder, K.H. Schleifer (eds.). 1992. The Prokaryotes. A handbook on the biology of bacteria: ecophysiology, isolation, identification, applications. Springer, New York.

Vobis, G. and E.E. Chaia. 1998. El rol de los actinomycetes en el suelo. En: Actas XVI Congreso Argentino de la Ciencia del Suelo, (AACS, ed.). Villa Carlos Paz, Córdoba.

Vobis, G., J. Schäfer and P. Kämpfer. Genus III. *Actinoplanes*, 122. pp. 1058–1088. *In*: M. Goodfellow, P. Kämpfer, H.-J. Busse, M.E. Trujillo, K. Suzuki, W. Ludwig, W.B. Whitmann (eds.). 2012. Bergey's Manual of Systematic Bacteriology, 2nd ed., Vol. 5., The Actinobacteria, Part B. Springer, New York .

Waksman, S.A. 1950. The Actinomycetes, their nature, occurrence, activities, and importance. Waltham, Massachusetts, USA.

Wall, L.G. 2000. The actinorhizal symbiosis. J. Plant Growth Regul. 19: 167–182.

Wall, L.G. and A.M. Berry. Early interactions, infection and nodulation in actinorhizal symbiosis. pp. 147–166. *In*: K. Pawlowski, W.E. Newton (eds.). 2008. Nitrogen-fixing Actinorhizal Symbioses. Springer. Dordrecht.

Wenzel, W.W. 2009. Rhizosphere process and management in plant-assisted bioremediation (phytoremediation) of soils. Plant Soil. 321: 385–408.

Werner, D. 1987. Pflanzliche und mikrobielle Symbiosen. Thieme, Stuttgart New York.

Young, J.M., L.D. Kuykendall, E. Martínez-Romero, A. Kerr and H. Sawada. 2001. A revision of *Rhizobium* Frank 1889, with an emended description of the genus, and the inclusion of all species of *Agrobacterium* Conn 1942 and *Allorhizobium undicola* de Lajudie et al. 1998 as new combinations: *Rhizobium radiobacter, R. rhizogenes, R. rubi, R. undicola* and *R. vitis*. Int. J. Syst. Evol. Microbiol. 51: 89–103.

Yuan, W.M. and D.L. Crawford. 1995. Characterization of *Streptomyces lydicus* WYEC108 as a potential biocontrol agent against fungal root and seed rots. Appl. Environ. Microbiol. 61: 3119–3128.

Zhao, K., P. Penttinen, T. Guan, J. Xiao, Q. Chen, J. Xu, K. Lindström, L. Zhang, X. Zhang and G.A. Strobel. 2011. The diversity and anti-microbial activity of endophytic actinomycetes isolated from medicinal plants in Panxi Plateau, China. Curr. Microbiol. 62: 182–19.

CHAPTER 2

Cultural Factors Affecting Heavy Metals Removal by Actinobacteria

Villegas Liliana Beatriz,[1,*] Rodríguez Analia,[1] Pereira Claudia Elizabeth[1] and Abate, Carlos Mauricio[1,2,3]

Introduction

Heavy metals are a subset of elements defined weakly that exhibit metallic properties. They are "the most toxic inorganic pollutants which occur in soils" and can be of natural or anthropogenic origin. This group mainly includes transition metals, some metalloids, lanthanides and actinides. Many definitions have been formulated in terms of weight or atomic mass which leads to the periodic table, chemical classification traditionally more solid and scientifically informative about elements; others have been based on the density and some of the chemical properties or degree of toxicity. However, the density or specific gravity is not of great importance in connection with the reactivity of a metal. It is important to note that this term has been applied even to semimetals (metalloids) such as arsenic,

[1]Planta Piloto de Procesos Industriales y Microbiológicos (PROIMI), CONICET, Av. Belgrano y Pasaje Caseros, 4000 Tucumán, Argentina.
[2]Facultad de Ciencias Naturales e Instituto Miguel Lillo, Universidad Nacional de Tucumán, Miguel Lillo 205, Tucumán.
[3]Facultad de Bioquímica, Química y Farmacia, Universidad Nacional de Tucumán, Ayacucho 471, Tucumán.
*Corresponding author: lbvilleg@hotmail.com

presumably due to the hidden assumption that "heaviness" and "toxicity" are in some way identical. This example further illustrates the confusion surrounding the term. For this reason "heavy metal" has been termed a "misinterpretation" by International Union of Pure Applied Chemistry (IUPAC) as a result of contradictory definitions and its lack of a "coherent scientific basis" (Duffus 2002, Bradl 2004).

From a physiological point of view, heavy metals fall into two main categories: (a) those essential as trace elements to maintain the metabolism of mammal, plant or human bodies, but at high concentrations they can lead to poisoning, like, iron (Fe), copper (Cu), zinc (Zn), cobalt (Co), nickel (Ni), selenium (Se) and others, (b) without physiological function and so, toxic even at low concentrations, as arsenic (As), lead (Pb), cadmium (Cd) and mercury (Hg). Heavy metals pose a critical concern to human health and environmental issues due to their high occurrence as a contaminant, low solubility in biota and the classification of several heavy metals as carcinogenic and mutagenic (Diels et al. 2002).

In general, soils have heavy metals as a result of geological and edaphogical processes. The natural content of chemical elements in soil is called local geochemical background or background level. Geochemical background was defined by Hawkes and Webb (1962 cited in Nakic et al. 2007) as the normal abundance of an element in barren earth material. Natural background levels reflect the usual processes unaffected by human activities. Many authors used this term as a synonym for natural background concentrations to describe natural variations in element concentration in the surficial environment although others are in disagree with this concept (Baize and Sterckman 2001, Nakic et al. 2007). Heavy metal pollution of soils derives from an increase in their concentration several times to the background level, arising from the implementation of certain human activities.

Heavy Metal Contamination

The Industrial Revolution was a historical period that took place in the nineteenth century mind, where changes in agriculture, manufacturing, mining, transportation and technology had a profound effect on the social, economic and cultural conditions of the times. This event in question began in the United Kingdom and it brought about changes that in the long run eventually revolutionized, not just the British economy but that of the rest of Western Europe, North America and gradually, much of the rest of the world.

While this industrialization brought many benefits and socio-economic progress to mankind, this was not accompanied by a criterion of preservation of the environment for the reason that the auto remediation capacity of the land was overestimated. Industrial activities, now considered synonymous with modernity and progress, generate a series of environmental hazards due to their effluents or emanations. Heavy metals are present in most industrial effluents; consequently they are increasing in the environment. Moreover, heavy metals cannot be degraded, so when they are released into the environment, they remain there indefinitely.

The accumulation of heavy metal, especially in urban areas, generates dispersion in soil, water and air and they subsequently seep into groundwater and the drinking water reservoir. Moreover, heavy metals could have long-term hazardous impacts on the health of soil ecosystems and adverse influences on the biological process of the soil. They also enter the food chain with great ease and accumulate, becoming increasingly concentrated (and hence dangerous) as one moves up the food chain.

In some developing countries, the pollution is produced, among other things, as a result of the gradual increase in population and rapid industrial development, which is aggravated by the lack of infrastructure regarding health facilities and inadequate treatment of waste.

Mobility and Bioavailabilty of Heavy Metals in Soil

The action of heavy metals on the environment depends on their mobility and bioavailability and these phenomena depend on total metal concentration and also on their association with the solid phase to which they are bound. The term mobility refers to a biological or chemical contaminating ability to move within the soil or ground water over a period of time. A contaminant may move under the influence of gravity as with light or dense non-aqueous phase liquids or under the influence of ground water flow as with dissolved constituents. The mobility of heavy metals is difficult to determine due to the many reactions inorganic compounds can have with various soil components. The most important factors which affect their mobility are pH, sorbent nature, presence and concentration of organic and inorganic ligands, including humic and fulvic acids, root exudates and nutrients. Furthermore, redox reactions, both biotic and abiotic, are of great importance in controlling the state of oxidation and thus, the mobility and toxicity of many elements (Violante et al. 2010).

On the other hand, bioavailability is the proportion of total metals that are available for incorporation into biota; this effect is known as bioaccumulation (John and Leventhal 1995). The free metal ion is, in general, the most bioavailable and toxic form of the metal. Therefore, total metal concentrations do not necessarily correspond with metal bioavailability because heavy metal speciation (physical-chemical forms) affects the mobility, the bioavailability and in consequence, the toxicity of the heavy metal. It is known that several metals exist in soils in more than one state of oxidation. One of the best examples is the chromium that exists under several states of oxidation ranging from Cr(–II) to Cr(VI). However, the trivalent [Cr(III)] and hexavalent [Cr(VI)] states are the most stable. Cr(III) occurs as a cation and hydroxide complex, in these conditions chromium is relatively immobile in soil, being strongly adsorbed by soils and readily forming insoluble precipitates, and its toxicity is low. On the other side, Cr(VI) occurs as oxyanions, chromate and dichromate. All these oxyanion forms are very soluble, relatively mobile in soils, being only weakly adsorbed by soils. Cr(VI) is also extremely toxic and a known carcinogen (Barceloux 1999a, Das and Mishra 2008). The fate of Cr in soil is partly dependent on the redox potential and the pH of the soil. Under reducing conditions, Cr(VI) can be reduced to Cr(III) by redox reactions in the presence of inorganic species, electron transfers at mineral surfaces, reaction with nonhumic organic substances such as carbohydrates and proteins, or reduction by humic substances in the soil. The reduction of Cr(VI) to Cr(III) in soil increasing with lower pH values (Zayed and Terry 2003). Although the Cr(III) is immobile, the additions of organic acids of low molecular weight (carboxylic and amino acids) to soil, significantly increased chromium accumulation from Cr(III) treated plants in the presence of increasing concentrations of organic acid, suggesting the existence of an interaction between Cr(III) and organic ligands leading to the formation of organically bound mobile Cr(III) (Srivastava et al. 1999 cited by Zayed and Terry 2003). Cr(III) generating reactive oxygen species (ROS), if present in high concentration, can cause toxic effects due to its ability to coordinate various organic compounds resulting in inhibition of some metalloenzyme systems (Shanker et al. 2005).

Papafilippaki et al. (2007) who estimate the heavy metal pollution due to agrochemical contamination in agricultural soils, reported that the relative availability and comparative mobility followed the order Cu>Pb>Zn>Cr. According to the outlook of these researchers, organic matter in the soil showed a strong relationship with this order due to the formation of soluble and insoluble complexes with the metals. Humified organic matter seemed to be involved in the formation of soluble complexes with Cu and Zn, while Cr(VI) was reduced to Cr(III).

Recent work showed that salinity increases the mobilization of heavy metals in soils. The mobilization of Pb, Cd, Cu and Zn appeared to be regulated by several or all of the following mechanisms; (i) competition with Mg and/or Ca for adsorption sites, (ii) complexation with chlorides, (iii) complexation with sulfates. Therefore, these results can be used for the evaluation of heavy metal mobility in saline soils, which is a growing problem in arid and semiarid regions where salinization and urbanization/industrialization are the main processes causing land degradation (Acosta et al. 2011).

Environmetal Remediation

In recent decades, environmental protection has become a relevant interest in society. Rapid human population growth, with consequent urbanization, global trend related to industrialization, ecological transition and change of land use, shows the urgent need to preserve the environment, because the damage and disruption are a serious danger to the diversity, health and human welfare. Through scientific and technological development, developed countries confront this trend with the help of planning, management and implementing economic-ecological policies that address environmental pollution, promoting the quality of urban centers and ecosystems that provide ecological services to them (such as watershed conservation, air purification), propelling sustainable development.

Among the techniques actually used in the removal of heavy metals include physicochemical methods (soil washing, excavation, ion exchange columns, treatment with alkaline solutions, flocculating agents, precipitation, ion exchange, reverse osmosis, etc.), in which all have the common denominator of high cost and are not applicable to large areas of contaminated soils and sediments. These techniques also have other methodological disadvantages like incomplete metal removal, higher reagent and energy supplies and hazardous waste generation, which demand extreme care to manipulate. In general *in situ* remediation processes should not be applied because it is impossible to treat a particular metal due to the competition that exists for the presence of others. These disadvantages are accentuated even more when metal concentrations to remove are very low (Silóniz et al. 2002, Viti et al. 2003).

For that reason, it is necessary to provide, through new biotechnological proposals, the development of techniques for removal at low cost, which should reduce the concentration of pollutants below regulatory levels, in order to protect human health and provide a permanent solution, either to extract the metals or to stabilize them in non-toxic or less toxic chemical species. Remedying an environment contaminated with biological methods (bioremediation), offers high specificity in removal of the metal of interest with operational flexibility both *in situ* and *ex situ* systems. It has several advantages in relation to the traditional technique, it is often less expensive and site disruption is minimal. It eliminates waste permanently, eliminates long-term liability and has greater public acceptance, with regulatory encouragement and can be coupled with other physical or chemical treatment methods. However, the bioremediation methods depend on having the right microbes, which have the physiological and homeostasis mechanisms effective, in the precise place with the correct environmental factors for degradation to occur. The success of these methods is dependent on an interdisciplinary approach, involving disciplines as microbiology, engineering, ecology, geology, and chemistry (Boopathy 2000).

Soil Microorganisms and Heavy Metal

The soil, like any other ecosystem, is an important habitat to thousands of organisms like a wide variety of fungi, algae, protozoa and different types of bacteria that vary in physiology. Microorganisms are an essential part of living soil and of outmost importance for soil health; they appear to be excellent indicators of soil health because they respond to changes in the soil ecosystem quickly (Nielsen and Winding 2002). Despite their small volume in soil, microorganisms are key players in the cycling of nitrogen, sulphur, and phosphorus, and the decomposition of organic residues.

The presence of pollutants as heavy metal has negative effects on the microflora. Some studies indicated that not all species have evolved resistance or tolerance mechanisms for metals. Many sensitive species might have been eliminated by the pollutants and their place is taken by the resistant species which have different ecological role.

Terms such as "resistance" and "tolerance" are arbitrary, often used interchangeably without clear distinction in literature but their meanings are different. According to Gadd (1992) "resistance" indicates the ability of microorganisms to survive the toxic effects of heavy metals through a detoxification mechanism occurring in direct response to the metal species in question and "tolerance" refers to the capacity of a microorganism to survive metal toxicity using the intrinsic properties and/or environmental modification of the toxicity

The resistant microorganisms have often failed to perform specific ecological functions (Ansari and Malik 2010). Several studies have demonstrated that microbial parameters may be useful as indicators of changing soil conditions caused by chemical pollution (Malik and Ahmad 2002).

Resistant microorganisms have a wide variety of answers or interactions with heavy metal. In general terms the interaction of bacteria and heavy metals causes their mobilization or immobilization. Some such interactions are mentioned below:

1. **Biotransformation:** As has been cited earlier, the mobility and toxicity of the heavy metal forms are different. Microorganisms exhibit a

number of enzymatic activities that transform certain metal species through oxidation, reduction, methylation and alkylation (Valls and Lorenzo 2002).

2. **Biosorption:** The terms of sorption and adsorption are used when the natural affinity of biological compounds is exploited for metallic elements, a process also known as passive bioaccumulation. May be involved cells dead or live. This approach depends on the components on the cell surface and the spatial structure of the cell wall on one side and on the other, the physico-chemical conditions of the surroundings where the cell is developed (Ledin 2000, Wang and Chen 2009). Earlier studies on heavy metals biosorption by microorganisms have shown that pH is the single most important parameter affecting the biosorption (Brady and Duncan 1994, Dönmez and Aksu 2001). Now it is recognized that ionic strength, temperature, and the presence of other metals and organic compounds, also have an important role in this process (Violante et al. 2010).
3. **Uptake or active bioaccumulation**: The term "uptake" is used when metabolism-dependent intracellular transport is implied during the accumulation and many features of a living cell, like intracellular sequestration followed by localization within cell components, metallothionein binding and complex formation can occur (Gadd 2004, Malik 2004). The ability of microorganisms to accumulate toxic metals varies with cell age.

Also it is important to mention that microorganisms participate in the cycling of carbon and thereby influence the amount and character of organic matter. This phenomenon can be of substantial importance for metal mobility, because organic compounds may bind metals. Microbial degradation of the metal–organic complex can change the speciation of the metal. However, metal binding in various organic substances may decrease the microbial degradation of the organic compound (Krantz-Rülcker et al. 1995). Also, microorganisms may produce or release organic compounds such as pigments or siderophores in the presence of heavy metals that can indirectly reduce or increase the mobility of heavy metals. On the other hand microorganisms also produce inorganic compounds such as sulfur that reduces the mobility of many metals and even causes them precipitate. In addition to these processes, microorganisms can influence metal mobility since they affect pH, redox potential (Eh), etc. All these parameters should be kept in mind while studying the influence of microbial metal accumulation on metal mobility (Ledin 2000).

For example, it is known that extreme pH generally decreases the rate of metal uptake: while at low pH the ions H^+ battle with the metallic ions for the combination sites of the cells, at high pH the metallic ions can rainfall for the formation of hydroxides or insoluble oxides, as well as to a wide range of hydroxy complexes. Most hydroxides are insoluble in water, except the hydroxides of alkaline metals, ammonium and barium. These compounds reduce the concentration of free ions in solutions and therefore their bioavailability and it is the general principle of heavy metal removal by chemical precipitation (Naja and Volesky 2010). Most previous studies, reported in literature for the metal removal by the majority of the microorganisms, pH between 4.0 and 8.0 is the range most accepted as the ideal (Fourest and Roux 1992, Brady and Duncan 1994). Later studies with yeasts, reported that in the absence of pH adjustment, the optimum initial pH for the removal of three cations: Cu(II), Cd(II) and Pb(II), was in the 4.5–5.5 range (Marques et al. 1999). However, a gradual pH increase was observed during the removal process, up to a final equilibrium value of 7.0 and 8.0. In other work, for optimum copper uptake by yeast adjustment pH at 5.0 was necessary (Villegas et al. 2005). Instead, a collection strain of fungi showed ability for heavy metals accumulation, but the optimum pH changed with the metal in study (Dursun et al. 2003).

Concerning at the effects of temperature on removal of heavy metals, these are restricted to metabolism-dependent accumulation of metals or biotransformation processes where the enzyme activities are involved. At low temperature (0–5°C) little or no metal is sequestered through metabolic processes by viable biomass. Most laboratory experiments are carried out in the temperature range 20–30°C, which has been reported optimal for metal accumulation and for the growth of the majority of the microorganisms (Brady and Duncan 1994). One clear example is finding different microbial groups that have different values of slope in different seasons indicating that the resistant population of microorganisms is variable with seasons (Ansari and Malik 2010).

Actinobacteria

Actinobacteria constitute a morphologically diverse group of Gram-positive bacteria ubiquitous in soils, where they are usually present in numbers of 10^5–10^6 colony-forming units per gram of soil. They play an important role in the recycling of organic carbon and are able to degrade complex polymers (Goodfellow and Williams 1983).

Some of actinobacteria were initially considered as minute fungi from their mycelium-like growth, which is why these bacteria were previously called "actinomycetes"; this name derives from the Greek "actys" (Ray) and "mykes" (fungus) (Ensign 1992). This form of growth favours the colonization of soil particles and the high guanosine plus cytosine content in their genomes distinguishes it from other Gram-positive bacteria

(Lacey 1997). Most of them are aerobic and neutrophilic, as many other soil microorganisms are mesophilic with an optimum growth between 25 and 30°C. Actinobacteria are mainly distributed in a wide variety of environments, such as terrestrial and marine. In general, they carry out their activities in the soil at a pH between 5 and 9, with an optimum near neutral (Johnston and Cross 1976, McBride and Ensign 1987). They have capacity of resistance to desiccation which has proven to be an important factor for their survival in soils that are often extremely dry. Some of them survive in soil as spores between bursts of vegetative growth when certain environmental conditions are favourable, such as nutrient supply, humidity and temperature or physiological interactions with other microorganisms. Scientists have long known that actinobacteria somehow kept soil bacteria populations in balance (Raja and Prabakarana 2011).

Probably most of the interest in this group of microorganisms lies in their ability to produce secondary metabolites that show bioactivities as antibiotics, enzyme inhibitors, immunosuppressors and others (Lam 2006, Manivasagan et al. 2009, Raja and Prabakarana 2011). Metabolites with antibiotic activity were isolated mainly from *Streptomyces* species, representing some 70% to 80% of the all isolated compounds. This genus is the largest, comprised 80% to 90% of the total of actinobacteria population, they are found predominantly in soil and they are noticeable for the odor of tilled soil which is attributed to geosmins, a group of their secondary metabolites (Raja and Prabakarana 2011). Members of the actinobacteria group were found to be capable of heavy metals removal, which is very important for potential use in bioremediation processes (Amoroso et al. 1998, Amoroso et al. 2001, Polti et al. 2007, Colin et al. 2012). With the aim of surviving in contaminated environments, they have to adjust or acquire tolerance or resistance to heavy metals. Natural bioremediation of the soils could be achieved *in situ* based on the activity of the indigenous soil microflora. Interestingly, it is just the study of these adaptation mechanisms to heavy metal presence that provides the useful tools to achieve a successfully bioremediation process.

Influence of Culture Medium Composition on Heavy Metal Clean up

Heavy metals are directly and/or indirectly involved in all aspects of microbial growth, metabolism and differentiation. Some heavy metals are essential as trace nutrients but toxic in high concentrations. Consequently, bacteria have developed sophisticated and specific cellular machinery consisting of an extensive network of specialized proteins, transporters and the regulation of gene expression that respond to either heavy metal deprivation or overload (Ma et al. 2009, Solioz et al. 2010). These mechanisms of cells to keep the optimal concentration of heavy metal bioavailability, mediated by balancing the heavy metal uptake, efflux and intracellular trafficking within compartments and storage processes so that the needs of the cell to the metal ion is satisfied, is called "homeostasis". Currently little is known about the transformation of heavy metals and their homeostasis mechanisms in an ecologically very important group of actinobacteria.

In general, microorganisms-heavy metal interactions are affected by numerous parameters as organic or inorganic composition of medium where the bacteria will grow, the pH and the temperature. When liquid culture media are employed, the availability of the nutrients and of the toxic substance, such as heavy metals, is better than in solid culture media. In artificial laboratory culture, soil bacteria show extreme sensitivity to heavy metals due to the high solubility of metal salts in the growth media. For this reason, these experimental systems provide representative information on the homeostasis mechanisms of the bacteria as well as how it affects the mobility and bioavailability of metals by changing the composition of the medium, pH and other factors. In this way, in heavy metals removal studies, minimal medium must be used as basal condition, because the complex media are inadequate due to the high concentration of organic components that absorb the metal ions. As was mentioned earlier, some heavy metal have high affinity to organic matter diminishing their bioavailability (Laxman and More 2002, Polti et al. 2007, Villegas et al. 2011).

Accordingly, the composition of the culture medium used may directly or indirectly influence the heavy metals removal by bacteria. At the same time, it depends on specific or nonspecific mechanisms that bacteria possess for removal or transformation of heavy metal and physicochemical characteristics of heavy metals in question.

Below is briefly show some results in relation to the state of knowledge about more common heavy metals removal by actinobacteria.

Copper (Cu) is an essential as well as toxic element on bacteria and other organisms, depending on concentrations. The excess copper avidly binds to many biomolecules such as proteins, lipids, and nucleic acids, regardless of its valence state. Furthermore, copper toxicity is exploited in agriculture as commercial fungicides and algicides (Barceloux 1999b). The major toxic effect of copper has frequently been claimed to be due to the generation of toxic reactive oxygen species in a Fenton-type reaction leading to the generation of hydroxyl radicals (•OH), hydrogen peroxide (H_2O_2), and superoxide (O^{2-}). But in contrast to other toxic metals, copper is also an essential trace nutrient because it serves as a cofactor for many enzymes due to its biologically suitable redox potential. This behaviour leads to harmful bacteria preventing Cu toxicity through buffering Cu in the cytosol strongly (Solioz et al. 2010). Briefly, this mechanism consists in the Cu(II) reduction to Cu(I) by metalloreductases at the cell surface and later transported across the plasma membrane through by the CopA Cu(I)-ATPase. Subsequently it is routed in the cytoplasm by small carrier soluble and specific proteins known as "copper-chaperones". The chaperone expression is highest at 0.75 mM media copper and declines at higher copper concentrations (Magnani and Solioz 2005). Little results have been reported regarding the resistance to copper in actinobacteria. Among this group, the genus *Streptomyces* has been the most studied mainly in *Streptomyces coelicolor* and *S. lividans* which showed Cu dependence in morphological development, being more pronounced in the last (Vijgenboom and Keijser 2002). Several studies show copper bioaccumulative capacity in this group of bacteria. It is important to mention that in an *Amycolpatosis* strain copper accumulative ability decreased with the presence of Cr(VI) (Achín Vera 2008, Colin et al. 2012). However, it is not clear at present as to the homeostasis mechanisms involved.

Other heavy metal with essential function in a number of enzymes is nickel (Ni), which like copper, has to be homeostatically regulated. Nickel is part of the active center of a nickel-containing superoxide dismutase found in *Streptomyces*, this enzyme plays a major role in defending the cell against oxidative stress (Ji-Sook et al. 2000).

Streptomycetes, are excellent producers of secondary metabolites such as antibiotics and pigments, and they modify the production in different growth media. Schmidt et al. (2009) studied seven *Streptomyces* nickel resistant strains and they try to predict the different molecular strategies that the strains employ to gain nickel resistance. Surprisingly they found that two different strain of *S. mirabilis* presented different behaviours as regards the resistance of nickel and production of pigments. One of them produced a higher amount of pigment and seemingly was not able to down-regulate the production, whereas the other one stopped the production at very high Ni(II) concentrations. Thus, these authors propose that the complex media stimulate pigment production and this interferes with Ni(II) resistance. This observation suggests the presence of two different mechanisms of resistance to nickel.

Both strains were shown to be more resistant to nickel when it was placed as $NiSO_4$ rather than $NiCl_2$, indicating that the use of different salts can lead to different results perhaps due to the ability of dissociation of these salts in the culture medium used. On the other hand, these authors also emphasize the finding that all strains of *S. mirabilis* resistant nickel also showed resistance to zinc [Zn(II)], so that these metals may share the mechanisms responsible for detoxification.

Chromium hexavalent [Cr(VI)], meanwhile, is present as chromate and because it has structural similarity to sulphate ions, some authors have suggested that chromate is able to use their transport routes to penetrate microbial cells (Cervantes and Campos-García 2007 as cited in Ramirez-Diaz et al. 2008). Inside the cell, Cr(VI) is readily reduced to Cr(III) by the action of various enzymatic or nonenzymatic activities; Cr(III) is unable to cross cellular membranes because it is positively charged under physiological condition. On the other hand, the Cr(III) generated may then exert diverse toxic effects in the cytoplasm (Cervantes et al. 2001, Ramirez-Diaz et al. 2008). However, other researchers have reported that the presence of sulphate ions in culture medium have no effect on Cr(VI) removal in several genera of gram negative bacteria (Wang and Xiao 1995, Zakaria et al. 2007) and gram positive bacteria (Liu et al. 2006) including bacteria of *Streptomyce*s genus (Laxman and More 2002). Recently a study shows that the addition of sulphate or phosphate anions to the culture medium increased specific Cr(VI) removal by *Streptomyces* sp. MC1. Nevertheless, the extracellular of total chromium concentrations did not change significantly under these culture conditions. Therefore, hexavalent chromium reduction in *Streptomyces* sp. MC1 may occur largely on the cell surface and this ion, as chromium oxyanions does not use the sulphate or phosphate transporter to penetrate into the cell in *Streptomyces* sp. MC1 (Pereira 2011). These results are inconsistent with those obtained previously by Poopal and Laxman (2009), while sulphate, nitrate, chloride and carbonate have no effect on chromate reduction during growth of *S. griseus*. Also it is important to mark that *Streptomyces* sp. MC1, was able to accumulate chromium under other culture conditions, and the chromium deposits was detected by transmission electron microscopy (Polti et al. 2010).

Recent studies in *Anthrobacer* sp. SUK 1201, isolated from chromute mine, showed that chromate reduction rate increased with Cr(VI) concentration and cell density but decreased with incubation due to Cr(VI) toxicity and reduction in cell number. This Cr(VI) reduction activity was effective in a narrow range of pH with an optimum of pH 7.0 and the optimum temperature was 35°C (Dey and Paul 2012).

Some actinobacteria require an electron donor as glucose, glycerol, acetate or lactose for Cr(VI) reduction and no reduction was observed in the absence of added electron donor (Laxman and More 2002, Ferro Orozco et al. 2007, Poopal and Laxman 2009). Recent work reported that specific consumption of glucose by *Streptomyces* sp. MC1 decreased in the presence of Cr(VI) in culture medium. However, no correlation between specific glucose consumption and specific Cr(VI) reduction was detected (Pereira 2011). In relation to chromate reductase activity, it was found and associated with the cell of different actinobacteria belonging to the genus *Streptomyces* and *Arthrobacter* (Desjardin et al. 2002, Megharaj et al. 2003, Poopal and Laxman 2009, Polti et al. 2010, Dey and Paul 2012). Evidently, Cr(VI)-reductase in these bacteria is associated mainly with the soluble fraction of the enzyme. *S. thermocarboxydus* NH50 and *Anthrobacer* sp.

SUK 1201 showed that Cr(VI) reduction was more rapid in the presence of glycerol than in the presence of glucose suggesting the existence of a soluble activity repressed partially by glucose (Desjardin et al. 2002, Dey and Paul 2012). However, *Streptomyces* sp. MC1 was exactly the opposite: these results suggest that Cr(VI) reduction in *Streptomyces* sp. MC1 could be an enzymatic activity regulated by glucose. In this strain chromate reductase showed activity within a broad pH (5.0 to 9.0) and temperature range (20 to 40°C) (Polti et al. 2010). This chromate reductase activity was increased significantly with NAD(P)H in addition to the cell-free extract (Megharaj et al 2003, Poopal and Laxman 2009, Polti et al. 2010). Unexpectedly, the result obtained by Desjardin et al. (2002) showed that the reducing activity is contained in the supernatant of cultures. This Cr(VI) reduction was enhanced drastically by the presence of a small concentration of cupric ions as $CuCl_2$ or $CuSO_4$. The presence of cations as Ni(II) or Cd(II) ions induced a slight reduction of Cr(VI). These authors suggest that the substance responsible for Cr(VI) reduction is not an enzyme because proteinase K has no effect on Cr(VI)-reduction. For this reason, the authors suggest that Cu(II) serve as a catalyst. Studies in *Amycolaptosis tucumanesis*, a copper resistant strain, also showed the ability to remove Cr(VI) in the bimetal system [Cu(II) and Cr(VI)]. Under this condition, the bacterium decreased the Cu(II) uptake but increased the Cr(VI) reduction, in relation to the simple metal system (Achín Vera 2008). The presence of 1 mM Cu(II) also enhanced the Cr(VI) reduction in *Anthrobacer* sp. SUK 1201 although growth slowed, while in other actinobacteria (*S. griseus*), the presence of this cation showed inhibitory effect on the Cr(VI) reduction (Poopal and Laxman 2009, Dey and Paul 2012). Both works reported that Cd, Ni, Co and Zn showed an inhibitory effect on this activity in varying degrees. It is interesting to mention that, in studies with cell free extracts of a gram negative bacteria, *Bacillus* sp., was able to Cr(VI) reduction under aerobic conditions, using NADH as an electron donor and produced a soluble Cr(VI)-reducing enzyme stimulated by the presence of Cu(II) (Camargo et al. 2003). Although the presence of Cu(II) favors the removal of Cr(VI) under some conditions, it is not yet clear as to the mechanisms by which it could positively regulate the action of the chromate reductase activity in some actinobacteria but it is important to note that the mechanisms of resistance in this group of bacteria would mainly consist of the external Cr(VI) reduction.

Cadmium (Cd) is an element of unknown biological function therefore the bacteria do not possess a specific homeostasis mechanism for this metal. Some works evidence that the actinobacteria can accumulate Cd, mainly by biosorption process. The Cd concentration and pH play an important role but were not observed to change with different temperatures (Puranik and Paknikar 1999, Yuan et al. 2009). Lebeau et al. (2002) showed that Cd biosorption increased when the microorganisms were immobilized. Other studies showed that one strain of *Streptomyces* genus (*Streptomyces tendae* F4) grow in the presence of Cd(II) concentration and this heavy metal was found mainly in the cell wall and the cytosolic fraction. Electron microscope images of this bacterium showed cytoplasm with dark granulate appearance (Siñeriz et al. 2009). This actinobacterium can produce a variety of hydroxamate siderophores simultaneaously. The siderophores, from the Ancient Greek nouns meaning "iron carrier", are small, high-affinity iron chelating compounds secreted by grasses and microorganisms such as bacteria and fungi. In addition to iron, siderophores can also form stable complexes with other heavy metal such as Cd, Cu, Pb and Zn, increasing the soluble metal concentration. For example, siderophores produced by *S. tendae* F4, enhanced the bioavailability of Cd and surprisingly their production is upregulated by the metal in the absence or presence of iron (Fe) concentration (Dimkpa et al. 2008, 2009). The interesting information is that a number of studies have demonstrated that metal resistance is often associated with antibiotic resistance due to physical linkage of the genes.

Obviously the successful removal of heavy metals by actinobacteria depends mainly on the specific interaction between actinobacteria and heavy metal. In this respect there is a lot of information but it is arbitrary, which requires more study and organization of existing results.

This chapter aims to highlight the importance of studying the main factors that influence the removal of metals. In some results it may be seen that a small variation in the composition of the culture medium affects the removal of metals negatively or positively. Whether this happens in one way or another depends on the specific or nonspecific heavy metal homeostasis mechanisms that bacteria possess, mechanisms not clearly understood yet.

The composition of the culture medium affects the speciation of metal and therefore their mobility and bioavailability, indirectly affecting the heavy metal removal of bacteria. All these variations can be easily studied in liquid minimal medium grown using factorial design with molecular tools such as proteomics and genomics approach will help to gather a little more knowledge of regulatory mechanisms. The identification and optimization or the cultural conditions that affect the heavy metal removal, represent key points for the development of a cost-competitive process. Once known these factors can be extrapolated to other media such as soil or sediment.

Remember that many of these contaminated sites contain more than one contaminant simultaneously. Then one can select the most suitable bacteria in each case and define the conditions that the bacteria can work in bioremediation process.

References Cited

Achín Vera, L.O. 2008. Remoción de Cu(II) y Cr(VI) por actinomycetes aislados de ambientes contaminados con metales pesados en sistemas de bimetal, Undergraduate thesis., Faculty of Biochemistry, Chemistry and Pharmacy, National University of Tucumán, Argentina.

Acosta, J.A., B. Jansen, K. Kalbitz, A. Faz and S. Martínez-Martínez. 2011. Salinity increases mobility of heavy metals in soils. Chemosphera. 85: 1318–1324.

Amoroso, M.J., R.G. Castro, F.J. Carlino, N.C. Romero, R.T. Hill and G. Oliver. 1998. Screening of heavy metal-tolerant actinomycetes isolated from the Salí River. J. Gen. Appl. Microbiol. 44: 129–132.

Amoroso, M.J., R. G. Castro, A. Durán, O. Peraud, G. Oliver and R.T. Hill 2001. Chromium accumulation by two *Streptomyces* spp. isolated from river sediments. J. Ind. Microbiol. Biotechnol. 26: 210–215.

Ansari, M.I. and A. Malik. 2010. Seasonal variation of different microorganisms with nickel and cadmium in the industrial wastewater and agricultural soils. Environ Monit Assess. 167: 151–163.

Baize, D. and T. Sterckeman. 2001. On the necessity of knowledge of the natural pedo-geochemical background content in the evaluation of the contamination of soils by trace elements. Sci. Total Environ. 264: 127–139.

Barceloux, D.G. 1999a. Chromium. Clin. Toxicol. 37: 173–194.

Barceloux, D.G. 1999b. Copper. Clin. Toxicol. 37: 217–230.

Boopathy, R. 2000. Factors limiting bioremediation technologies. Bioresour. Technol. 74: 63–67.

Bradl, H.B. 2004. Adsorption of heavy metal ions on soils and soils constituents. J. Colloid Interface Sci., 277: 1–18.

Brady, D. and J.R. Duncan 1994. Bioaccumulation of metal cations by *Saccharomyces cerevisiae*. Appl. Microbiol. Biotechnol. 41: 149–154.

Camargo, F., B.C. Okeke, F. Bento and W.T. Frankenberger. 2003. *In vitro* reduction of hexavalent chromium by a cell-free extract of *Bacillus* sp. ES 29 stimulated by Cu^{2+}. Appl. Microbiol. Biotechnol. 62: 569–573.

Cervantes, C. and J. Campos-García. 2007. Reduction and efflux of chromate by bacteria. *In:* D.H. Nies, S. Silver (eds.). Molecular Microbiology of Heavy Metals. Springer-Verlag, Berlin, pp. 407–420. Quoted in Ramirez-Diaz et al. 2008.

Cervantes C., J. Campos-García, S. Devars, F. Gutiérrez-Corona, H. Loza-Tavera, J.C. Torres-Guzmán and R. Moreno-Sánchez. 2001. Interactions of chromium with microorganisms and plants. FEMS Microbiol Reviews. 25: 335–347.

Colin, V.L., L.B. Villegas and C.M. Abate. 2012. Indigenous microorganisms as potential bioremediators for environments contaminated with heavy metals. Review. Int. Biodeterior. Biodegradation. 69: 28–37.

Das, A. and S. Mishra. 2008. Hexavalent Chromium (VI): Health hazards and Environmental Pollutant. J. Environ. Res. Dev 2: 386–92.

Desjardin, V., R. Bayard, N. Huck, A. Manceau and R. Gourdon. 2002. Effect of microbial activity on the mobility of chromium in soils. Waste Management. 22: 195–200.

Dey, S. and Paul A.K. 2012. Optimization of cultural conditions for growth associated chromate reduction by *Arthrobacter* sp. SUK 1201 isolated from chromite mine overburden. J. Hazard. Mater. 213: 200–206.

Diels, L., N. Van der Lelie and L. Bastiaens. 2002. New development in treatment of heavy metal contaminated soils. Rev. Environ. Sci. Biotechnol. 1: 75–82.

Dimkpa, C., A. Svatos, D. Merten, G. Büchel and E. Kothe. 2008. Hydroxamate siderophores produced by *Streptomyces acidiscabies* E13 bind nickel and promote growth in cowpea (*Vigna unguiculata L.*) under nickel stress. Can. J. Microbiol. 54: 163–72.

Dimkpa, C., D. Merten, A. Svatos, G. Büchel and E. Kothe. 2009. Siderophores mediate reduced and increased uptake of cadmium by *Streptomyces tendae* F4 and sunflower (*Helianthus annuus*), respectively. J. Appl. Microbiol. 107: 1687–96.

Dönmez, G. and Z. Aksu. 2001. Bioaccumulation of copper(II) and nickel(II) by the non-adapted and adapted growing *Candida* sp. Wat. Res. 35: 1425–1434.

Duffus, J.H. 2002. Heavy metals-a meaningless term? Pure Appl. Chem. 74: 793–807.

Dursun, A.Y., G. Uslu, Y. Cuci and Z. Aksu. 2003. Bioaccumulation of copper(II), lead(II) and chromium(VI) by growing *Aspergillus niger*. Process Biochem. 38: 1647–1651.

Ensign, J.C. 1992. Introduction to the Actinomycetes. In The Prokaryotes. A handbook on the biology of bacteria: ecophysiology, isolation, identification, applications. Edited by A. Ballows, H.G. Trüper, M. Dworkin, W. Harder & K.-H. Schleifer. Springer-Verlag. New York. pp. 811–815.

Ferro Orozco, M., E.M. Contreras, N.C. Bertola and N.E. Zaritzky. 2007. Hexavalent chromium removal using aerobic activated sludge batch systems added with powdered activated carbon. Water SA. 33: 239–244.

Fourest, E. and J.C. Roux. 1992. Heavy metal biosorption by fungal mycelial by-products: mechanisms and influence of pH. Appl. Microbiol. Biotechnol. 37: 399–403.

Gadd, G.M. 1992. Metals and microorganisms: A problem of definition. FEMS Microbiol. Lett. 100: 197–204.

Gadd, G.M. 2004. Microbial influence on metal mobility and application for bioremediation. Geoderma, 122: 109–119.

Goodfellow, M. and S.T. Williams. 1983. Ecology of actinomycetes. Ann. Rev. Microbiol. 37: 189–215.

Hawkes, H.E. and J.S. Webb. 1962. Geochemistry in Mineral Exp loration. New York: Harper.

Ji-Sook, H., O. So-Young, K.F. Chater You-Hee and R. Jumg-Hye. 2000. H_2O_2-sensitive Fur-like Represor CatR Regulating the major catalase gene in *Streptomyces coelicolor*. J. Biol. Chem. 275: 38254–38260.

John, D.A. and J.S. Leventhal.1995. Bioavailability of Metals. In Preliminary Compilation of Descriptive Geoenvironmental Mineral Deposit Models, ed. E. du Bray, pp. 10–18. USGS, Denver.

Johnston, D. and T. Cross. 1976. Actinomycetes in lake muds: dormant spores or metabolically active mycelium? Freshwater Biology. 6: 465–470.

Krantz-Rülcker, C., E. Frändberg and J. Schnürer. 1995. Metal loading and enzymatic degradation of fungal cell walls and chitin. BioMetals. 8: 12–18.

Lacey, J. 1997. *Actinomycetes* in compost. Ann. Agric. Environ. Med. 4: 113–121.

Lam Kin, S. 2006. Discovery of novel metabolites from marine actinomycetes. Cur. Op. Microbiol. 9: 245–251.

Laxman, R. and S. More. 2002. Reduction of hexavalent chromium by *Streptomyces griseus*. Miner. Eng. 15: 831–837.

Lebeau, T., D. Bagot, K. Jézéquel and B. Fabre. 2002. Cadmium biosorption by free and immobilised microorganisms cultivated in a liquid soil extract medium: effects of Cd, pH and techniques of culture. Sci. Total Environ. 291: 73–83.

Ledin, M. 2000. Accumulation of metals by microorganisms—processes and importance for soil systems. Earth-Science Reviews. 51: 1–31.

Liu, Y.G., W.H. Xu, G.M. Zeng, X. Li and H. Gao. 2006. Cr(VI) reduction by *Bacillus* sp. isolated from chromium landfi ll. Process Biochem. 41: 1981–1986.

Ma, Z., F.E. Jacobsen and D.P. Giedroc. 2009. Metal Transporters and Metal Sensors: How Coordination Chemistry Controls Bacterial Metal Homeostasis. Chem. Rev. 109: 4644–4681.

Magnani, D. and M. Solioz. 2005. Copper chaperone cycling and degradation in the regulation of the cop operon of *Enterococcus hirae*. BioMetals. 18: 407–12.

Malik A. 2004. Metal bioremediation through growing cells. Environment International. 30: 261–278.

Malik, A. and M. Ahmad. 2002. Seasonal variation in bacterial flora of the wastewater and soil in the vicinity of industrial area. Environ. Monit. and Assess. 73: 263–273.

Manivasagan, P., S. Gnanam, K. Sivakumar, T. Thangaradjou, S. Vijayalakshmi and T. Balasubramanian. 2009. Antimicrobial and Cytotoxic Activities of an Actinobacteria (*Streptomyces* sp. PM–32) Isolated from an Offshore Sediments of the Bay of Bengal in Tamilnadu. Adv. Biol. Res. 3: 231-236.

Marques, P.A., H.M. Pinheiro, J.A.Teixeira and M.F. Rosa. 1999. Removal efficiency of Cu^{2+}, Cd^{2+} and Pb^{2+} by waste brewery biomass: pH and cation association effects. Desalination. 124: 137–144.

McBride, M.J. and J.C. Ensign. 1987. Metabolism of endogenous trehalose by *Streptomyces griseus* spores and by spores or cells of other actinomycetes. J. Bacteriol. 169: 5002–7.

Megharaj, M., S. Avudainayagam and R. Naidu. 2003. Toxicity of hexavalent chromium and its reduction by bacteria isolated from soil contaminated with tannery waste. Curr. Microbiol. 47: 51–54.

Naja, G.M. and B. Volesky. 2010. Treatment of Metal-Bearing Effl uents: Removal and Recovery. *In:* L.K. Wang J.P. Chen, Y.T. Hung, N.K. Shammas (eds.). Handbook on heavy metals in the environment. Taylor & Francis, Boca Raton, pp. 247–291.

Nakic Zoran, K. Posavec and A. BacÏaniv. 2007. A Visual Basic Spreadsheet Macro for Geochemical Background Analysis. Ground water. 45: 642–647.

Nielsen, M.N. and A. Winding. 2002. Microorganisms as indicators of soil health, NERI Technical Report No. 388, National Environmental Research Institute, Copenhagen, Denmark.

Papafilippaki, A., D. Gasparatos , C. Haidouti and G. Stavroulakis. 2007. Total and bioavailable forms of Cu, Zn, Pb and Cr in agricultural soils: A study from the hydrological basin of Keritis, Chania, Greece, Global Nest. J. 9: 201–206.

Pereira, C.E. 2011. Estudio de la influencia de iones sulfatos y fosfatos en la remoción de Cr(VI) por *Streptomyces* sp. MC1, Undergraduate thesis, Faculty of Biochemistry, Chemistry and Pharmacy, National University of Tucumán, Argentina.

Polti, M., M.J. Amoroso and C. Abate. 2010. Chromate reductase activity in *Streptomyces* sp. MC1 J. Gen. Appl. Microbiol. 56: 11–18

Polti, M.A., M.J. Amoroso and C.M. Abate. 2007. Chromium(VI) resistance and removal by actinomycete strains isolated from sediments. Chemosphere. 67: 660–667.

Poopal, A.C. and R.S. Laxman. 2009. Studies on biological reduction of chromate by *Streptomyces griseus*. J. Hazard. Mater. 169: 539–545.

Puranik, P.R. and K.M. Paknikar. 1999. Biosorption of Lead, Cadmium, and Zinc by Citrobacter Strain MCM B-181: Characterization Studies. Biotechnol. Prog. 15: 228–237.

Raja, A. and P. Prabakarana. 2011. Actinomycetes and Drug-An Overview. Amer. J. Drug Disc. Develop. 1: 75–84.

Ramírez-Díaz, M.I., C. Díaz-Pérez, E. Vargas, H. Riveros-Rosas, J. Campos-García and C. Cervantes. 2008. Mechanisms of bacterial resistance to chromium compounds. BioMetals. 21: 321–332.

Schmidt, A., G. Haferburg, A. Schmidt, U. Lischke, D. Merten, F. Ghergel, G. Büchel and E. Kothe. 2009. Heavy metal resistance to the extreme: Streptomyces strains from a former uranium mining area. Chemie der Erde. 69: 35–44.

Shanker, A.K., C. Cervantes, H. Loza-Taverac and S. Avudainayagam. 2005. Chromium toxicity in plants. Environ. Internat. 31: 739–753.

Silóniz, M., L. Balsolobre, C. Alba, M. Valderrama and J. Peinado. 2002. Feasibility of copper uptake by the yeast *Pichia guilliermondii* isolated form sewage sludge. Res. Microbiol. 153: 173–180.

Siñeriz Louis, M., E. Kothe and C.M. Abate. 2009. Cadmium biosorption by *Streptomyces* sp. F4 isolated from former uranium mine. J. Basic Microbiol. 49: 55–62.

Solioz, M., H.K. Abicht, M. Mermod and S. Mancini. 2010. Response of Gram-positive bacteria to copper stress. J. Biol. Inorg. Chem. 15: 3–14.

Srivastava, S., S. Prakash and M.M. Srivastava. 1999. Chromium mobilization and plant availability—the impact of organic complexing agents. Plant Soil. 212: 203–208.

Valls, M. and V. Lorenzo. 2002. Exploiting the genetic and biochemical capacities of bacteria for the remediation of heavy metal pollution. FEMS Microbiol. Rev. 26: 327–328.

Vijgenboom, E. and B. Keijser. 2002. Copper and the Morphological development of *Streptomyces,* Handbook of copper pharmacology and toxicology, Humana Press.

Villegas, L.B., M.J. Amoroso and L.I.C. Figueroa. 2005. Copper tolerant yeasts isolated from polluted area of Argentina. J. Basic Microbiol. 45: 381–391.

Villegas, L.B., M.J. Amoroso and L.I.C. Figueroa. 2011. Interaction of Copper or Chromium with Yeasts: Potential Application on Polluted Environmental Clean Up. Chapter 6, In Bioremediation: Biotechnology, Engineering and Environmental Management. Alexander C. Mason (Ed.). Nova Science Publisher. New York.

Violante, A., V. Cozzolino, L. Perelomov, A.G. Caporale and M. Pigna. 2010. Mobility and biovailability of heavy metals and metalloids in soil environmments. Soil. Sci. Plant Nutr. 10: 268–292.

Viti, C., A. Pace and L. Giovannetti. 2003. Characterization of Cr (VI)-resistant bacteria isolated from chromium-contaminated soil by tannery activity. Curr. Microbiol. 46: 1–5.

Wang, J.L. and C. Chen. 2009. Biosorbents for heavy metals removal and their future. Biotechnol. Adv. 27: 195–226

Wang, Y.T. and C. Xiao. 1995. Factors affecting hexavalent chromium reduction in pure cultures of bacteria. Water Res. 29: 2467–2474.

Yuan, H.P., J.H. Zhang, Z.M. Lu, H. Min and C. Wu. 2009. Studies on biosorption equilibrium and kinetics of Cd^{2+} by *Streptomyces* sp. K33 and HL-12. J. Hazard. Mater. 164: 423–431.

Zayed, A.M and N. Terry. 2003. Chromium in the environment: factors affecting biological remediation. Plant and Soil. 249: 139–156.

CHAPTER 3

Morphological Changes and Oxidative Stress in Actinobacteria during Removal of Heavy Metals

Verónica L. Colin,[1,2,*] Liliana B. Villegas,[1,3] Claudia E. Pereira,[1] María J. Amoroso[1,3,4] and Carlos M. Abate[1,4,5]

Introduction

Heavy metal pollution is one the most serious environmental problems facing our planet today and immediate solutions are needed to tackle it. Heavy metals such as copper (Cu) and chromium (Cr) play an important role as trace elements in biochemical reactions but these metals are toxic at higher concentrations. Cu is a very versatile heavy metal, able to cycle between two redox states, oxidized Cu(II) and reduced Cu(I), with Cu(I) being highly

[1]Planta Piloto de Procesos Industriales y Microbiológicos (PROIMI), CONICET, Av. Belgrano y Pje. Caseros, 4000, Tucumán, Argentina.
[2]Universidad de San Pablo-Tucumán, Argentina.
[3]Facultad de Ciencias de la Salud. Universidad del Norte Santo Tomás de Aquino-Tucumán, Argentina.
[4]Facultad de Bioquímica, Química y Farmacia, Universidad Nacional de Tucumán, 4000 Tucumán, Argentina.
[5]Facultad de Ciencias Naturales e Instituto Miguel Lillo, Universidad Nacional de Tucumán, 4000 Tucumán, Argentina.
*Corresponding author: veronicacollin @yahoo.com.ar

unstable. All living organisms require Cu as a catalytic cofactor for basic biological process such as the respiration (Puig and Thiele 2002); however, both prokaryotic and eukaryotic cells possess homeostatic mechanisms to regulate the concentration of Cu and minimize the toxic effects produced by excessive levels of the same.

Cr is the seventh most abundant element on earth and occurs in diverse oxidation states, but only Cr(VI) and Cr(III) are ecologically important (Cefalu and Hu 2004). Although Cr(III) is an essential trace element, at high concentrations it has negative effects on cellular structures. On the other hand, Cr(VI) which is more mobile and soluble in water than Cr(III), is always toxic to living organisms and has been listed as a priority pollutant and a human carcinogen by the US Environmental Protection Agency.

In our region (Tucumán-Argentina), mining and industrial activities have led to large scale Cu and Cr contamination (Viti et al. 2003, Cefalu and Hu 2004, Albarracín et al. 2005, Cheung and Gu 2007). At present, the conventional technologies used to remove heavy metals from the environment involve physicochemical processes, which are costly and require large amounts of energy and specialized equipment. Microbe-based heavy metals removal is now considered to be an effective alternative method to the conventional processes. In this connection, actinobacteria, a group of free-living saprophytes microorganisms found predominantly in soils (Kavitha and Vijayalakshmi 2007, Mohan and Vijayakumar 2008), have been identified as potential bioremediators of environment contaminated with heavy metal (Colin et al. 2012). Because of their filamentous nature, actinobacteria are considered an intermediate group between bacteria and fungi (Pandey et al. 2004). Their metabolic diversity, particular growth characteristics, mycelial form and relatively rapid ability to colonize selective substrates, make them well-suited for use as agents for bioremediation of both inorganic and organic compounds (Albarracín et al. 2005, 2007, 2010a,b, Polti et al. 2007, 2009, 2010, 2011, Dávila Costa et al. 2011, Benimeli et al. 2003, 2008, Fuentes et al. 2010, 2011).

Copper and chromium are redox-active metals, which can act as catalysts in the formation of oxygen reactive species (ROS). Oxidative stress caused by the presence of ROS, which the cell is unable to counterbalance, could result in damage to one or more biomolecules including DNA, RNA, proteins and lipids. It is clear that a more lucid understanding of the mechanisms of metals toxicity on living cells might lead to the development of novel technologies to mitigate such toxicity. To this end, microorganisms are useful models for the study of various aspects of oxidative stress. It is due to that the principles of cellular defense are very similar in all types of organisms (Poljsak et al. 2010). Antioxidant ability of many actinobacteria has been referring in terms of their activity levels of antioxidant enzymes such as superoxide dismutase (SOD) and catalase (CAT), as well as of other enzymes and proteins that act to decrease ROS concentrations (Villegas et al. 2007, Dávila-Costa et al. 2011). Likewise, studies related to the inexorable morphological alterations in diverse microorganisms have been associated with the oxidative stress caused by the presence of heavy metals (Ackerley et al. 2006, Chourey et al. 2006, Villegas et al. 2009, Francisco et al. 2010). These morphological changes are frequently considered an adaptive response of microorganisms against stress conditions. The present chapter discusses both morphological alterations and antioxidant enzymes levels changes in native actinobacteria of our region, during oxidative stress caused by Cu(II) and/or Cr(VI).

Heavy Metals and Indigenous Actinobacteria Morphology

Cellular morphology is frequently considered as one of the key factors of physiological studies of filamentous microorganisms (Martins et al. 2004). In this connection, morphological characterization of actinobacteria grown either in the presence or absence of heavy metals could give proportionate relevant information about their physiological state. Morphological changes in two strains of our collection culture as *Streptomyces* sp. MC1 resistant to Cr(VI) (Polti et al. 2007) and *Amycolatopsis tucumanensis* DSM 45259^T resistant to Cu(II) (Albarracín et al. 2010a) under heavy metal exposure were therefore, widely studied.

Morphology of *Streptomyces* sp. MC1 under Cr(VI)-exposure

Streptomyces sp. MC1 showed, in previous studies, the ability to reduce Cr(VI) to less toxic species (Pereira 2010). This biological reduction seems to occur largely on the cell surface. In fact, chromate reductase activity was effectively detected in all cellular fractions of this strain by Polti et al. (2010). Based upon these findings, morphology of this strain was exanimate under Cr(VI)-exposure. Scanning Electron Microscopy (SEM) of this strain showed the typical branching mycelia of the genus *Streptomyces* that fragments into rod-like elements (Polti et al. 2011). However, these authors noted a slight difference in morphology between Cr(VI) exposed and non-exposed *Streptomyces* sp. MC1 filaments, since the exposed were rounder and shorter than non-exposed ones. Recently, Pereira (2010) demonstrated that the addition of sulphate or phosphate ions to the liquid minimal medium (MM) at final concentrations of 5 mM, significantly increase the ability to remove Cr(VI) by this strain from the minimal medium (MM). In this connection, digital image analysis was used to quantify the morphology of *Streptomyces* sp. MC1 cultivated in MM (called control C) and also in MM supplemented with sulphate (S) or phosphate (P). Cells cultivated under the same conditions, but with Cr(VI) to a final concentration of 20 mg L^{-1}, labeled as CCr, SCr, and PCr respectively, were also analyzed.

Microscopic morphology was quantified by estimation of two microscopic parameters such as hyphal diameter (D) and hyphal growth unit length (L/N) (Trinci 1974). Lengths of filaments were obtained by using the Nikon Eclipse Net software package version 1.20 with a pre-set calibration in µm. The samples were analyzed at a total magnification of 2500×. Finally, macroscopic morphology was analyzed by determining the average diameters of bacterial floccules. Although the morphological parameters as D and L/N were not significantly altered in S and P, compared to the C (Table 1), the presence of Cr(VI) in the MM encouraged alteration inexorable in cellular morphology, mainly on branching degree (Fig. 1). Thus, L/N was decreased between a 41 to 50% in CCr(VI), SCr(VI) and PCr(VI), compared to their respective controls (C, S, and P) (Table 1). Branching length was also decreased about 70% in cells Cr(VI)-exposed (data not shown). Finally, D values, were not modified by the supplement of Cr(VI) to the MM.

Oxidative stress faced by a cell exposure to the Cr(VI) produce inevitable alterations of the growth and microscopic morphology. As a result, the morphological changes undergone by the cells can be linked to the stress level that causes metal. Ackerley et al. (2006), for example, reported on morphological alteration of *Escherichia coli* K-12 cells exposed to chromate, which exhibited extreme filamentous morphology. Cells that had been preadapted by overnight growth in the presence of chromate were less stressed than non-adapted controls. In the same years, Chourey et al. (2006) studied cellular morphology of Cr(VI)-exposed *Shewanella oneidensis* MR-1 cells, compared to the non-exposed control cells. In contrast to the untreated control cells, Cr(VI)-exposed cells formed apparently aseptate, non-motile

Table 1. *Streptomyces sp.* MC1 morphological parameters determined in minimal medium (MM).
(D): Hyphal diameter. (L): Total hyphal length. (N): Number of tips. (L/N): Hyphal growth unit length. Results are presented as the mean ± standard deviation calculated from thirty independent experiments. The values with different letters are significantly different (P<0.05).

Culture Conditions	Morphological Parameters	
	D (µm)	L/N (µm)
C	0.77 ± 0.04 a	43.0 ± 3.5 a
CCr	0.76 ± 0.10 a	60.5 ± 7.8 b
S	0.79 ± 0.06 a	40.5 ± 6.2 a
SCr	0.82 ± 0.07 a	58.3 ± 8.2 b
F	0.76 ± 0.10 a	40.1 ± 3.3 a
FCr	0.77 ± 0.11 a	60.0 ± 8.9 b

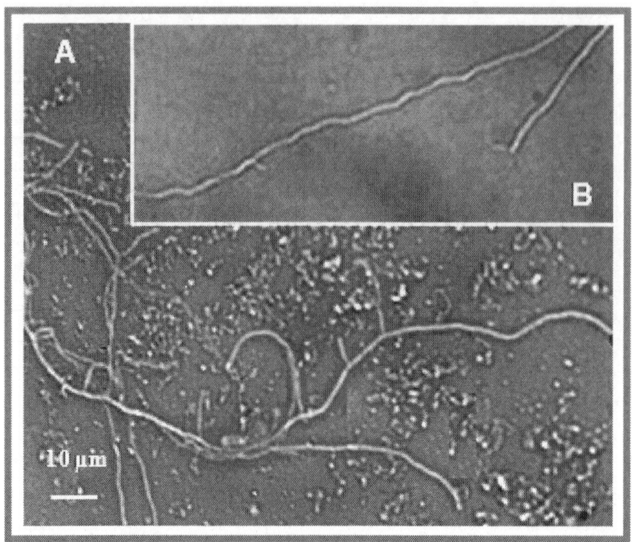

Figure 1. Differential interference contrast images of *Streptomyces* MC1 sp. cells cultivated in minimal medium (MM) in the presence (A) as well as in absence of Cr(VI) (B). Both panels are at the same magnification as (A), where the bar represents 10 μm (Pereira 2010).

filaments that tended to aggregate. Recently, Francisco et al. (2010) studied the cellular morphology of *Ochrobactrum tritici* grown in a minimal medium in the presence of chromate or dichromate, compared to control cells without the addition of chromium. These authors found that while the chromate-exposed cells did not show morphological alterations compared to control, the dichromate-exposed cells showed inexorable morphological alterations. In our particular case, a decrease in hyphal branching ability was observed in the presence of Cr(VI), which was reflected in an increase in L/N. The branching process is a dynamic and complex mechanism that involves the coordinated action of numerous enzymes and cytoplasm precursors (Barreteau et al. 2008, Bouhss et al. 2008), where the initiation of new branching could become challenging for the cells under inhospitable conditions. Therefore, a decrease in the branching degree observed in Cr(VI)-exposed cells, could represent an adaptive response by stressed cells.

In terms of macroscopic morphology of this strain, the reference cultures (C, S, and P) did not exhibit significant differences. In all three of these conditions, bacterial flocculation was observed, creating structures that ranged from "compact pellets" to loose mycelia aggregates that could best be described as "clumps" (Fig. 2A). In contrast, only compact pellets of a homogeneous size were observed in the presence of Cr(VI): CCr, SCr, and PCr (Fig. 2B).

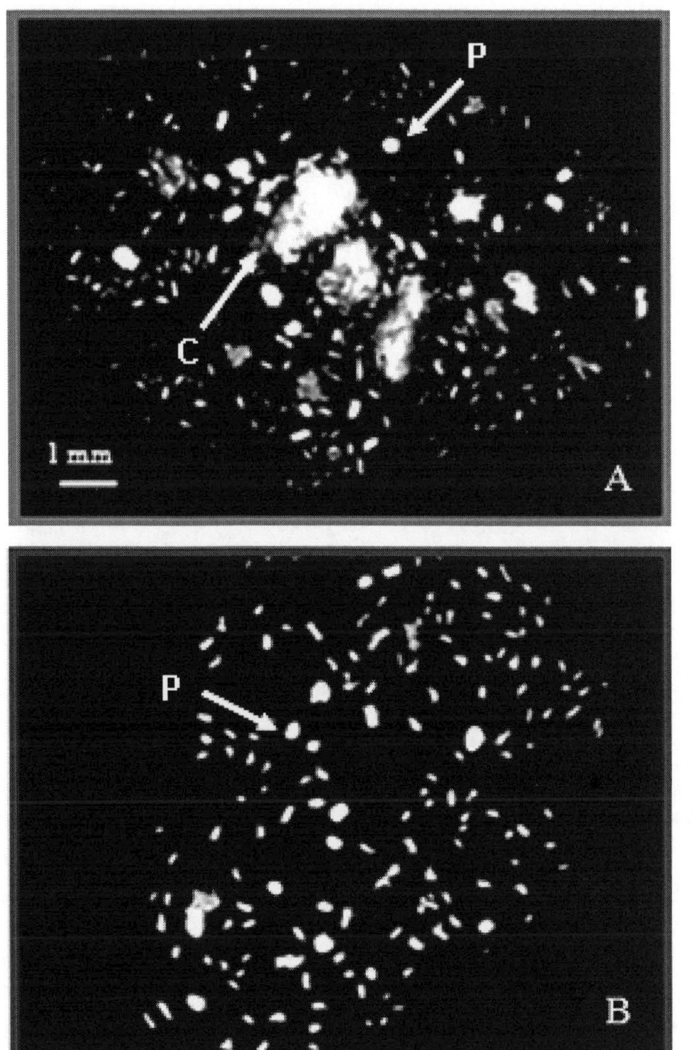

Figure 2. Macroscopic morphology of *Streptomyces* sp. MC1 cultivated in minimal medium (MM) either in absence (A), or in the presence of Cr(VI) (B). **C**: Clumps. **P**: Pellets. Both panels are at the same magnification as (B), where the bar represents 1 mm (Pereira 2010).

The quantification of mycelial aggregation of the filamentous microorganisms grown in submerged cultures can also be performed (Papagianni 2006). In order to characterize the mycelial aggregation in *Streptomyces* sp. MC1 liquid media, either in the presence or absence of Cr(VI), histograms were designed based upon the frequency distributions of the largest mean diameter of the bacterial floccules (Fig. 3). In the absence

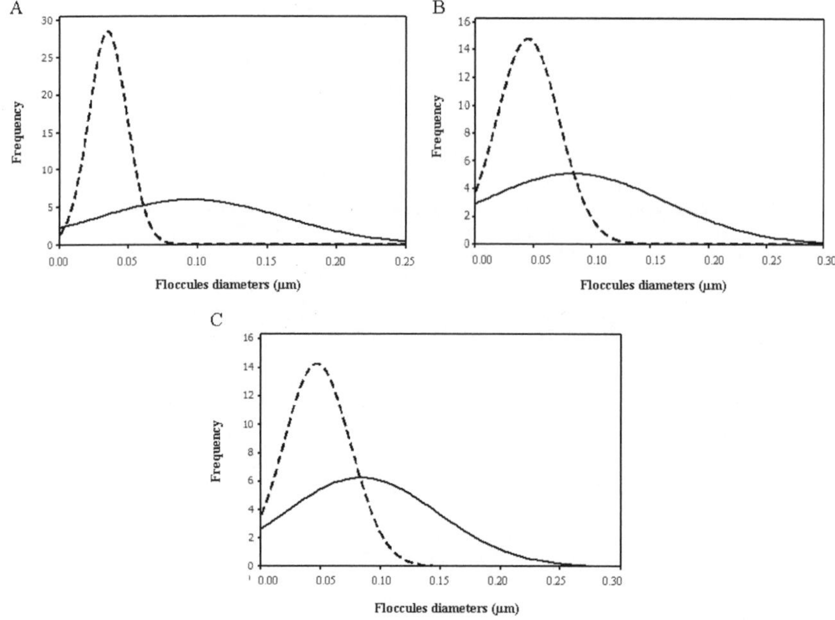

Figure 3. Histograms showing the distribution of the largest mean diameter of bacterial floccules (n=100) from *Streptomyces* sp. MC1, obtained in minimal medium (MM) in the absence (—) and presence of 20 mg L^{-1} Cr(VI) (– –). (A) Without added sulphate and phosphate. (B) In the presence of 5mM sulphate. (C) In the presence of 5mM phosphate (Pereira 2010).

of metal, a mixture of small pellets (0.042 ± 0.006) and prominent clumps (0.084 ± 0.006) was observed. In the presence of Cr(VI), the degree of bacterial flocculation increased and only compact pellets of a homogeneous size (0.042 ± 0.006) were observed. On the basis of these observations, it has been concluded that bacterial floccules obtained in the absence of metal were generally more loose and larger than those developed in the presence of Cr. This was reflected in the 95% confidence intervals for the largest mean diameter of these structures, as well as in the relative positions of the peaks for the fitted normal distributions.

Amycolatopsis tucumanensis DSM 45259^T morphology upon Cu(II)-exposure

Concerning the *A. tucumanensis* DSM 45259^T, a study the effect of Cu(II) on the morphological characteristics of this Cu(II)-resistant strain and of a non Cu(II)-resistant strain as *Amycolatopsis eurytherma* DSM 44348^T was carried out by Dávila-Costa et al. (2011). *A. tucumanensis* DSM 45259^T and *A. eurytherma* DSM 44348^T were cultured on agar-MM with increasing concentrations of Cu(II) (0 to 30 mg L^{-1}). These authors note that both strains growing in the absence of Cu(II), produced branched substrate mycelium as well as a profuse aerial mycelium, both with types of hyphae fragmented into squarish rod-shaped elements. In contrast, it has been pointed out that when Cu(II) is added to the MM, *A. tucumanensis* DSM 45259^T morphology was slightly modified, while *A. eurytherma* DSM 44348^T displayed a patent morphological alteration. Indeed, it was observed that a wide occurrence of hyphae undergo cellular degeneration and these hyphae could be distinguished by the aberrant shapes they displayed. These results, clearly demonstrated the inexorable morphological alteration in the presence of Cu(II) in the sensitive strain whereas in the resistant strain, the morphology was not affected practically.

Heavy metals and antioxidant enzymes

Actinobacteria, like other aerobic organisms, generate ROS endogenously as by products of metabolic processes. These forms of oxygen are highly damaging to cellular constituents, including DNA, lipids and proteins (Imlay 2003). ROS can also be formed by exposure of cells either to ionizing radiation, redox-cycling chemicals present in the environment or by exposure to heavy metals. Through these mechanisms, all aerobically growing organisms are continuously exposed to reactive oxidants, under physiological conditions; therefore, endogenous antioxidant cellular defence systems are necessary to maintain ROS at basal, unharmful level and repair damages. Oxidative molecules are mainly detoxified via SOD, which in two steps dismutase the superoxide to O_2 and H_2O_2. Subsequently, the H_2O_2 is detoxified by CAT. These enzymes create the first and the most important line of antioxidant defence (Lushchak and Gospodaryov 2005). However, when ROS production is accelerated or there is a defect in the antioxidant system ROS, an oxidative stress results (Costa and Moradas-Ferreira 2001). Based on these antecedents, the effects of Cr(VI) and Cu(II) on the SOD and CAT activity levels from *Streptomyces* sp. MC1 and *A. tucumanensis* DSM 45259^T was evaluated by Villegas et al. (2007). Exponential phase cells were broken in a French press at 20,000 psi, and the supernatant or cellular extract, was then used for enzymatic activity determination. CAT activity was determinated according to (Izawa et al. 1996) while gel staining and quantification SOD activity assays were performed as described by Beauchamp and Fridovich (1971). Proteins in the cellular extracts were detected by Bradford reagent (Bradford 1976). The quantification assays in native PAGE revealed the number as well as the intensity of the bands with SOD activity (Fig. 4). Thus, whereas *Streptomyces* sp. MC1 presented only one band with SOD activity (Fig. 6A), which was inhibited in presence of H_2O_2 (lane 7 and 9) but not by NaN_3 (lane 8 and 10); two bands with SOD

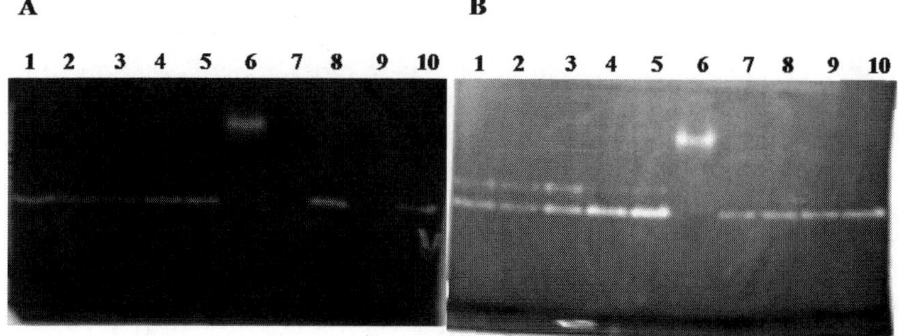

Figure 4. SOD activity of exponential phase cells from *Streptomyces* sp. MC1 (A) and *A. tucumanensis* DSM 45259^T (B) grown in MM with 10 and 20 ppm of Cu(II) or Cr(VI). Control cells (Lane 1); cells incubated with 10 ppm of Cr(VI) (Lane 2); cells incubated with 20 ppm of Cr(VI) (Lane 3); cells incubated with 10 ppm of Cu(II) (Lane 4); cells incubated with 20 ppm of Cu(II) (Lane 5); bacterial origin Mn-SOD (Lane 6); cells incubated with 20 ppm of Cr(VI) + H_2O_2 (Lane 7); cells incubated with 20 ppm of Cr(VI) + NaN_3 (Lane 8); cells incubated with 20 ppm of Cu(II) + H_2O_2 (Lane 9); and cells incubated with 20 ppm of Cu(II) + NaN_3 (Lane 10) (Villegas et al. 2007).

activity were observed for *A. tucumanensis* DSM 45259^T (Fig. 6B), one of which was inhibited by H_2O_2 (lane 7 and 9) as well as by NaN_3 (lane 8 and 10). However, no inhibition was observed for the other band with H_2O_2 and NaN_3 (lane 7 to 10).

In terms of CAT activity, this increase in time-dependent manner parallels the increase in the concentration of the metals (Fig. 5) (Villegas et al. 2007). However, these strains showed important differences between them. *Streptomyces* sp. MC1 showed, for example, values of CAT activity highly increased when it is cultivated in the presence of Cu(II) compared to control cells without metal. CAT activity against Cr(VI) was, however, similar to the control cells without metal (Fig. 5B).

Concerning *A. tucumanensis* DSM 45259^T, it showed the same behavior as *Streptomyces* sp. MC1 but against Cr(VI), with values of CAT activity highly increased in the presence of metal compared to control cells. In addition, the values of CAT activity were similar for both assayed concentrations (Fig. 5A).

Dávila Costa et al. (2011) performed comparative studies on ROS production as well as SOD and CAT activities of *A. tucumanensis* DSM 45259^T and one strain sensitive as *A eurytherma* DSM 44348^T against Cu(II). These authors noted that the increase in ROS production from the basal level to the stress conditions was lesser in the Cu(II)-resistant strain than in the metal-sensitive strain. In addition, detected levels of antioxidants enzymes as SOD and CAT in *A. tucumanensis* DSM 45259^T were significantly high relative to *A. eurytherma* DSM 45259^T. On the other hand, levels of thioredoxin reductase enzyme and metallothioneins, which are essential

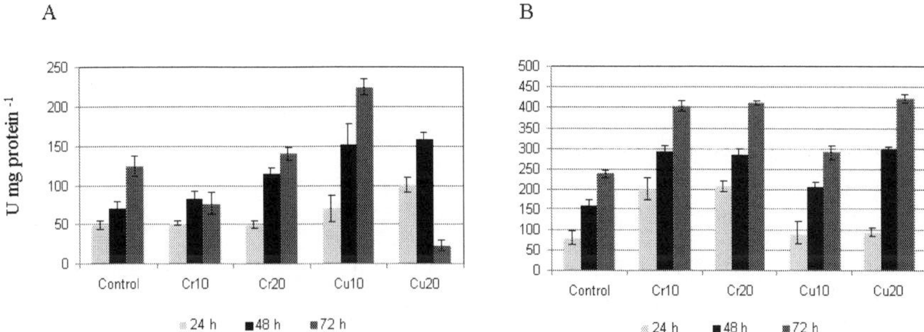

Figure 5. CAT activity of exponential phase cells from *Streptomyces* sp. MC1 (A) and *A. tucumanensis* DSM 45259^T (B), grown in minimal medium (MM) with 10 and 20 ppm of Cu(II) or Cr(VI) (Villegas et al. 2007).

for maintaining a cytoplasmatic reductive environment (Attarian et al. 2009), were found to be greater in Cu(II)-resistant strain than in sensitive strain. Finally, as mentioned above, the drastic raise of ROS production in *A. eurytherma* DSM 45259^T was accompanied by inexorable morphological alteration, in contrast to the observed in *A. tucumanensis* DSM 45259^T whose morphology, was not practically modified.

Conclusion

As discussed in the present chapter, antioxidant ability of native actinobacteria such as *Streptomyces* sp. MC1 and *Amycolatopsis tucumanensis* DSM 45259^T has been mainly referred in terms of SOD and CAT activities levels. Important evidence on the efficiency of the antioxidant defense system, which permits to these microorganisms to survive in environment polluted with heavy metals, has been widely reported by our research group. In addition, whereas inexorable morphological alterations were observed as result of heavy metal toxicity on sensitive microorganisms, resistant-strains were not significantly affected, or less, discrete morphological changes were detected. These studies seem to be highly promising in terms of the creation of platforms that encourage the development of bioremediation processes based on metal-resistant microorganisms. The most important characteristic of this approach is that the system is environment friendly. However, more studies could be required to develop future applications of native strains for cleaning up our environment and we continue to work on this subject.

Acknowledgments

We gratefully acknowledge the financial assistance of CIUNT D/401, and CONICET, Argentina.

References

Ackerley, D.F., Y. Barak, S.V. Lynch, J. Curtin and A. Matin. 2006. Effect of chromate stress on *Escherichia coli* K-12. J. Bacteriol. 188: 3371–3381.

Albarracín, V.H., M.J. Amoroso and C.M. Abate. 2005. Isolation and characterization of indigenous copper-resistant actinomycete strains. Chem. Erde-Geochem. 65: 145–156.

Albarracín, V.H. 2007. Estudios de los aspectos fisiológicos, bioquímicos y moleculares de la resistencia al cobre en cepas de actinomycetes, PhD thesis, Faculty of Biochemistry, Chemistry and Pharmacy, National University of Tucumán, Argentina.

Albarracín, V.H., P. Alonso-Vega, M.E. Trujillo, M.J. Amoroso and C.M. Abate. 2010a. *Amycolatopsis tucumanensis* sp. nov., a novel copper resistant actinobacterium isolated from polluted sediments in Tucumán, Argentina. Int. J. Syst. Evol. Micr. 60: 397–401.

Albarracín, V.H., M.J. Amoroso and C.M. Abate. 2010b. Bioaugmentation of copper polluted soil microcosms with *Amycolatopsis tucumanensis* to diminish phytoavailable copper for *Zea mays* plants. Chemosphere. 79: 131–137.

Attarian, R., C. Bennie, H. Bach and Y. Av-Gay. 2009. Glutathione disulfide and S-nitrosoglutathione detoxification by *Mycobacterium tuberculosis* thioredoxin system. FEBS Lett. 583: 3215–3220.

Barreteau, H., A. Kovac, A. Boniface, M. Sova, S. Gobec and D. Blanot. 2008. Cytoplasmic step of pectidoglycan biosynthesis. FEMS Microbiol. Rev. 32: 168–207.

Beauchamp, C. and I. Fridovich. 1971. Superoxide dismutase: improved assays and an assay applicable to acrylamide gels. Anal. Biochem. 44: 276–287.

Benimeli, C.S., M.J. Amoroso, A.P. Chaile and G.R. Castro. 2003. Isolation of four aquatic streptomycetes strains capable of growth on organochlorine pesticides. Bioresource Technol. 89: 133.138.

Benimeli, C., M.S. Fuentes, C.M. Abate and M.J. Amoroso. 2008. Bioremediation of lindane-contaminated soil by *Streptomyces* sp. M7 and its effects on Zea mays growth. Int. Biodeter. Biodegradation. 61: 233–239.

Bouhss, A., A.E. Trunkfield, D.T. Bugg and D. Mengin-Lecreulx. 2008. The biosynthesis of peptidoglycan lipid-linked intermediates. FEMS Microbiol. Rev. 32: 208–233.

Bradford, M.M. 1976. A rapid and sensitive method for the quantification of microgram quantities of protein utilizing the principle of protein-dye binding. Anal. Biochem. 72: 248–254.

Cefalu, W.T. and F.B. Hu. 2004. Role of chromium in human health and in diabetes. Diabetes Care. 27: 2741–2751.

Cheung, K.H. and J.D. Gu. 2007. Mechanism of hexavalent chromium detoxification by microorganisms and bioremediation application potential: a review. Int. Biodeter. Biodegradation. 59: 8–15.

Chourey, K., M.R. Thompson, J. Morrell-Falvey, N.C. VerBerkmoes, S.D. Brown, M. Shah, J. Zhou, M. Doktycz, R.L. Hettich and D.K. Thompson. 2006. Global Molecular and Morphological Effects of 24-Hour Chromium(VI) Exposure on *Shewanella oneidensis* MR-1. Appl. Environ. Microbiol. 72: 6331–6344.

Colin, V.L., L.B. Villegas and C.M. Abate. 2012. Indigenous microorganisms as potential bioremediators for environments contaminated with heavy metals. Int. Biodeter. Biodegradation. 69: 28–37.

Costa, V. and P. Moradas-Ferreira. 2001. Oxidative stress and signal transduction in *Saccharomyces cerevisiae*: insights into ageing, apoptosis and diseases. Mol. Aspects Med. 22: 217–246.

Dávila Costa, J.S., V.H. Albarracín and C.M. Abate. 2011. Responses of environmental *Amycolatopsis* strains to copper stress. Ecotox. Environ. Safe. 74: 2020–2028.

Francisco, R., A. Moreno and P.V. Morais. 2010. Different physiological responses to chromate and dichromate in the chromium resistant and reducing strain *Ochrobactrum tritici* 5bvl1. Biometals. 23: 713–725.

Fuentes, M.S., C.S. Benimelli, S. Cuozzo and M.J. Amoroso. 2010. Isolation of pesticide-degrading actinomycetes from a contaminated site: Bacterial growth, removal and dechlorination of organochlorine pesticides. Int. Biodeter. Biodegradation. 64: 434–441.

Fuentes, M.S., J.M. Saez, C.S. Benimelli and M.J. Amoroso. 2011. Lindane biodegradation by defined consortia of indigenous *Streptomyces* strains. Water Air Soil Poll. 22: 217–231.

Imlay, J.A. 2003. Pathways of oxidative damage. Annu. Rev. Microbiol. 57: 395–418.

Izawa, S., Y. Inoue and A. Kimura. 1996. Importance of catalase in the adaptive response to hydrogen peroxide: analysis of acatalasaemic *Saccaromyces cerevisiae*. Biochem. J. 320: 61–67.

Kavitha, M. and M. Vijayalakshmi. 2007. Studies on Cultural, Physiological and Antimicrobial Activities of *Streptomyces rochei*. J. Appl. Sci. Res. 3: 2026–2029.

Lushchak, L. and D. Gospodaryov. 2005. Catalases protect cellular proteins form oxidative modification in *Saccharomyces cerevisiae*. Cell Biol. Int. 29: 187–192.

Martins, A.M.P., K. Pagilla, J.J. Heijnen and M.C.M. van Loosdrecht. 2004. Filamentous bulkingsludge: a critical review. Water Research. 38: 793–817.

Mohan, R. and R. Vijayakumar. 2008. Isolation and characterization of marine antagonistic Actinomycetes from West coast of India. Facta Universitatis 15: 13–19.

Pandey, B. 2004. Studies on the antimicrobial activity of actinomycetes isolated from Khumbu region of Nepal, PhD thesis, Tribhuvan University, Kathmandu, Nepal.

Papagianni, M. 2006. Quantification of the fractal nature of mycelial aggregation in *Aspergillus niger* submerged cultures. Microb. Cell Fact. 5: 3, DOI: 10.1186/1475-2859-5-5.

Pereira, C.E. 2010. Estudio de la influencia de iones sulfatos y fosfatos en la remoción de Cr(VI) por *Streptomyces* sp. MC1, Undergraduate thesis, Faculty of Biochemistry, Chemistry and Pharmacy, National University of Tucumán, Argentina.

Poljsak, B., I. Pócsi, P. Raspor and M. Pesti. 2010. Interference of chromium with biological systems in yeast and fungi: a review. J. Basic Microbiol. 50: 21–36.

Polti, M.A., M.J. Amoroso and C.M. Abate. 2007. Chromium(VI) resistance and removal by actinomycete strains isolated from sediments. Chemosphere. 67: 660–667.

Polti, M.A., R.O. García, M.J. Amoroso and C.M. Abate. 2009. Bioremediation of chromium (VI) contaminated soil by *Streptomyces* sp. MC1. J. Basic Microbiol. 49: 285–292.

Polti, M.A., M.J. Amoroso and C.M. Abate. 2010. Chromate reductase activity in *Streptomyces* sp. MC1. J. Gen. Appl. Microbiol. 56: 11–18.

Polti, M.A., M.J. Amoroso and C.M. Abate. 2011. Intracellular chromium accumulation by *Streptomyces* sp. MC1. Water, Air Soil Poll. 214: 49–57.

Puig, S. and D.J. Thiele. 2002. Molecular mechanisms of copper uptake and distribution. Curr. Opin. Chem. Biol. 6: 171–180.

Trinci, A.P.J. 1974. A study of the kinetics of hyphal extension and branch initiation of fungal mycelia. J. Gen. Appl. Microbiol. 81: 225–236.

Villegas, L.B., M.J. Amoroso and C.M. Abate. 2007. Effect of Cu(II) and Cr(VI) on oxidative stress actinomycetes strains isolated from polluted area. XLIII Reunion Annual of SAIB, Mar del Plata-Argentina.

Villegas L.B., M.J. Amoroso and L.I. Castellanos de Figueroa. 2009. Responses of *Candida fukuyamaensis* RCL-3 and *Rhodotorula mucilaginosa* RCL-11 to copper stress. J. Basic Microbiol. 49: 395–403.

Viti, C., A. Pace and L. Giovannetti. 2003. Characterization of Cr (VI)-resistant bacteria isolated from chromium-contaminated soil by tannery activity. Curr. Microbiol. 46: 1–5.

CHAPTER 4

Activation of Silent Genes in Actinobacteria by Exploiting Metal Stress

Götz Haferburg* and Erika Kothe

Introduction

The microorganism is always right, your friend, and a sensitive partner; there are no stupid microorganisms; microorganisms can and will do anything; they are smarter, wiser, more energetic than chemists, engineers, and others; if you take care of your microbial friends, they will take care of your future (Perlman 1980). These five laws of the distinguished biotechnologist David Perlman seem to be meant as an amusing humanization of microbes to elicit from the audience a sympathetic sentiment for microbiology and industrial microbiology in particular. However we find the settled belief in the potency of the microbial cell. It could also be understood as an advice to the microbiologist, never to stop seeking the appropriate approach to make use of the inconceivable spectrum of capabilities concealed in the microbial world.

This chapter contains both basic principles of gene activation and potential applications, with special focus on metal stress in screening programs for new antibiotics from actinobacteria. The secondary metabolite pattern of a potential producer strain can be affected by the type of metal

Friedrich-Schiller University Jena, Institute of Microbiology, Microbial Communication, Neugasse 25, 07743 Jena, Germany.
*Corresponding author: goetz.haferburg@posteo.de

supplemented and applied metal concentration. It has been shown how the pattern of secondary metabolites is altered in a number of strains if the culture contains traces (or large quantities) of metals. Thus, metal stress seems exploitable in antibiotic screenings. Not only does metal stress induce the production of metabolites linked to metal-dependent metabolism like metal chelators. Rather, the entire secondary metabolite pattern can be changed if the cultures are spiked with small metal concentrations. Yet, unknown secondary metabolites with a medical potential are thought to be discovered by the utilization of metals for awakening sleeping gene clusters.

Silent, sleeping and cryptic genes

What are these genes: sleeping, or silent? Not all genes of all biosynthetic clusters within a microbial genome are expressed at any time. Quite on the contrary, it seems that most of these clusters are silent or show a very low expression level if standard laboratory conditions are applied (Chiang et al. 2011). They are simply not induced because of the missing environmental signal. For metabolome changes, a systems biology approach is being developed for some model organisms, for instance, for the best investigated actinobacteria *Streptomyces coelicolor* is exposed to salt stress (Kol et al. 2010). However, the information available for inducible secondary metabolism gene clusters is scarce and the inducing conditions for many secondary metabolites potentially produced by surveying genome information are mostly lacking. But gene mining in actinobacteria genomes frequently results in surprising new findings as was shown *in silico* for metallothioneins and metallohistins distribution in the phylum which helps to explain resistance mechanisms in actinobacteria organisms thriving in metalliferous habitats (Schmidt et al. 2010).

Genomes of many organisms contain gene clusters for secondary metabolites; well studied producers besides actinobacteria are bacilli, cyanobacteria and myxobacteria. Secondary metabolites are "products which are not formed under all circumstances and [with] no obvious metabolic function" (Bu'Lock 1965). This means that gene clusters encoding synthesis of secondary metabolites can be well hidden genetic entities which await arousal by induction. The way of induction is hardly (if at all) to be correlated to the delivered product. A good example is the detection of the first antibiotic from an anaerobic bacterium formerly believed to be too energy-limited to produce secondary metabolites at all. In *Clostridium cellulolyticum*, a search for potential non-ribosomal peptide synthases and polyketide synthetases revealed potential clusters which, upon induction by environmental factors substituted in the media from earth extracts, produced a new class of metabolites, the closthioamides (Lincke et al. 2010).

It might be important to differentiate between silent and cryptic genes, since the focus here is on arousing silent genes. Cryptic genes also encode specific functions not expressed under any known circumstances. But in contrast to silent genes, activation of cryptic genes seems to need a mutational event or other genetic mechanisms like recombination or insertion (Tamburini and Mastromei 2000). Hence, regulation and signal transduction is different for silent gene clusters which may be induced by environmental factors.

Physiology of secondary metabolism was studied first in plants. One of the founding fathers of secondary metabolite research was the botanist Ernst Stahl (1848 - 1919) who held the chair for botany in Jena. Studies on defensive mechanisms of plants against herbivorous snails, published in 1888, was probably his most influential publication for later research on secondary plant compounds. It took a while before the research field was expanded to microorganisms. The cornerstone of natural compound research in microbiology was laid by Alexander Fleming's (1881–1955) accidental discovery of the antibiotic activity of penicillin from a contaminating fungus on a *Staphylococcus* culture in 1928. Since the times of Stahl and Fleming, a tremendous amount of knowledge on biochemistry of secondary metabolites has accumulated. From the chemical point of view, secondary metabolites are natural products of comparably low molecular weight (< 3000 Daltons) characterized by a high structural variety (polyketides, non-ribosomal peptides, terpenes, aminoglycosides, bacteriocins, beta-lactams, butyrolactones, siderophores, melanins, etc.). But still, there is no uniformly accepted explanation why such metabolites are produced. What ecological role do they play and how do organisms benefit from the production of these energetically expensive compounds? Hypotheses reach from "ballast for the cells" to "waste detoxification" or "ecological fitness". In contrast to the term secondary metabolite, the definition of antibiotics seems to be far less problematic. An antibiotic is "inhibiting the growth or the metabolic activities of bacteria and other microorganisms by a chemical substance of microbial origin" (Waksman 1947).

Antibiotics producing actinobacteria

The number of increasingly antibiotic-resistant pathogenic bacteria is rising steadily. Both, Gram-positive (especially *Staphylococus aureus, Streptococcus pneumoniae* and several enterococci) and Gram-negative (especially *Acinetobacter baumannii, Klebsiella pneumoniae* and *Pseudomonas aeruginosa*) pathogenic bacteria have the capacity to acquire and spread multiple antibiotic resistances, driving the antibacterial drug development (Theuretzbacher 2011). Involved with an increase of the average age of the population in most of the industrial countries is the susceptibility to infectious diseases. Examples for an increased risk potential regarding spread of infectious diseases can be seen in growing numbers of surgeries and immunocompromised patients. In addition, the acceleration of mobility in a globalized world leads to a faster spread of pathogens and improper use and application of antibiotics is one of the major reasons why drug resistance could expand across various groups of microorganisms. Horizontal gene transfer is a dynamic process turning resistance recipients into a new type of multipliers (Cantón and Morosini 2011). Resistance genes are acquired by plasmids, transposable elements or phages and transmitted to the descendants of the recipient colony. Furthermore they can act as new donors to susceptible cells; thus transmitting resistance genes over time and various prokaryotic phyla.

Antibiotics as feed supplement for fast muscle build-up in stockbreeding were in use for more than four decades. Use of antibiotics in animal feed was stopped in some European countries in the 1980s (Sweden and Denmark). Still, across Europe this practice was not banned before 2006 although the danger of resistance spread and ways of acquired multiple antibiotic resistances were known (for review on antimicrobial resistance in agriculture see: Silbergeld et al. 2008). For these reasons it is obvious that new antibiotics, preferably directed at new targets in the bacterial or fungal cell, are urgently needed. However, during the past decades, screening programs for new antibiotics have been down-scaled and development of drugs into a stage III trial is steadily decreasing. The graph for methicillin-resistant *Staphylococcus aureus* or vancomycin-resistant bacterial infections follows a reverse tendency. This has led to serious problems, especially in hospitals. Less new antibiotics become available, but at the same time, more resistant bacteria appear in medical centers (Fig. 1a, b).

It is in particular the phylum *Actinobacteria* that comprises a tremendous amount of antibiotic producers (Table 1). Antibiotics are produced by other large groups of prokaryotes like myxobacteria and bacilli as well. But about two thirds (7900) of all antibiotics known, originate from actinobacteria and here, with 80%, first and foremost from the genus *Streptomyces* (Kieser et al. 2000). The overall number of antibiotics expected from actinobacteria depends on the authors and their calculation, but ranges from about 5000 (Demain and Fang 2000) to more than four times this number (Berdy 2005) can be found. Many eukaryotes, like fungi or plants also produce a great number of antibiotics, with 2600 and 5000, respectively (Kieser et al. 2000). Nevertheless, actinobacteria hold a special position not only for the high numbers and structural variety of antibiotics, but also because

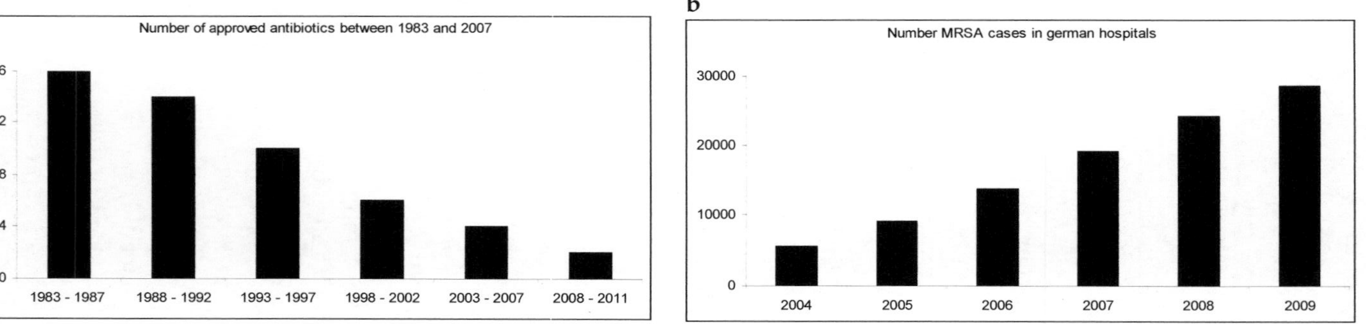

Figure 1. a. Decrease of antibiotics approved in the last three decades (source: Infectious Disease Society of America, IDSA, and Demain 2009); b. Increase of MRSA infections in Germany in the 2000s (following Gastmeier 2011).

Table 1. Antibiotics widely used today (compiled from Singh and Greenstein 2000, Davies 2006, Sosa 2007).

Antibiotic class	Year of discovery	Cellular/molecular target	Producer organism	Resistance reported
Sulfonamides	1937	Folate synthesis	Synthetically produced	1960s
Penicillins	1928	Bacterial cell wall	*Penicillium, Streptomyces* also produced synthetically	1940s
Polymyxin	1947	Cell membrane	*Bacillus polymyxa*	1940s
Chloranphenicol	1949	23S rRNA of ribosome	*Streptomyces venezuelae* also produced synthetically	1950s
Tetracyclines	1953	30S subunit of ribisomes	*Streptomyces aureafaciens Streptomyces rimosus*	1950s
Cephalosporins	1953	Bacterial cell wall	*Cephalosporium*	1970s
Aminoglykosides	1943	30S subunit of ribisomes targets also other molecules	*Streptomyces griseus* and *Micromonospora*	1950s
Vancomycin	1958	Bacterial cell wall	*Amycolatopsis orientalis*	1980s
Lincosamide	1966	23S rRNA of ribosomes	Semisynthetic, derivative of lincomycin by *S. lincolnensis*	1980s
Rifamycin	1957	RNA polymerase	*Amycolatopsis mediterranei* also produced synthetically	
Trimethopim/ Sulphamethoxazole	1973	Dihydrofolate reductase	Synthetically produced	1970s
Carbapenems	1976	Bacterial cell wall	*Streptomyces cattleya*	1990s
Monobactams	1982	Bacterial cell wall	(Semi)synthetically produced	
Linezolid	1987	23S rRNA of ribosomes	Synthetically produced	2000s
Daptomycin	1987	Cell membrane	*Streptomyces roseosporus*	2000s

of their comparatively easy screening and well established fermentation techniques (for search and discovery strategies see: Bull et al. 2000, Donadio et al. 2002).

As new antibiotics are needed to fight multi-resistant pathogens, all resources need to be re-evaluated which should also include environmental activation of secondary metabolites, e.g., by metal stress. Some streptomycete secondary metabolites, including antibiotics, have already been shown to bind metals, for example actinorhodin (a chelator of iron produced by *S. coelicolor;* Coisne et al. 1999), isatin (scavenging heavy metals from the medium of *S. albus;* Gräfe and Radics 1986), or melanin (involved in heavy metal tolerance in *S. scabies;* Beausejour and Beaulieu 2004). The use of actinobacteria with their already proven ability to produce different classes of metabolites can thus be envisioned to provide a specifically rich source of new antibiotics in metal-induced screening routines.

Heavy metal resistance of actinobacteria

Besides the possession of a secondary metabolism, actinobacteria seem to be, remarkably adaptive microorganisms from an ecological perspective. Actinobacteria have conquered a vast number of metalliferous habitats. A great variety of different actinobacterial genera has been isolated from naturally or anthropogenically metal (and metalloid) rich habitats. Many of the isolates are able to resist remarkably high concentrations of different metals in soils and sediments. In geogenically nickel enriched serpentine soils of Tuscany, Italy (Mengoni et al. 2001) or in New Caledonia (Hery et al. 2003), high numbers of adapted, nickel tolerant or nickel resistant strains were found, including *S. yatensis* from which a high number of antimicrobial and antitumor compounds could be identified (Saintpierre et al. 2003).

The case-dependent heavy metal influence on secondary metabolism has been analyzed using *S. galbus*. While low concentrations of cadmium or chromium stimulated pigment production, nickel and mercury showed a negative effect (Raytapadar et al. 1995). In *S. rishiriensis,* nickel had a positive effect on the production of the antibiotic coumermycin A1 (Claridge et al. 1966). In *S. coelicolor,* the production of actinorhodin is sensitive to mercury, cadmium, copper, nickel and lead, while chromium slightly stimulated expression (Abbas and Edwards 1989, 1990). The influence of heavy metals on the metabolic profile of metal resistant actinobacteria was also observed with respect to substrate spectra utilized (Sprocati et al. 2006).

Plant interactions may also be involved in adaptation of microbes to environmental stress factors. Interestingly, endophytic metal resistant actinobacteria also inhabit hyperaccumulating plants and act as growth promoting bacteria (Rajkumar et al. 2009, Sun et al. 2010). The rhizosphere of nickel hyperaccumulator plants like, e.g., *Sebertia accuminata* and *Thlaspi goesingense* was shown to be specifically rich in nickel resistant strains (Stoppel and Schlegel 1995, Pal et al. 2007). Additionally, cadmium resistant *Microbacterium* and *Arthrobacter* strains were described as predominant endophytes in metal resistant *Solanum nigrum* plants growing on a mine tailing (Luo et al. 2011).

Isolation from metal contaminated sites yielded strains which can survive high metal loads and, at the same time, can be checked for the metallome, the transcriptome, proteome and metabolome changes associated with (heavy) metal exposition (Haferburg and Kothe 2010). Mercury resistant *Streptomyces* isolates were derived from the metal polluted sediment of Baltimore's Inner Harbor in Chesapeake Bay (Ravel et al. 1998a, b). An extreme nickel and zinc resistance could be found among *Streptomyces* strains of the former uranium mining area in Thuringia, Germany (Schmidt et al. 2009). Some of these strains occurring in a post-mining area do not only survive, but also grow in presence of more than 100 mM nickel or zinc. Lin and colleagues isolated a zinc resistant strain, now named *Streptomyces zinciresistens* sp. nov., from a copper and zinc mine in Shaanxi in the north-western mining province China (Lin et al. 2011). A metal resistant strain of the genus *Amycolatopsis* was isolated from copper contaminated sediments in the province of Tucumán, Argentina (Albarracín et al. 2010). Strains belonging to the rare actinobacteria genera *Kitasatospora* and *Lentzea* were shown to cope well with high nickel concentrations (Haferburg et al. 2009). The antibiotic resistance of several *Mycobacterium* species like, the human pathogens *M. tuberculosis* and *M. fortuitum,* correlates with resistance against a huge set of metal cations (Agranoff and Krishna 2004). Metalloids like arsenic are often found as a result of geogenic or anthropogenic contamination. The biotechnologically important *Corynebacterium glutamicum* serves as model for arsenic resistance (Mateos et al. 2006). An additional option for resistance factors is the expression of efflux pumps. This has been found mainly in Gram-negatives; however, for *S. acidiscabies* a partial sequence of such a metal efflux system has been identified (Amoroso et al. 2000).

The mechanisms conferring resistance should be viewed, before a decision is made on whether they might impact induction of silent gene clusters or not. These mechanisms can be extracellular or intracellular sequestration and storage (for thorough review see: Haferburg and Kothe 2007). Metal biochemistry, influences metal detoxification not only via different resistance mechanisms, but directly and indirectly by sequestration of cations and metal stress response, respectively (Fig. 2).

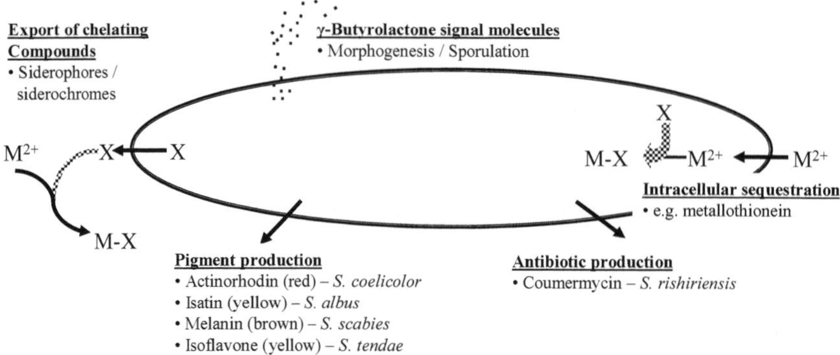

Figure 2. Overview of influences of metals on the secondary metabolism of streptomycetes.

Genetic control of antibiotic production

Already in the middle 1960s, the frequency of product rediscovery could be shown to drive screening programs with up to 10,000 new strains of *Streptomyces* necessary to identify one new antimicrobial principle (Gräfe 1992). It thus seems logical to review the general control mechanisms in actinomycetes to find a general mechanism that may be exploited. In addition, the genetic modification by controlled expression of regulatory genes may present a way to produce compounds encoded in the genome, but not expressed under any known laboratory conditions, as seen above.

In actinomycetes, biosynthesis of antibiotics is commonly encoded chromosomally in clusters of up to 30 genes. Antibiotic resistance genes are an essential part of these clusters and unavoidable for survival of the producer. There are very few examples for extrachromosomal elements coding for antibiotic biosynthesis, one example being methylenomycin (Kinashi et al. 1992). Also biosynthesis clusters usually contain regulatory genes for the activation of transcription. These act pathway specific, but are in turn controlled by pleiotropic regulators. They may also include two component systems, protein kinases, quorum sensing γ-butyrolactones, the alarmone ppGpp or alternative σ-factors and their regulatory networks (Chater and Bibb 1997). By genome mining, almost 1000 genes, including 65 σ-factors, have been identified from the *S. coelicolor* genome with a potential regulatory function in secondary metabolism (Bentley et al. 2002). Specifically σ-factor cascades, just like those known for sporulation in bacilli and clostridia, are seen with up to four different serially connected promoter recognition sequences (Missiakas and Raina 1998).

In most cases analyzed so far, the onset of antibiotic production is linked to late exponential or early stationary phase when at least one of the nutritional sources, nitrogen, phosphorous or carbon, starts to become depleted. At this point of time, morphological differentiation is initiated as well. Additional activation may occur when vegetative mycelium breaks down and "fatal attraction" takes place (Shi and Zusman 1993, Challis and Hopwood 2003). Thus, a common regulator for antibiotic production and morphogenesis can be postulated for streptomycetes. The A-factor, a γ-butyrolactone of *S. griseus* is a key molecule in signaling. During growth, A-factor is continuously synthesized and accumulated. This is perfectly in line with its function as quorum sensing signal. If a certain concentration is reached, allosteric binding to the receptor protein ArpA leads to inactivation of this repressor, liberating the promoter region of *adpA* and expressing the activator AdpA. AdpA activates the transcription of a number of genes, among them regulators for antibiotic production and morphogenesis. It is believed that in almost all streptomycetes, γ-butyrolactones occur as regulators for secondary metabolism and morphological differentiation in transition to stationary phase and spore formation.

Regulators that are controlled by external signals via butyrolactones (and their corresponding binding receptors) have been grouped as SARPs, *Streptomyces* antibiotic regulatory proteins (Wietzorrek and Bibb 1997). The SARP protein family comprises of most known pathway dependent transcriptional regulators which usually act as positive regulators. The mode of action (and flow of information) has been termed *pyramidal* control strategy (Martín and Liras 2010). Additionally, antibiotic synthesis may be controlled by two-component systems which act in linking external stress signals to antibiotic biosynthesis. While single two-component systems may target antibiotic synthesis directly, a closer look reveals intense cross talk between different global regulators in *Streptomyces*. This mode of action can be seen as a *network* of global regulators. To give an example, in the *S. coelicolor* genome more than 60 two-component systems were found. Most of them are not integrated into antibiotic gene clusters, but do show an influence on antibiotic production. In addition, primary and secondary metabolisms are strongly interwoven by the action of a complex network of *c*ross *t*alking *g*lobal *r*egulators. This mechanism has been called CTGR (Martin and Liras 2010).

Studies on the regulation of primary metabolism using *S. coelicolor* and *S. clavuligerus* have revealed a growth rate dependent precursor channeling for antibiotic synthesis. In case of unbalanced growth, e.g., in periods of weak growth, precursors accumulate. This is due to the lack of feed back inhibition. Subsequently, the pool of precursors is emptied by biosynthesis of secondary metabolites like antibiotics (Hood et al. 1992, Khetan et al. 1999). These results display the complex and multiple superimposed regulation circuits in place of control of secondary metabolism. A re-arrangement of pathways within primary metabolism is likely to lead to a shift in secondary metabolism and hence may lead to awakening silent gene clusters, including those for antibiotics production. One factor known to exert such a pleiotropic effect is the presence of heavy metals in sublethal concentrations, i.e., metal stress. Changes in precursor generation and accumulation likely to occur in metal supplemented cultures may thus be expected to result in different patterns of secondary metabolite formation.

In contrast to the well investigated culture conditions necessary to lead to antibiotic production, far less is known about the factors determining type and quantity of the produced compounds. For instance, the genomes of *S. avermitilis* and *S. coelicolor* are predicted to encode at least 25 and 22 different secondary metabolites, respectively (Omura et al. 2001, Bentley et al. 2002). However, only a small set of these metabolites is produced under given culture conditions. For a thorough screening, a variation of culture conditions may thus lead to the production not only of new, but also of more of the same secondary metabolites. Approaches using the addition of inducers including butyrolactones, metals or other stressors, or new media compositions may be supplemented in future by molecular approaches targeting regulatory networks (for review see: Hopwood 1989).

Activation of sleeping genes and potential of metal-based screening

It is believed that only 10 to 20% of all existing natural antibiotics are known (Gräfe 1992). Antibiotic producing strains commonly produce a good deal more than one secondary metabolite of biological activity. Hence, it would be a waste of labor and screening efforts if strains once isolated and incorporated into strain collections are considered too early as weak producers or producers of already known compounds. The OSMAC concept (One Strain–Many Compounds) represents the opposite pole of high throughput screening and explains the benefits of modified culture conditions for a more complete spectrum of secondary metabolites very well (Bode et al. 2002). Up to twenty different metabolites belonging to diverse product families can be detected from a single producer strain by changing screening parameters like, media composition, aeration, culture vessel and addition of enzyme inhibitors.

The main challenge is to trace yet undiscovered, but potentially applicable new compounds. The structural and functional diversity of compounds produced by streptomycetes is remarkably broad, but the concentration of synthesized product is commonly very low. This is in line with current findings that sub-inhibitory concentrations of antibiotics already exert quite remarkable changes in the transcriptome of co-occurring microbes. Whether or not antibiotics should be regarded as signaling molecules in interspecies microbial communication, is currently discussed. This new finding might explain, why extremely low concentrations of antibiotics can often be measured, during fermentation even under optimized production conditions. However, for feasible industrial application, the antibiotic screening has to be accompanied by the development of production schemes for increased product yield. Usually, undirected mutagenesis has been employed before molecular genetic tools were available. Even now, transposon mutagenesis as an undirected approach is used to increase the rate of production of a certain metabolite. However, the use of direct approaches becomes more and more prominent, especially with producer strains, the genomes of which have been sequenced. The decreasing cost of genome sequencing is helpful with respect to targeted alteration of biosynthesis genes leading to production of an entire library of custom made variants of a certain antibiotic class. However, genetic manipulation, as well as genome sequencing in actinobacteria, is slowed down by certain biological properties. The genome sizes are usually very large, comparable to those of typical higher fungi, with approximately 10 times the genome size of a "typical" bacterium. In addition, the high GC content in the genome leads to problems with shot gun sequencing applying short reads. The modification restriction systems even of model streptomycetes are rather effective so as not to allow integration of DNA produced for example, in standard *E. coli* cloning strains. The transformation systems available are tedious and often necessitate large spans of homologous sequence in order to allow for integration into the genome of the targeted strains. Transformation of new isolates involves difficulties in many instances (for review of the molecular cloning in streptomycetes please see: Kieser et al. 2000).

The solution to the problems reviewed above might focus more directly on streptomycete biology. As discussed above, metals might be exploited to discover new antibiotics. At the same time it is feasible and might also lead to an increase in production of a certain antibiotic. In a very limited study of only 10 stains derived from a heavy metal contaminated environment, two could be identified to produce new compounds after being exposed to sub-inhibitory nickel and cadmium concentrations (Haferburg et al. 2009). One of these strains, *S. tendae* F4, even yielded a new class of isoflavones, the formation of which was analyzed to yield a new, formerly unknown chemical mechanism (Ndejouong et al. 2009, Ueberschaar et al. 2011). Generally, the induction of antimicrobial activity against several human pathogens, including *Candida albicans*, was shown to occur in 50% of the metal supplemented cultures in this study (Table 2).

Judging from this very limited investigation, the use of metal stress for both screening and production seems a promising and easy strategy for

Table 2. Screening of actinobacteria for metal prompted induction of bioactive secondary metabolites (Haferburg et al. 2009).

Strain	Isolated from	Metal Resistance	Biologycal activity in metal induced culture against
Streptomyces acidiscabies E13	Metal rich former mining site	5 mMNi	*S. aureus* 134/94, *M. smegmatis* 987
Streptomyces tendae F4	Metal rich former mining site	1 mMCd/1 mM Ni	*C. albicans*
Streptomyces ciscaucasicus PT1	Metal rich former mining site	1 mM Ni	
Streptomyces sp. PT5	Metal rich former mining site	1 mM Ni	
Streptomyces aureus PT13	Metal rich former mining site	1 mM Ni	
Streptomyces purpurascens Tosca 2	Ultramafic soil	1 mM Ni	*E. coli* 458
Streptomyces lincolnensis Tosca 3	Ultramafic soil	10 mM Ni	
Lentzea waywayandensis Tosca 4	Ultramafic soil	1 mM Ni	
Kitasatospora sp. JE12	Non-metalliferous habitat	5 mM Ni	*S. aureus* 134/94, *M. smegmatis* 987
Streptomyces sp. WiP14	Non-metalliferous habitat	1 mM Ni	*E. coli* 458

application in the programs for discovery of new drugs. It could also show that the aspect of basic science, discovery of new biosynthesis routes and regulatory circuits and networks, is still awaiting many new discoveries in the versatile organism group of actinobacteria. For the induction of antibiotic genes, raising the concentration of metals in the fermentation media by 10 to 100 fold has been implied (Iwai and Omura 1982). Thus, a threshold of production can be overcome. However, such concentrations generally require tolerance, since these can be considered inhibiting to the growth of sensitive strains. Still, a stress response is associated with a switch in metabolism. This metabolic shift is associated with the formation of new secondary metabolites or an increased compound production.

Conclusions

The use of actinobacteria, for the discovery and production of antibiotics specifically streptomycetes and rare actinomycetes, already has quite a long history. The current situation of higher prevalence for multiresistances among pathogens and the decrease of detection of new antibiotics and new drug targets have led to a situation where merely screening programs for new strains does not seem well suited for the necessary drug development. Hence, new strategies are needed to provide us with the urgently needed new compounds. Here, we discussed the use of metal stress to allow for expression of genes not expressed under standard cultivation and screening conditions.

In order to develop this approach, a better understanding of different regulatory pathways and networks seems essential. With respect to metal induced transcriptional and translational regulation as well as changes in the metabolome, the term metallomics has been introduced (Szpunar 2005, Mounicou et al. 2009, Haferburg and Kothe 2010). However, specifically for streptomycetes and actinobacteria, the results are still not sufficient to get a holistic view on the regulation involved. Thus, the current knowledge on antibiotics production, regulation of secondary metabolism and heavy metal resistance were reviewed to allow for the introduction of a new, metallomics driven approach to be introduced into current and future screening programs for new drugs. This quest has, however to be, accompanied by a development of targeted approaches, which might need more time for drug discovery, due to the comparatively slow development of molecular tools for new isolates.

The introduction of sub-inhibitory metal concentrations into screening programs for new, metal resistant strains as well as the re-screening of established strain collections might provide an alternative. The idea for this proposal comes from detailed investigation of microbial communication, which has started only recently. The investigation of pure strains does not reflect the natural situation in which evolution has led to the development of secondary metabolites. Hence, a better understanding of metal induced molecular communication channels needs to be exploited in antibiotic screening, as in the presented proposition.

Acknowledgements

This work was supported with projects performed under the EU-FP 7 project UMBRELLA and within the frame of the DFG-funded research training group "Alteration and element mobility at microbe-mineral interphases" included in the "Jena School for Microbial Communication", a graduate school financed by the German Federal and State governments through DFG.

References

Abbas, A. and C. Edwards. 1989. Effects of Metals on a Range of *Streptomyces* Species. Appl. Environ. Microbiol. 55: 2030–2035.

Abbas, A. and C. Edwards. 1990. Effects of Metals on *Streptomyces coelicolor* Growth and Actinorhodin Production. Appl. Environ. Microbiol. 56: 675–80.

Agranoff, D. and S. Krishna. 2004. Metal ion transport and regulation in *Mycobacterium tuberculosis*. Front Biosci. 9: 2996–3006.

Albarracín, V.H., P. Alonso-Vega, M.E. Trujillo, M.J. Amoroso and C.M. Abate. 2010. *Amycolatopsis tucumanensis* sp. nov., a copper-resistant actinobacterium isolated from polluted sediments. Int. J. Syst. Evol. Microbiol. 60: 397–401.

Amoroso, M.J., D. Schubert, P. Mitscherlich, P. Schumann and E. Kothe. 2000. Evidence for high affinity nickel transporter genes in heavy metal resistant *Streptomyces* spec. J. Basic Microbiol. 40: 295–301.

Beausejour, J. and C. Beaulieu. 2004. Characterization of *Streptomyces scabies* mutants deficient in melanin biosynthesis. Can. J. Microbiol. 50: 705–709.

Bentley, S.D., K.F. Chater, A.M. Cerdeno-Tarraga, G.L. Challis, N.R Thomson, K.D. James, D.E. Harris, M.A. Quail, H. Kieser, D. Harper, A. Bateman, S. Brown, G. Chandra, C.W. Chen, M. Collins, A. Cronin, A. Fraser, A. Goble, J. Hidalgo, T. Hornsby, S. Howarth, C.H. Huang, T. Kieser, L. Larke, L. Murphy, K. Oliver, S. O'Neil, E. Rabbinowitsch, M.A. Rajandream, K. Rutherford, S. Rutter, K. Seeger, D. Saunders, S. Sharp, R. Squares, S. Squares, K. Taylor, T. Warren, A. Wietzorrek, J. Woodward, B.G. Barrell, J. Parkhill and D.A. Hopwood. 2002. Complete genome sequence of the model actinomycete *Streptomyces coelicolor* A3(2). Nature. 417: 141–147.

Berdy, J. 2005. Bioactive microbial metabolites. A personal view. J. Antibiot. 58: 1–26.

Bode, H.B., B. Bethe, R. Höfs and A. Zeeck. 2002. Big effects from small changes: possible ways to explore nature's chemical diversity. Chembiochem. 3: 619–27.

Bull, A.T., A.C. Ward and M. Goodfellow. 2000. Search and discovery strategies for biotechnology: the paradigm shift. Microbiol. Mol. Biol. Rev. 64: 573–606.

Bu'Lock, J.D. 1965. The biosynthesis of natural products: an introduction to secondary metabolism, page 2, McGraw-Hill publishing, 1965, University of Michigan.

Cantón, R. and M.I. Morosini. 2011. Emergence and spread of antibiotic resistance following exposure to antibiotics. FEMS Microbiol. Rev. 35: 977–991.

Challis, G.L. and D.A. Hopwood. 2003. Synergy and contingency as driving forces for the evolution of multiple secondary metabolite production by *Streptomyces* species. Proc. Natl. Acad. Sci. USA. 100: 14555–14561.

Chater, K.F. and M.J. Bibb. Regulation of Bacterial Antibiotic Production. Biotechnology, Vol 7: Products of Secondary Metabolism. *In:* Kleinkauf, H. and H. von Dohren (Eds.). 1997. VCH, Weinheim, Germany, pp. 57–105.

Chiang, Y.M., S.L. Chang, B.R. Oakley and C.C. Wang. 2011. Recent advances in awakening silent biosynthetic gene clusters and linking orphan clusters to natural products in microorganisms. Curr. Opin. Chem. Biol. 15: 137–143.

Claridge, C.A., V.Z. Rossomano, N.S. Buono, A. Gourevitch and J. Lein. 1966. Influence of Cobalt on Fermentative Methylation. Appl. Microbiol. 14: 280–283.

Coisne, S., M. Bechet and R. Blondeau. 1999. Actinorhodin production by *Streptomyces coelicolor* A3(2) in iron-restricted media. Lett. Appl. Microbiol. 28: 199–202.

Davies, J. 2006. Where have All the Antibiotics Gone? Can. J. Infect. Dis. Med. Microbiol. 17: 287–290.

Demain, A.L. and A. Fang. 2000. The natural functions of secondary metabolites. Adv. Biochem. Eng. Biotechnol. 69: 1–39.

Demain, A.L. 2009. Antibiotics: Natural products essential to human health. Med. Res. Rev., 29: 821–842.

Donadio, S., P. Monciardini, R. Alduina, P. Mazza, C. Chiocchini, L. Cavaletti, M. Sosio, A.M. Puglia. 2002. Microbial technologies for the discovery of novel bioactive metabolites. J. Biotechnol. 99: 187–198.

Gastmeier, P. 2011. Zur Entwicklung nosokomialer Infektionen im Krankenhausinfektions-Surveillance-System (KISS). Epidemiol. Bull. 5: 35–37.

Gräfe U. and L. Radics. 1986. Isolation and structure elucidation of 6-(3′-methylbuten-2′-yl) isatin, an unusual metabolite from *Streptomyces albus*. J. Antibiot. 39: 162–163.

Gräfe, U. 1992. Biochemie der Antibiotika. Struktur, Biosynthese, Wirkmechanismus. Spektrum, Heidelberg.

Haferburg, G., I. Groth, U. Möllmann, E. Kothe and I. Sattler. 2009. Arousing sleeping genes: shifts in secondary metabolism of metal tolerant actinobacteria under conditions of heavy metal stress. Biometals. 22: 225–234.

Haferburg, G. and E. Kothe. 2007. Microbes and metals: interactions in the environment. J. Basic. Microbiol. 47: 453–467.

Haferburg, G. and E. Kothe. 2010. Metallomics: lessons for metalliferous soil remediation. Appl. Microbiol. Biotechnol. 87: 1271–1280.

Hery, M., S. Nazaret, T. Jaffre, P. Normand and E. Navarro. 2003. Adaptation to nickel spiking of bacterial communities in neocaledonian soils. Environ. Microbiol. 5: 3–12.

Hood, D.W., R. Heidstra, U.K. Swoboda and D.A. Hodgson. 1992. Molecular genetic analysis of proline and tryptophan biosynthesis in *Streptomyces coelicolor* A3(2): interaction between primary and secondary metabolism. Gene. 115: 5–12.

Hopwood, D.A. 1989. Antibiotics: opportunities for genetic manipulation. Philos. Trans. R. Soc. Lond. B. Biol. Sci. 324: 549–62.

Iwai, Y. and S. Omura. 1982. Culture conditions for screening of new antibiotics. J. Antibiot. 35: 123–141.

Khetan, A., L.H. Malmberg, Y.S. Kyung, D.H. Sherman and W.S. Hu. 1999. Precursor and cofactor as a check valve for cephamycin biosynthesis in *Streptomyces clavuligerus*. Biotechnol. Prog. 15: 1020–1027.

Kieser, T., M.J. Bibb, M.J. Buttner, K.F. Chater, D.A. Hopwood. Preparation and analysis of genomic and plasmid DNA. *In:* T. Kieser, M.J. Bibb, M.J. Buttner, K.F. Chater and D.A. Hopwood (eds.). 2000. Practical Streptomyces Genetics. The John Innes Foundation, Norwich, UK, pp. 161–210.

Kinashi, H., M. Shimaji-Murayama and T. Hanafusa. 1992. Integration of SCP1, a giant linear plasmid, into the *Streptomyces coelicolor* chromosome. Gene. 115: 35–41.

Kol, S., M.E. Merlo, R.A. Scheltema, M. de Vries, R.J. Vonk, N.A. Kikkert, L. Dijkhuizen, R. Breitling and E. Takano. 2010. Metabolomic characterization of the salt stress response in *Streptomyces coelicolor*. Appl. Environ. Microbiol. 76: 2574–2581.

Lin, Y.B., X.Y. Wang, H.F. Li, N.N. Wang, H.X. Wang, M. Tang and G.H. Wei. 2011. *Streptomyces zinciresistens* sp. nov., a zinc-resistant actinomycete isolated from soil from a copper and zinc mine. Int. J. Syst. Evol. Microbiol. 61: 616–620.

Lincke, T., S. Behnken, K. Ishida, M. Roth and C. Hertweck. 2010. Closthioamide: an unprecedented polythioamide antibiotic from the strictly anaerobic bacterium *Clostridium cellulolyticum*. Angew. Chem. Int. Ed. Engl. 49: 2011–2013.

Luo, S.L., L. Chen, J.L. Chen, X. Xiao, T.Y. Xu, Y. Wan, C. Rao, C.B. Liu, Y.T. Liu, C. Lai and G.M. Zeng. 2011. Analysis and characterization of cultivable heavy metal-resistant bacterial endophytes isolated from Cd-hyperaccumulator *Solanum nigrum* L. and their potential use for phytoremediation. Chemosphere. 85:1130–1138.

Martín, J.F. and P. Liras. 2010. Engineering of regulatory cascades and networks controlling antibiotic biosynthesis in *Streptomyces*. Curr. Opin. Microbiol. 13: 263–273.

Mateos, L.M., E. Ordóñez, M. Letek and J.A. Gil. 2006. *Corynebacterium glutamicum* as a model bacterium for the bioremediation of arsenic. Int. Microbiol. 9: 207–215.

Mengoni, A., R. Barzanti, C. Gonnelli, R. Gabbrielli and M. Bazzicalupo. 2001. Characterization of nickel-resistant bacteria isolated from serpentine soil. Environ. Microbiol. 3: 691–698.

Missiakas, D. and S. Raina. 1998. The extracytoplasmic function sigma factors: role and regulation. Mol. Microbiol. 28: 1059–1066.

Mounicou, S., J. Szpunar and R. Lobinski. 2009. Metallomics: the concept and methodology. Chem Soc. Rev. 38: 1119–1138.

Ndejouong, B.S., I. Sattler, H.M. Dahse, E. Kothe and C. Hertweck. 2009. Isoflavones with unusually modified B-rings and their evaluation as antiproliferative agents. Bioorg. Med. Chem. Lett. 19: 6473–6476.

Omura, S., H. Ikeda, J. Ishikawa, A. Hanamoto, C. Takahashi, M. Shinose, Y. Takahashi, H. Horikawa, H. Nakazawa, T. Osonoe, H. Kikuchim, T. Shiba, Y. Sakaki and M. Hattori. 2001. Genome sequence of an industrial microorganism *Streptomyces avermitilis*: deducing the ability of producing secondary metabolites. Proc. Natl. Acad. Sci. USA. 98: 12215–12220.

Pal, A., G. Wauters and A.K. Paul. 2007. Nickel tolerance and accumulation by bacteria from rhizosphere of nickel hyperaccumulators in serpentine soil ecosystem of Andaman, India. Plant Soil. 293: 37–48.

Perlman, D. 1980. Some problems on the new horizons of applied microbiology. Dev. Indust. Microbiol. 21: 15–23.

Rajkumar, M., N. Ae and H. Freitas. 2009. Endophytic bacteria and their potential to enhance heavy metal phytoextraction. Chemosphere. 77: 153–160.

Ravel, J., M.J. Amoroso, R.R. Colwell and R.T. Hill. 1998a. Mercury-resistant actinomycetes from the Chesapeake Bay. FEMS Microbiol. Lett. 162: 177–184.

Ravel, J., H. Schrempf and R.T. Hill. 1998b. Mercury resistance is encoded by transferable giant linear plasmids in two chesapeake bay *Streptomyces* strains. Appl. Environ. Microbiol. 64: 3383–3388.

Raytapadar, S., R. Datta and A.K. Paul. 1995. Effects of some heavy metals on growth, pigment and antibiotic production by *Streptomyces galbus*. Acta Microbiol. Immunol. Hung. 42: 171–177.

Saintpierre, D., H. Amir, R. Pineau, L. Sembiring and M. Goodfellow. 2003. *Streptomyces yatensis* sp. nov., a novel bioactive streptomycete isolated from a New-Caledonian ultramafic soil. Antonie Van Leeuwenhoek. 83: 21–26.

Schmidt, A., G. Haferburg, A. Schmidt, U. Lischke, D. Merten, G. Gherghel, G. Büchel and E. Kothe. 2009. Heavy metal resistance to the extreme: *Streptomyces* strains from a former uranium mining area. Che. Erde–Geochem. 29: 35–44.

Schmidt, A., M. Hagen, E. Schütze, A. Schmidt and E. Kothe. 2010. *In silico* prediction of potential metallothioneins and metallohistins in actinobacteria. J. Basic. Microbiol. 50: 562–569.

Shi, W. and D.R. Zusman. 1993. Fatal attraction. Nature. 366: 414–415.

Silbergeld, E.K., J. Graham and L.B. Price. 2008. Industrial food animal production, antimicrobial resistance, and human health. Annu. Rev. Public. Health. 29: 151–169.

Singh, M.P. and M. Greenstein. 2000. Antibacterial leads from microbial natural products discovery. Curr. Opin. Drug. Discov. Devel. 3: 167–176.

Sosa, A. 2007. The Threat of Antibiotic-resistant Bacteria and the Development of New Antibiotics. In Antimicrobial Resistance in Bacteria (Amábile-Cuevas, C.F., ed.), pp. 7–24, Horizon Bioscience.

Sprocati, A.R., C. Alisi, L. Segre, F. Tasso, M. Galletti and C. Cremisini. 2006. Investigating heavy metal resistance, bioaccumulation and metabolic profile of a metallophile microbial consortium native to an abandoned mine. Sci. Total Environ. 366: 649–658.

Stoppel, R.D. and H.G. Schlegel. 1995. Nickel resistant bacteria from anthropogenically nickel polluted and naturally nickel percolated ecosystems. Appl. Environ. Microbiol. 61: 2276–2285.

Sun, L.N., Y.F. Zhang, L.Y. He, Z.J. Chen, Q.Y. Wang, M. Qian and X.F. Sheng. 2010. Genetic diversity and characterization of heavy metal-resistant-endophytic bacteria from two copper-tolerant plant species on copper mine wasteland. Bioresour. Technol. 101: 501–509.

Szpunar, J. 2005. Advances in analytical methodology for bioinorganic speciation analysis: metallomics, metalloproteomics and heteroatom-tagged proteomics and metabolomics. Analyst. 130: 442–465.

Tamburini, E. and G. Mastromei. 2000. Do bacterial cryptic genes really exist? Res. Microbiol. 151: 179–182.

Theuretzbacher, U. 2011. Resistance drives antibacterial drug development. Curr. Opin. Pharmacol. 11: 433–438.

Ueberschaar, N., B.S. Ndejouong, L. Ding, A. Maier, H.H. Fiebig and C. Hertweck. 2011. Hydrazidomycins, cytotoxic alkylhydrazides from *Streptomyces atratus*. Bioorg. Med. Chem. Lett. 19: 5839–5841.

Waksman, S.A. 1947. What is an antibiotic or an antibiotic substance? Mycologia. 39: 565–569.

Wietzorrek, A. and M. Bibb. 1997. A novel family of proteins that regulates antibiotic production in streptomycetes appears to contain an OmpR-like DNA-binding fold. Mol. Microbiol. 25: 1181–1184.

CHAPTER 5

Overview of Copper Resistance and Oxidative Stress Response in *Amycolatopsis tucumanensis*, a Useful Strain for Bioremediation

José S. Dávila Costa,[1,*] Erika Kothe,[2] María J. Amoroso[1,4] and Carlos M. Abate[1,3,4]

Introduction

Heavy metals and remediation strategies

Heavy metal pollution is one of the most important environmental problems today. Various industries, such as mining and smelting of metalliferous, surface finishing industry, energy and fuel production, fertilizer and pesticide industry and application, leatherworking, photography, aerospace

[1]Pilot Plant of Industrial and Microbiological Processes (PROIMI), CONICET. Av. Belgrano y Pasaje Caseros, 4000 Tucumán, Argentina.
[2]Institute of Microbiology, Friedrich-Schiller-University, Neugasse, 07743 Jena, Germany.
[3]Natural Sciences College and Miguel Lillo Institute, National University of Tucumán, 4000 Tucumán, Argentina.
[4]Biochemistry, Chemistry and Pharmacy College, National University of Tucumán, 4000 Tucumán, Argentina.
*Corresponding author: jsdavilacosta@gmail.com

and atomic energy installation, etc.; produce and discharge wastes containing different heavy metals into the environment. There is a growing shortage of metals as a kind of resource and this also brings about serious environmental pollution, threatening human health and ecosystem. Three kinds of heavy metals are considered: toxic metals (such as Hg, Cr, Pb, Zn, Cu, Ni, Cd, As, Co, Sn, etc.), precious metals (such as Pd, Pt, Ag, Au, Ru, etc.) and radionuclides (such as U, Th, Ra, Am, etc.) (Wang and Chen 2006).

Conventional physicochemical treatment methods, which have been practiced for several decades for the removal of toxic heavy metals from waste water, may be ineffective or very expensive. These disadvantages become more pronounced at metal concentration less than 100 mg L^{-1}. It is well known that heavy metal contamination is a widespread phenomenon. This is especially true in developing countries where high-cost remediation technology is not affordable (Tsui et al. 2006).

The potential of using bioaccumulation or biosorption, which have been used for the removal of heavy metal ions by microorganisms, has become an attractive subject over the past decade (Haferburg et al. 2007). It is the pragmatic goal of current bioprocess research on metal removal from treatable sources to identify species of microorganisms that are capable of efficient uptake environmentally and economically important metals such as copper. Therefore, screening for microbes with high accumulation capacities and studying their stable resistance characteristics is an inevitable part of any remediation strategy (Vidali 2001).

Bioremediation as sustainable strategy for cleaning up the environment

Environmental biotechnology is not a new field; composting and waste water treatments are familiar examples of old environmental biotechnologies. However, numerous studies in molecular biology and ecology offer opportunities for more efficient biological processes. Notable accomplishments of these studies include the clean-up of polluted water and land areas (Vidali 2001).

Bioremediation can be defined as any process that uses biological agents, such as bacteria, fungi, or green plants, to remove or neutralize contaminants in polluted soil or water. Bacteria and fungi generally work by breaking down contaminants into less harmful substances. Plants can be used to aerate polluted soil and stimulate microbial action. They can also absorb contaminants such as salts and metals into their tissues, which are then harvested and disposed of.

There are a number of advantages of bioremediation, which may be employed in areas which cannot be reached easily without excavation. For example, hydrocarbon spills (or more specific: gasoline) may contaminate groundwater well below the surface of the ground. Injecting the right organisms, in conjunction with oxygen-forming compounds, might significantly reduce concentrations after a period of time.

Potential microorganisms for bioremediation processes can be isolated from almost any environmental conditions. Microbes will adapt and grow at subzero temperatures, as well as extreme heat, desert conditions, in water, with an excess of oxygen, and in anaerobic conditions, with the presence of hazardous compounds or on any waste stream. The main requirements are an energy source and a carbon source. Because of the adaptability of microbes and other biological systems, these can be used to degrade or remediate environmental hazards. Probably, the main advantage of studying environmental strains for application in bioremediation technologies is that afterwards these can be used for bioaugmentation of polluted soil, without any considerable environmental impact (Benimeli et al. 2003, Polti et al. 2007, Albarracín et al. 2010a, Dávila Costa et al. 2011b).

Amycolatopsis genus

In 1986, Lechevalier et al. proposed the formation of the genus *Amycolatopsis* (family *Pseudonocardiaceae*; Zhi et al. 2009) to accommodate nocardioform actinobacteria that contain meso-diaminopimelic acid, arabinose and galactose in their cell wall peptidoglycan and lack mycolic acids. It currently contains 47 species with validly published names (http://www.bacterio.cict.fr/a/amycolatopsis.html, accessed December 2011) and their representatives have been studied thoroughly because of their important secondary metabolism (e.g., vancomycin and rifamycin); nevertheless, this genus had never been studied in the field of bioremediation until the time that *Amycolatopsis tucumanensis* was isolated from copper-polluted sediments (Albarracín et al. 2005).

Copper duality

In the primordial, anaerobic world, copper existed only in Cu(I) state as insoluble sulphides. However, the ensuing oxygen evolution of the early atmosphere by microorganisms, triggered the oxidation of insoluble Cu(I) to soluble and hence more bioavailable Cu(II) (Crichton and Pierre 2001). Copper has a dual nature as an essential yet dangerous metal ion; it is a necessary cofactor for many proteins and enzymes in a variety of key biological processes, but it can also be highly toxic, partly due to its affinity for binding sites that should be occupied by other metals. Copper resistance mechanisms normally include (1) efflux transporters, (2) intracellular copperbinding chaperones, or (3) unspecific chelators such as glutathione or mycothiol (Stoyanov et al. 2001, Rensing and Grass 2003); Cu(II) handling is estimated to involve almost half of the copper proteome of a given organism (Bertini et al. 2010).

Since copper is a redox-active metal, it is a potential candidate for engaging as a catalyst in the formation of Reactive Oxygen Species (ROS), causing serious damage cellular components. The production of these oxidants as well as the protection against them is intrinsic to every living cell; however, in conditions of stress, normal capacities of these mechanisms are insufficient, triggering cells to increase and expand their antioxidative network (Georgopoulus et al. 2002). In most living organisms superoxide dismutases (SOD) and catalases (CAT) are the main enzymes involved in antioxidant mechanisms (Cuypers et al. 2010); in addition, thioredoxin systems are essential for maintaining a cytoplasmatic reductive environment, appropriate DNA synthesis, and transcription regulation in cells (Attarian et al. 2009).

Amycolatopsis tucumanensis, a Strain with Bioremediation Skill

Copper resistance and cupric reductase activity

A. tucumanensis DSM 45259^T, the strain of a recently recognized novel species of the genus *Amycolatopsis*, has been studied for its remarkable copper resistance as well as for its ability to bioremediate copper polluted soil microcosms (Albarracín et al. 2005, 2010b, Dávila Costa et al. 2011a,b).

A. tucumanensis has remarkable copper resistance; its resistance strength was highlighted against other actinobacteria strains, both, resistant (*Streptomyces* sp. AB5A) and sensitive (*Streptomyces coelicolor* DSM 40783^T and *Amycolatopsis eurytherma* DSM 44348^T) as it showed the highest growth in Cu(II) amended media (Fig. 1) (Dávila Costa et al. 2011a).

Copper specific bioaccumulation in *A. tucumanensis* increased with incubation time up to a maximum level on day 5 when incubated in liquid medium amended with 32 mg L^{-1} of Cu(II). *A. tucumanensis* showed the highest bioaccumulation value for growing cells (25 mg g^{-1}) among other microorganisms previously proposed for bioremediation processes (Cabral 1992, Amoroso et al. 1998, Albarracin et al. 2005). The high uptake of copper in *A. tucumanensis* is initially associated with its ability to bind copper, intracellulary and also in its exopolymeric fraction (Albarracín et al. 2008, Dávila Costa et al. 2011a,b). The exopolymer which was characterized as an exopolysaccharide (EPS), is formed only by glucose and it has been shown to be green when cells were incubated in the presence of copper; under controlled conditions (without copper) the EPS is white or cream colored. This suggests a copper binding property of the EPS, as has been proposed

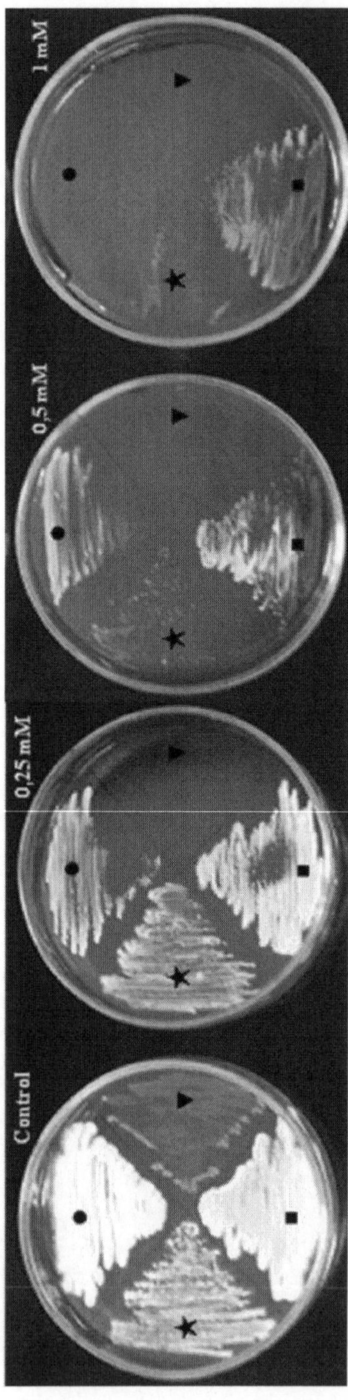

Figure 1. Copper resistance profiles of the selected strains on culture medium were added to Cu(II) at different concentrations (0.25, 0.50 and 1 mM). *A. tucumanensis* (filled square), *Streptomyces* sp. AB5A (filled star), *S. coelicolor* (filled inverse triangle), *A. eurytherma* (filled circle).

previously for other microorganisms. Regarding intracellular copper, it was found: in the cytosol (68%), cell wall (11%) and membranes (3%) (Albarracín et al. 2008, Ozturk et al. 2010).

As has been shown, copper exited only in Cu(I) state as insoluble sulphides in the anaerobic world. However, the presence of oxygen in the atmosphere triggered off the oxidation of insoluble Cu(I), to soluble Cu(II). As an ancient relict, Cu(I) is the preferred form for handling copper by the cell, copper ATPases or eukaryotic CtrI-type transporters appear to transport Cu(I) while copper chaperones bind Cu(I) for its delivery to cuproenzymes (Finney and O'Halloran 2003). In this sense, relationship between cupric reductase activity (RA_{Cu}) and copper resistance strength was assessed in *A. tucumanensis*. RA_{Cu} obtained for *A. tucumanensis* were on average 65% higher than those obtained to others copper-resistant and copper-sensitive actinobacteria strains; these results clearly demonstrated that RA_{Cu} was higher as higher the copper resistance strength of the strain, suggesting that these two abilities may be in a close relationship (Dávila Costa et al. 2011a). Besides, time is one of the parameters which could have influence in the cost/benefit relationship during a bioremediation process; as we expected, *A. tucumanensis* also showed the best values of percentage of total Cu(II) reduced per minute at different temperatures (Table 1) (Dávila Costa et al. 2011a).

Table 1. Percentage of total Cu(II) reduced per minute.

	A. tucumanensis	*Streptomyces* sp. AB5A (copper-resistant strain)	*A. eurytherma* (copper-sensitive strain)	*S. coelicolor* (copper-sensitive strain)
10°C	**1.82±0.09**	0.86±0.10	0.23±0.03	0.13±0.07
25°C	**2.09±0.07**	1.35±0.10	0.20±0.13	0.19±0.05
30°C	**2.21±0.02**	1.00±0.08	0.34±0.11	0.26±0.04
45°C	**2.12±0.19**	1.05±0.04	0.58±0.07	0.51±0.08

Bioremediation of copper-contaminated soil microcosms

As we mentioned above, development of promising new bioremediation strategies for heavy metal polluted areas, is of concern nowadays. Soil bioremediation constitutes a special challenge because of its heterogeneity and also because well adapted microorganisms are needed to bioremediate this particular environment. Hence, it is essential to research the application of microorganisms to experimentally heavy-metal polluted soil microcosms (Tabak et al. 2005). Consequently, taking into account the noteworthy copper-resistance strength of *A. tucumanensis,* it was assessed for its ability to clean up copper polluted soil microcosms and, to address the success of the bioremediation process, *Zea mays* plants grown were used in this bioaugmented soil microcosm as bioindicators.

Probably, the versatility of *A. tucumanensis* for growing well in uncommon culture media (including soil medium) with unusual carbon and nitrogen sources and wide pH and temperature ranges, was one of the most important detected features during these assays (Albarracín et al. 2010b, Dávila Costa et al. 2011a). This strain was able to abundantly colonized soil microcosms experimentally contaminated with 80 mg of copper kg^{-1} of soil, reaching a maximal population of 2.5 x 10^9 CFU g^{-1} of soil. The soil bioaugmentation microcosm assays were carried out at 30°C, a temperature that some soils in the Northwest of Argentina can reach during summer; nevertheless, an acceptable soil colonization of *A. tucumanensis* in soil microcosm was observed when the strain was kept at room temperature (22°C) as well (Albarracín et al. 2010b). Copper immobilization ability of *A. tucumanensis* on soil was assessed measuring the bioavailable copper in the soil solution extracted from soil microcosms supplemented with 80 mg of copper kg^{-1} of soil, by using chemical and physical methods and, thereby, 31% lower amounts of the metal were found in soil solution as compared to soil microcosm non-bioaugmented with *A. tucumanensis*. The results obtained when using *Z. mays* as bioindicator revealed, 20% and 17% lower tissue contents of copper in roots and leaves, respectively, after using *A. tucumanensis* to bioremediate soil microcosm. These results confirmed the efficiency of the bioremediation process of *A. tucumanensis* (Albarracín et al. 2010b).

Antioxidant response upon copper-induced stress

Several physiological and pathological states have been associated with high levels of Reactive Oxygen Species (ROS), a condition termed oxidative stress; ROS readily react with and damage vital cellular structures, among them lipids, DNA, and proteins (Leichert et al. 2007). So far, an efficient copper-induced oxidative stress response had been never studied as a possible resistance mechanism and within actinobacteria, some genera such as *Amycolatopsis* lack of any information (Dávila Costa et al. 2011b). Antioxidative network within actinobacteria was widely studied in *Mycobacterium* genus but mainly with the purpose of attacking its protection system, making these pathogenic cells more vulnerable (Newton et al. 2008).

ROS production in *A. tucumanensis* increased when increasing the copper specific biosorption (Table 2). In turn, it was not expected to find an outstandingly high basal level of ROS production in *A. tucumanensis*. On this fact, one should consider that *A. tucumanensis* was isolated from an extreme environment like riverside sediments with 600 mg Cu(II) kg^{-1} (normal

Table 2. Copper specific biosorption and ROS production of the cultured cells grown until exponential phase.

	Copper specific biosorption (mg Cu g^{-1} of cells)			
	0	6	10	15
ROS production (fluorescence intensity mg^{-1} protein)	310	524	743	900

Cu(II) concentration is up to 50 mg Cu(II) kg^{-1}) (Georgopoulus et al. 2002). The high basal level of ROS production found in *A. tucumanensis* might be inherent to this strain in order to keep well-adapted antioxidant machinery, which was essential to survive in a copper-challenged environment (Dávila Costa et al. 2011b).

Superoxide anion ($O_2^{\cdot-}$) is formed by a single electron transfer to molecular oxygen and it is further reduced to hydrogen peroxide either chemically or by the action of superoxide dismutase (SOD); subsequently hydrogen peroxide is detoxified by catalase (CAT). SOD activity in *A. tucumanensis* increased together with the concentration of copper in the culture medium in a range from 0 to 20 mg L^{-1} of Cu(II); however, no significant difference between activities from cells grown in presence of 20 and 30 mg L^{-1} of Cu(II) were observed. Regarding CAT activity in culture medium with 10 mg L^{-1} of Cu(II), *A. tucumanensis* displayed ca. 1.3 times higher levels when compared to the control without copper; nevertheless CAT activity decreased in the presence of 20 and 30 mg L^{-1} of Cu(II) (Dávila Costa et al. 2011b).

Several reports have highlighted the significance of the thioredoxin system [thioredoxin reductase (TrxR) and the redox active protein thioredoxin (Trx)] in the oxidative stress response, suggesting that loss of thioredoxin reductase results in a cellular damage (Ritz et al. 2000, Trotter and Grant 2002). The TrxR activity profiles obtained from *A. tucumanensis* were higher as higher the concentration of copper in the culture medium, moreover, TrxR activity in the presence of 30 mg L^{-1} of Cu(II) increased 2.5-fold relative to the control without copper.

Metallothioneins (MT) are proteins or polypeptides linked to protection against high concentration of heavy metals, metal homeostasis and oxidative stress (Palmiter 1998, Suhy et al. 1999, Robinson et al. 2001). *A. tucumanensis* grown in culture medium supplemented with 10, 20 and 30 mg L^{-1} of Cu(II) had significantly higher MT concentration than in the control without copper. In addition, levels of MT in untreated and copper-treated cells from *A. tucumanensis* were on an average 3.5-fold more elevated compared to another copper-sensitive *Amycolatopsis* strain (Dávila Costa et al. 2011b).

Morphological damage in copper-challenged cultures

Cell death has long been known to be a fundamental feature of animal development, but recently it has become a fashionable subject of general biological interest. Most studies on cell death have been carried out on animal cells. By contrast, little or no attention has been given to the process of cell death in prokaryotes, although they have been widely used as models for the study of many other basic cellular processes (Miguélez et al. 1999).

The morphological features of hyphal development on copper amended media in *A. tucumanensis* were analyzed by scanning electron microscopy. When growing on culture medium amended with 10, 20 and 30 mg L^{-1} of Cu(II), *A. tucumanensis* produced well-developed, branched substrate mycelium as well as a profuse aerial mycelium (Fig. 2); both types of hyphae fragmented into squarish rod shaped elements. The aerial mycelium displayed spore-like structures with smooth surface in long, straight to flexuous chains (Fig. 2). All these properties are consistent with the normal

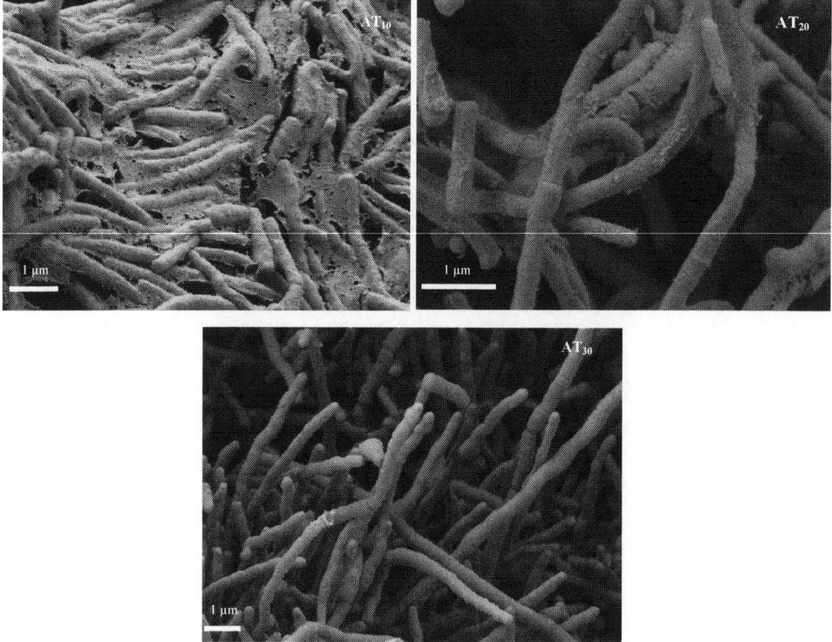

Figure 2. Changes in morphology on copper-increasing concentrations in *A. tucumanensis* (AT). 10, 20 and 30: concentration of copper (mg L^{-1}) in the culture medium. No degenerative hyphae were observed in *A. tucumanensis* at any copper concentrations.

morphological characteristics of the genus *Amycolatopsis*, meaning that the morphology of our strain was not appreciably affected by the toxic effects of copper (Dávila Costa et al. 2011b).

On the contrary, when the copper-sensitive strain *Amycolatopsis eurytherma* was grown in culture medium supplemented with high concentrations of copper (10, 20 and 30 mg L^{-1}), it displayed a patent morphological alteration. Only 10 mg L^{-1} of copper added to the media, caused symptoms of cellular degeneration such as discrete depressions along the wall together with hyphae shrinkage and collapse in the aerial as well as the substrate mycelium (Fig. 3). At higher concentrations (20 and 30 mg L^{-1}), it was observed that an intricate network of substrate hyphae was in different stages of cellular degeneration; a fraction of the hyphal population metamorphosed into chains of spores, while nonsporulating hyphae degenerated and died, displaying tubular-deflated structures (Fig. 3). Since many hyphae displayed aberrant morphologies, the toxic effect of copper in *A. eurytherma* morphology was irreversible (Dávila Costa et al. 2011b).

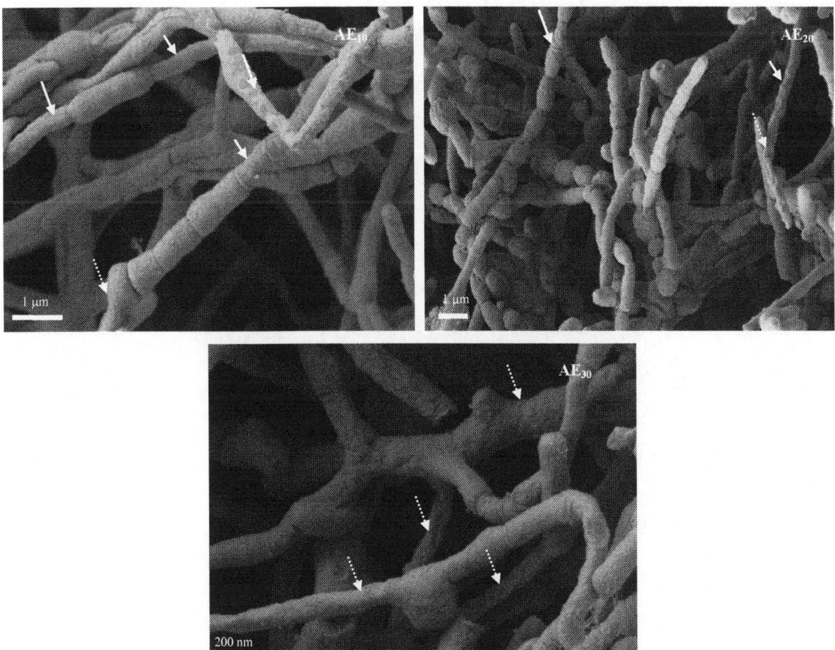

Figure 3. Changes in morphology upon copper-increasing concentrations in the copper-sensitive strain, *A. eurytherma* (AE). 10, 20 and 30: concentration of copper (mg L^{1}) in the culture medium. Depressions along the wall (short arrows). Degenerating hyphae (long arrows). Tubular-deflated structures (long dashed arrows).

Conclusion

In summary, it was clearly demonstrated that *A. tucumanensis* has remarkable copper resistance as well as efficient cupric reductase activity at different temperatures. Its copper resistance strength was highlighted against other actinobacteria strains as it showed the highest growth in Cu(II) amended media, and the highest copper-specific biosorption value (25 mg Cu g^{-1} of cells); the phenotype of the bioaccumulation of copper involved: metal biosorption by EPS and intracellular accumulation. It was also revealed that cupric reductase activity was higher as higher the copper resistance strength of the strain, suggesting that these two abilities may be in close relationship.

Biosorbed copper induced ROS production in *A. tucumanensis*, nonetheless, a noteworthy antioxidant mechanism given by high levels of antioxidant enzymes and metallothioneins was detected in this strain. This efficient antioxidant mechanism allowed *A. tucumanensis* to avoid serious damage in its morphology.

Over the past years several studies were carried out in *A. tucumanensis* in order to get a more detailed picture of its promising features; fortunately our results clearly contributed to enlarge the knowledge on the complex cupric-resistome of this novel strain. Further work in this direction will enable us to obtain new tools to modify this strain genetically to be used in large-scale soil bioremediation strategies.

References Cited

Albarracín, V.H., M.J. Amoroso and C.M. Abate. 2005. Isolation and characterization of indigenous copper resistant actinomycete strains. Chem. Erde-Geochem. 65: 145–156.

Albarracín, V.H., B. Winik, E. Kothe, M.J. Amoroso and C.M. Abate. 2008. Copper bioaccumulation by the actinobacterium *Amycolatopsis* sp. AB0. J. Basic Microbiol. 48: 323–330.

Albarracín, V.H., P. Alonso-Vega, M. Trujillo, M.J. Amoroso and C.M. Abate. 2010a. *Amycolatopsis tucumanensis* sp. nov., a novel copper resistant actinobacterium isolated from polluted sediments. Int. J. Syst. Evol. Microbiol. 60: 397–401.

Albarracín, V.H., M.J. Amoroso and C.M. Abate. 2010b. Bioaugmentation of copper polluted soil with *Amycolatopsis tucumanensis* to disminish phytoavailable copper for *Zea mays* plants. Chemosphere. 79: 131–137.

Amoroso, M.J., G.R. Castro, F.J. Carlino, N.C. Romero, R.T. Hill and G. Oliver. 1998. Screening of heavy metal-tolerant actinomycetes isolated from the Salí River. J. Gen. Appl. Microbiol. 44: 129–132.

Attarian, R., C. Bennie, H. Bach and Y. Av-Gay. 2009. Glutathione disulfide and S-nitrosoglutathione detoxification by *Mycobacterium tuberculosis* thioredoxin system. FEBS Lett. 583: 3215–3220.

Benimeli, C.S., M.J. Amoroso, A.P. Chaile and G.R. Castro. 2003. Isolation of four aquatic *Streptomycetes* strains capable of growth on organochlorine pesticides. Bioresour. Technol. 89: 133–138.

Bertini, I., G. Cavallaroa and K.S. McGreevya. 2010. Cellular copper management—a draft user's guide. Coordination Chemistry Reviews. 254: 506–524.

Cabral, J.P.S. 1992. Selective binding of metal ions to *Pseudomonas syringae* cells. Microbios. 71: 47–53.

Crichton, R.R. and J.L. Pierre. 2001. Old iron, young copper: from Mars to Venus. Biometals. 14: 99–112.

Cuypers, A., M. Plusquin, T. Remans, M. Jozefczak, E. Keunen, H. Gielen, K. Opdenakker, A. Ravindran Nair, E. Munters, T.J. Artois, T. Nawrot, J. Vangronsveld and K. Smeets. 2010. Cadmium stress: anoxidative challenge. Biometals. 23: 927–940.

Dávila Costa, J.S., V.H. Albarracín and C.M. Abate. 2011a. Cupric reductase activity in copper-resistant *Amycolatopsis tucumanensis*. Water Air Soil Poll. 216: 527–535.

Dávila Costa, J.S., V.H. Albarracín and C.M. Abate. 2011b. Responses of environmental *Amycolatopsis* strains to copper stress. Ecotoxicol. Environ. Saf. 74: 2020–2028.

Finney, L. and T. O'Halloran. 2003. Transition metal speciation in the cell: insights from the chemistry of metal ion receptors. Science. 300: 931–936.

Georgopoulus, P.G., A. Roy, R.E. Opiekun, M.J. Yonone-Lioy and P.J. Lioy. 2002. Environmental dynamics and human exposure to copper, Vol 1: Environmental dynamics and human exposure issues. International Copper Association Ltd. New York, USA.

Haferburg, G., M. Reinicke, D. Merten, G. Büchel and E. Kothe. 2007. Microbes adapted to acid mine drainage as source for strains active in retention of aluminium or uranium. J. Geochem. Expl. 92: 196–204.

Lechevalier, M.P., H. Prauser, D.P. Labeda and J.S. Ruan. 1986. Two new genera of nocardioform actinomycetes: *Amycolata* gen. nov. and *Amycolatopsis* gen. nov. Int. J. Syst. Bacteriol. 36: 29–37.

Leichert, L.I., F. Gehrke, H.V. Gudiseva, T. Blackwell, M. Ilbert, A.K. Walker, J.R. Strahler, P.C. Andrews and U. Jakob. 2007. Quantifying changes in the thiol redox proteome upon oxidative stress in vivo. PNAS. Doi 10.1073 pnas.0707723105.

Miguélez, E.M., C. Hardisson and M.B. Manzanal. 1999. Hyphal death during colony development in *Streptomyces* antibioticus: Morphological evidence for the existence of a process of cell deletion in a multicellular prokaryote. J. Cell Biol. 145: 515–525.

Newton, G.L., N. Buchmeier and R.C. Fahey. 2008. Biosynthesis and functions of mycothiol, the unique protective thiol of Actinobacteria. Microbiol. Mol. Biol. Rev. 72: 471–494.

Ozturk, S., B. Aslim and Z. Suludere. 2010. Cadmium(II) sequestration characteristics by two isolates of *Synechocystis* sp. in terms of exopolysaccharide (EPS) production and monomer composition. Bioresour. Technol. 101: 9742–8.

Palmiter, R.D. 1998. The elusive function of metallothioneins. Proc. Natl. Acad. Sci. USA. 95: 8428–8430.

Polti, M.A., M.J. Amoroso and C.M. Abate. 2007. Chromium (VI) resistance and removal by actinomycete strains isolated from sediments. Chemosphere 67: 660–667.

Rensing, C. and G. Grass. 2003. *Escherichia coli* mechanisms of copper homeostasis in a changing environment. FEMS Microbiol. Rev. 27: 197–213.

Ritz, D., H. Patel, B. Doan, M. Zheng, F. Aslund, G. Storz and J. Beckwith. 2000. Thioredoxin 2 is involved in the oxidative stress response in *Escherichia coli*. J. Biol. Chem. 275: 2505–2512.

Robinson, N., S. Whitehall and J. Cavet. 2001. Microbial metallothioneins. Adv. Microb. Physiol. 44: 183–213.

Stoyanov, J., J. Hobman and N. Brown. 2001. CueR (YbbI) of *Escherichia coli* is a MerR family regulator controlling expression of the copper exporter CopA. Mol. Microbiol. 39: 502–512.

Suhy, D.A., K.D. Simon, D.I. Linzer and T.V. O'Halloran. 1999. Metallothionein is part of a zinc-scavenging mechanism for cell survival under conditions of extreme zinc deprivation. J. Biol. Chem. 274: 9183–9192.

Tabak, H.H., P. Lens, E.D. Van Hullebusch and W. Dejonghe. 2005. Developments in bioremediation of soils and sediments polluted with metals and radionuclides. Microbial processes and mechanisms affecting bioremediation of metal contamination and influencing metal toxicity and transport. Rev. Environ. Sci. Biotechnol. 4: 115–156.

Tsui, M.T.K., K.C. Cheung, N.F.Y. Tam and M.H. Wong. 2006. A comparative study on metal sorption by brown seaweed. Chemosphere. 65: 51–57.

Vidali, M. 2001. Bioremediation An overview. Pure Appl. Chem. 73: 1163–1172.

Trotter, E.W. and C.M. Grant. 2002. Thioredoxins are required for protection against a reductive stress in the yeast *Saccharomyces cerevisiae*. Mol. Microbiol. 46: 869–878.

Wang, J.L. and C. Chen. 2006. Biosorption of heavy metals by *Saccharomyces cerevisiae*: a review. Biotechnol. Adv. 24: 427–451.

Zhi, X.Y., W.J. Li and E. Stackebrandt. 2009. An update of the structure and 16S rRNA gene sequence-based definition of higher ranks of the class Actinobacteria, with the proposal of two new suborders and four new families and emended descriptions of the existing higher taxa. Int J. Syst. Evol. Microbiol. 59: 589–608.

CHAPTER 6

Multi-metal Bioremediation by Microbial Assisted Phytoremediation

Martin Reinicke,[1] Frank Schindler,[1] Martin Roth[2] and Erika Kothe[1,*]

Introduction

The contamination of the mesosphere with heavy metals is a worldwide problem, mainly caused by anthropogenic activities like mining processes. Acid mine drainage (AMD) associated with this activity is a common problem, especially when sulphidic, pyrite containing ores are exploited (Johnson 2003). Through oxidation and erosion, accelerated by microbial pyrite oxidation and leaching, acid is generated which results in environmental conditions mobilizing heavy metals. Rainfall and rising groundwater levels lead to heavy metal contaminated waters, which are bioavailable for both plants and other organisms.

Heavy metals may be essential, but are toxic at higher concentrations and influence environmental nutrient cycles (Valsecchi et al. 1995). The impact of high heavy metal concentrations and salinity can be seen with high conductivity, low organic matter content of substrates, low microbial biomass (Brookes and McGrath 1984, Chander et al. 1995), reduced plant

[1]Friedrich Schiller University Institute for Microbiology Neugasse 25, 07743 Jena, Germany
[2]Leibniz Institute for Natural Product Research and Infection Biology, Hans Knöll Institute, Beutenbergstr. 11a, 07745 Jena, Germany.
*Corresponding author: Erika.kothe@uni-jena.de

growth and loss of soil functions (Adriano 2001, Lee et al. 2002, Speir 2008). All these changes result directly in a reduction of soil activity and soil fertility (Chaudri et al. 1992). The uptake of heavy metals into plants, especially concentrating in edible parts, invokes danger of contaminating the food web also including that which human beings consumption (Rogival et al. 2007, Zhuang et al. 2009). High loads of heavy metals in groundwater effect plant growth and microbial populations. Finally, the self regeneration of the soil is limited (Baath 1989) and an active intervention is required for remediation.

A problem for bioremediation of heavy metal pollution is the fact that metals are not degradable and hence persist over a long period of time (Oliveira and Pampulha 2006). The redox state of metals is mainly relevant for mobility and thus for the risk of metal contaminations. The type of heavy metal contamination is dependent on the geological, hydrogeological and mineralogical conditions, the soil texture and pollution history. In most areas, not just one single heavy metal, but a multi-metal mixture of different concentrations is prevalent, and because of their motilities in hydrogeological systems, Cu, Fe and Ni are dominant, representing an important part of multi-metal contaminations worldwide.

Microbial communities are affected dramatically by heavy metal contaminations associated with a reduction of microbial biomass (Yang et al. 2007), changes in community structure and resulting activities (Kandeler et al. 1996, Pennanen et al. 1998, Crowley 2008). Under the harsh environmental conditions, only tolerant or resistant microorganisms will survive, provided they have developed strategies to reduce the toxic effects of the prevailing heavy metals (Ellis and Weightman 2003).

Actinomycetes are ubiquitously distributed in soil habitats and associated mostly with organic soil constituents. This large group of GC-rich Gram-positive bacteria includes the genus *Streptomyces*, which are well known as producers of secondary metabolites like antibiotics (Hopwood 2006). The broad range of secondary metabolites of these actinobacteria as well as their production of exoenzymes plays an important role in soil function and nutrient cycles. Their filamentous growth characteristics and the ability to form exospores is a particularly advantageous trait to establish this group of bacteria in soil and even more so for survival in polluted areas. According to their spore formation, the die-off during semi-dry conditions is lower, improving their persistence in low humidity areas. In the rizosphere, the distribution of actinomycetes is heterogeneous with higher abundances in the root zone where secondary metabolite producers are enriched (Basil et al. 2004), which is attributed at least in part to the production of plant growth promoting factors like IAA (Khamna et al. 2009), or the production of metal binding or chelating compounds such as siderophores (Dimkpa et al. 2008,

Dimkpa et al. 2009). Therefore actinomycetes are potential candidates for microbial assisted phytoremediation of heavy metal contaminated sites.

For remediation of heavy metal polluted sites, several methods including techniques like excavation and *ex situ* treatment, installation of wetlands, or bioremediation have been proposed. The latter allows handling of the pollution *in situ* without disturbance of soil structure and movement of contaminated material.

In situ bioremediation approaches can be classified into bioattenuation, in which the natural decrease of contamination is monitored, biostimulation, enhancing of degradation processes by addition of nutrients, electron acceptors or substrates, and bioaugmentation, inoculation of bacteria with capabilities to detoxify the contaminated sites (Iwamoto and Nasu 2001). Phytoremediation should be extended to the influence of bacteria related to root bacteria and can be divided into approaches of phytoextraction where contaminants are mobilized and taken up into above-ground harvestable plant biomass, which is then composted or burned (Salt et al. 1995), or phytostabilization where non-contaminated plant material can be produced although the contamination in the ground persists, e.g., for renewable energy production.

Re-vitalisation through Soil Amendments

One area that is well suited for investigations relating to mining activities and metal contamination is found at the former mining site near Ronneburg, Germany. After the active mining period of over 40 years, the mining site was remediated by the removal of highly contaminated material and covering the site with up to 40 cm uncontaminated, sandy-silty backfill substrate.

At this site, a field trial was established to understand the interdependencies of heavy metals, microbial populations and plant growth essential for phytoremediation. Additions of 5 cm of compost or topsoil ploughed in for two plots were compared to a similar, unamended plot, each 12 x 12 m (Büchel et al. 2005). New contaminations through capillary transport occurred when fractions of mobile heavy metals were rising to the surface, resulting in concentrations exceeding the limits given by the German law. The area was characterized by measuring the physicochemical parameters and analysis of the hydrogeological situation (Carlsson and Büchel 2005, Grawunder et al. 2009).

A mix of *Melilotus albus* Med (white clover) and *Festuca rubra commutata* Wf (red fescue) was used as a seed crop. While *Festuca* germinated and grew well, *Melilotus* did not perform well under the harsh conditions of the test field site.

Changes of Abiotic Parameters

The substrate at the field site shows no horizons or signs of soil formation, with sandy, nutrient-poor, low organic matter and high heavy metal and sulphate load material of pH values ranging from 3.5 to 4.5 in high local heterogeneity. A strongly reduced plant growth is visible outside the test field, where in 2004 a mixture of monocotyl species and clover was seeded. The soil additions at the test field shifted the physico-chemical parameters of the substrate to a more neutral environment with lower contamination.

All samples contained 15% clay, with 20% silt and 65% sand in the control and compost plots, while the topsoil plot contained 30% silt and 55% sand. Through soil addition, the content of organic carbon as well as nitrogen content increased, with a more pronounced effect seen with compost as compared to topsoil addition (Table 1). As shift in pH results in a higher rate of metal sorption (García-Sánchez et al. 1999, Merdy et al. 2009) and the introduction of organic matter, especially with the compost treatment, increased metal ion complexation by functional groups of organic particles, like humic and fulvic acids (Harter and Naidu 1995, Violante et al. 2010), better plant growth was expected.

Table 1. Soil parameter changes upon addition of topsoil or compost.

Plot (12 x 12 m)	pH	carbon [g*kg^{-1}]	nitrogen [g*kg^{-1}]	sulphur [g*kg^{-1}]
control	4.07	1.9	n.d.	0.1
topsoil	5.97	11.7	0.8	0.1
compost	6.95	30.1	2.5	0.1

n.d. not detectable, below measurement limit

Microbial Biomass and Activity

Heavy metal polluted soils are mostly characterized by reduced soil activity and low microbial biomass. The observed number of colony forming units in soil samples from 5 to 10 cm increased on amendments resulting in hundred to thousand fold microbial biomasses (Fig. 1). After one season with growth of *Festuca rubra*, the number of cultivable bacteria dropped in the compost four-fold, a two fold increase was seen in the topsoil plot. The highest effect was visible in the control plot, which had received no amendment. The plant-derived root exudates thus provided conditions that lead to an increase of more than ten fold in the microbial community. Thus, a sustainable effect of the soil amendments is seen, but the most prominent change could be observed by the initiation of pedogenesis with a root zone on the non-amended control plot (see Fig. 1).

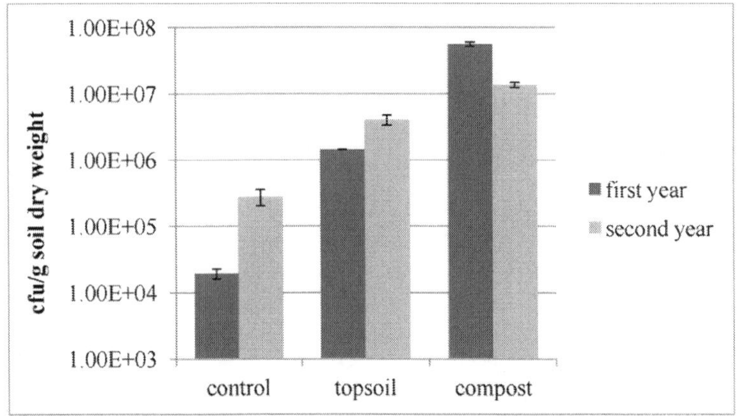

Figure 1. Microbial biomass after one year of addition of compost and topsoil examined by cultivation.

The influence of heavy metals on the number of exo- or endospores was tested by pre-treatment of soil and vegetative cells which were inactivated at 80°C for 1 hr. In contrast to our expectations, a low content of germinating spores (1% of CFU) in the control soil was measured while compost and topsoil heat treated soil samples contained 10 and 30% spores, respectively. This unexpected finding suggests that either the spores formed were limited in germination rates after additional stress had been applied during heat treatment, or else that vegetative Gram-positives with persistent cell walls preventing DNA extraction were present, something not seen in DNA-dependent community analyses.

Soil respiration is a basal parameter reflecting microbial activity (Isermeyer 1952). The increase in microbial biomass of the amended plots is reflected in soil respiration which was sustainable after one year. While in the control plot, 0.63 (± 0.12) µg CO_2 g^{-1} soil dry mass was formed, the topsoil plots and compost plots had 3.63 (± 0.14) and 22.83 (± 0.21) µg CO_2 g^{-1} soil dry mass formation.

Changes in Microbial Diversity

The composition of microbial communities is influenced by environmental conditions and soil properties. The functional diversity, in turn, affects the habitat through excretion of metabolites, degradation of organic matter and redox reactions. Contaminations, in the form of toxic, multi-metal and salt contaminations, generate a selective pressure with the consequence of loss of bacterial groups introduced with the amendments that do not share the ability to cope with these abiotic factors. In our case, microbial populations of the substrate without amendment were strongly dominated by Gram-positives, as determined by cultivation experiments, with the phylogenetic groups *Bacilli* and *Actinobacteria* dominating. Expected Gram-negative bacteria, like *Pseudomonas*, were not obtained by cultivation-dependent assessment of microbial diversity.

In comparison, cultivation-independent DNA-based methods revealed the presence of a large diversity of Gram-negatives. These data suggest that strong-walled Gram-positives as well as endo- or exospores might be strongly under-represented using DNA-dependent methods to access microbial diversity (Priha et al. 2004, Dauphin et al. 2009), which is not so obvious in normal soils. In the substrate of the test field, however, Gram-negatives which are obviously present, as evidenced by the sequencing strategy (Fig. 2), fail to grow in sufficient quantities to be detectable by cultivation.

The Gram-negatives identified in the DNA-based technology are, in large parts, reflecting the acidic and generally detrimental conditions, as 5 groups of acidobacteria represent a quarter of the community. The amendment with topsoil did not change this fraction of the community substantially (data not shown). In topsoil, however, Gamma-proteobacteria were absent, and bacilli replaced the clostridia. A few strains belonging to verrucomicrobia, planctomycetes and bacteroidetes were detected.

A much stronger effect is expected in the compost-amended plot where, with the compost layer about 10^8 cfu g^{-1} soil dry weight, typical soil and compost bacteria were added. The mixing resulted in approximately 30% compost and 70% original substrate, allowing us to predict a profile of the initially combined communities. The actual community structure observed after one year of adaption differed on the prediction in lower abundance of acidobacteria, Beta- and Gamma-proteobacteria and higher occurrence of verrucomicrobia, chloroflexi and planctomycetes (Fig. 2). This reflects a sustainable effect of soil treatment with small amounts of compost and a persistence of the influence over a longer time period.

Microbially Assisted Phytoextraction

To test the effect of microbial processes on metal element cycling, the control, topsoil and compost amended plots were divided into nine subplots, each 4m x 4m in size. To compare microbial inoculation effects on plant growth and metal uptake, the subplots were inoculated with 20 L pregrown culture of streptomycetes, or streptomycetes and mycorrhizal inoculum of *Glomus intraradices* Sy 167 at spore densities of around 700 spores m^{-2}, or left untreated, each treatment in three replicates (Schindler et al. 2012). The inoculated streptomycete strains *Streptomyces acidiscabies* E13 (Amoroso et al. 2000) and *S. tendae* F4 (Sineriz et al. 2009) had both been isolated from the same mining region.

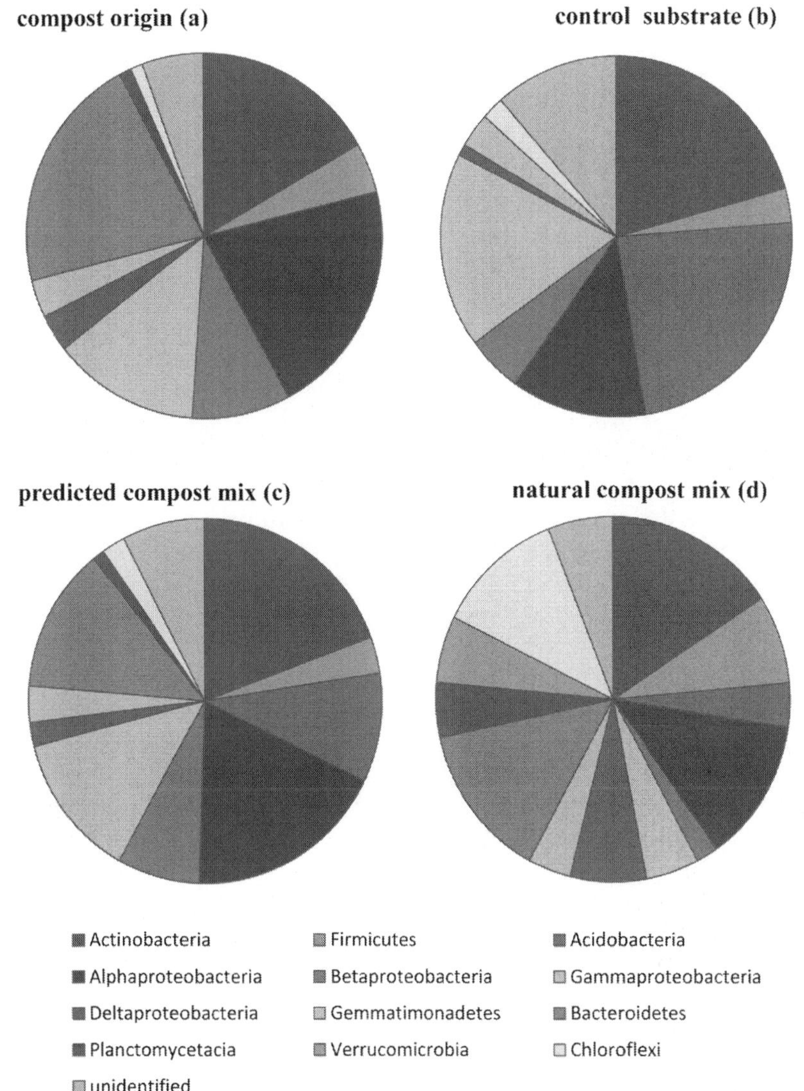

Figure 2. Composition of microbial communities observed in the control (unamended test field substrate; a) and the original compost used as amendment (b) were used to predict community structure after amendment (predicted compost mix; c). After one year of adaptation, the natural compost mix community structure was found (d).

In order to assess the effects, both total content and bioavailable content of metals need to be determined. The preferential uptake of elements into above-ground plant biomass can then be compared to the prevalence of metals in the substrate or water phase.

Element Availability

Sequential extractions (Zeien and Brümmer 1989) applicable to soil with low $CaCO_3$-content or $CaCO_3$-free soils (Grawunder et al. 2009) was used to determine seven fractions: (F1) mobile, (F2) specifically adsorbed, (F3) bound to Mn-oxides, (F4) bound to organic matter, (F5) bound to amorphous Fe-oxides, (F6) bound to crystalline Fe-oxides and (F7) the residual fraction. A sample from an undisturbed forest soil (Jenzig) was used as a control. Cu is common in various substrates and easily soluble in weathering processes, especially in acidic environments with mean levels from 13 to 24 mg/kg (Kabata-Pendias and Pendias 2001). Through the intensive use of fungicides, the concentration increased up to 1500 mg kg^{-1} in agricultural soils (Besnard et al. 2001). The major part of Cu is bound and not available for plants and microorganisms (Fig. 3). Due to the low pH, Cu is available to a somewhat larger extend in the untreated control substrate. Mixed with topsoil or compost, the organic fraction (F4) became more dominant, while Mn oxides decreased even though more than 50–60% is still tightly bound in the residual fraction. The inoculation had little effect on Cu bioavailability. Inoculation and treatment are therefore well suited for phytostabilization.

In general, Ni is siderophilic and attracted to metallic iron. It occurs mainly in organically bound forms in surface soil horizons, while Fe or Mn oxide-bound Ni seems to be the best source for plants (Norrish 1975). In our soil samples, Ni was inhomogeneously distributed with 40–55 µg g^{-1} still within the guidelines values (15–70 µg g^{-1}) of the German law (BBodSchV 1999). As most other metals, Ni was predominantly found in the crystalline Fe-bound and residual fractions (70–80%, Fig. 3). When compost was added to the substrate, the bioavailable Ni fraction, constituted less than 1% of the total content, while the organics-bound fraction was larger.

As seen with all soils, most of the iron is immobile in the test field substrate. However, due to siderophore production by strains used for augmentation (Dimkpa et al. 2008), a shift to more bioavailable Fe could be observed (Fig. 3).

Metal Uptake into Plant Biomass

The total concentrations of Cu and Ni in the soil range from 40 to 79 µg g^{-1}. The proportion of Fe in the soil is much higher and ranges between 42000 und 53000 µg g^{-1} (Fig. 4). Pearson's correlations with bilateral significances at the 0.01 level (s**) were calculated using IBM® SPSS® Statistics software (version 19 for Microsoft® Windows). These non-parametric correlations were calculated for all samples (element vs. element). Calculated Pearson index revealed no correlation between Cu, Fe and Ni for total contents.

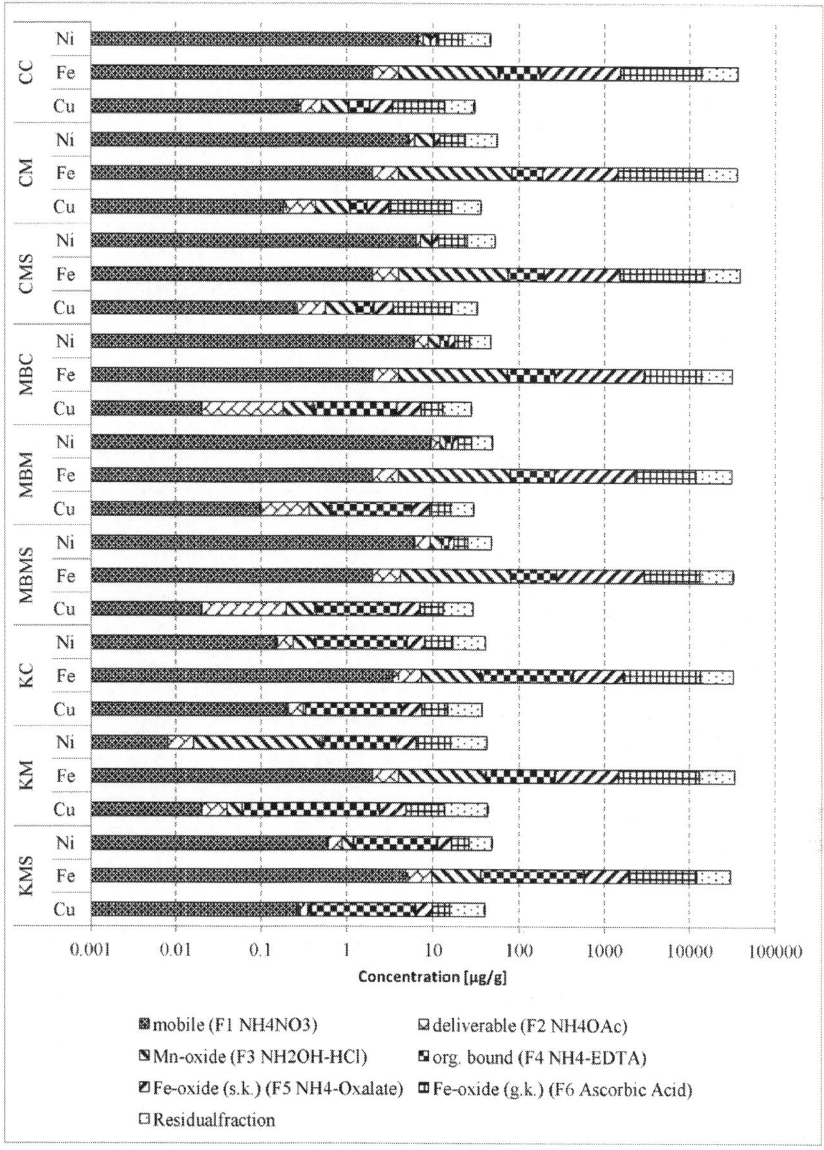

Figure 3. Sequential extraction profiles for Cu, Fe and Ni.

If the focus is on the bioavailable contents, Cu is almost constantly available in all plots, while Fe and Ni concentrations vary considerably between the substrates. The amendments increased the bioavailability of Fe and Ni. A correlation analysis using the Pearson index revealed a weak negative correlation between Cu or Fe and Ni (Table 2).

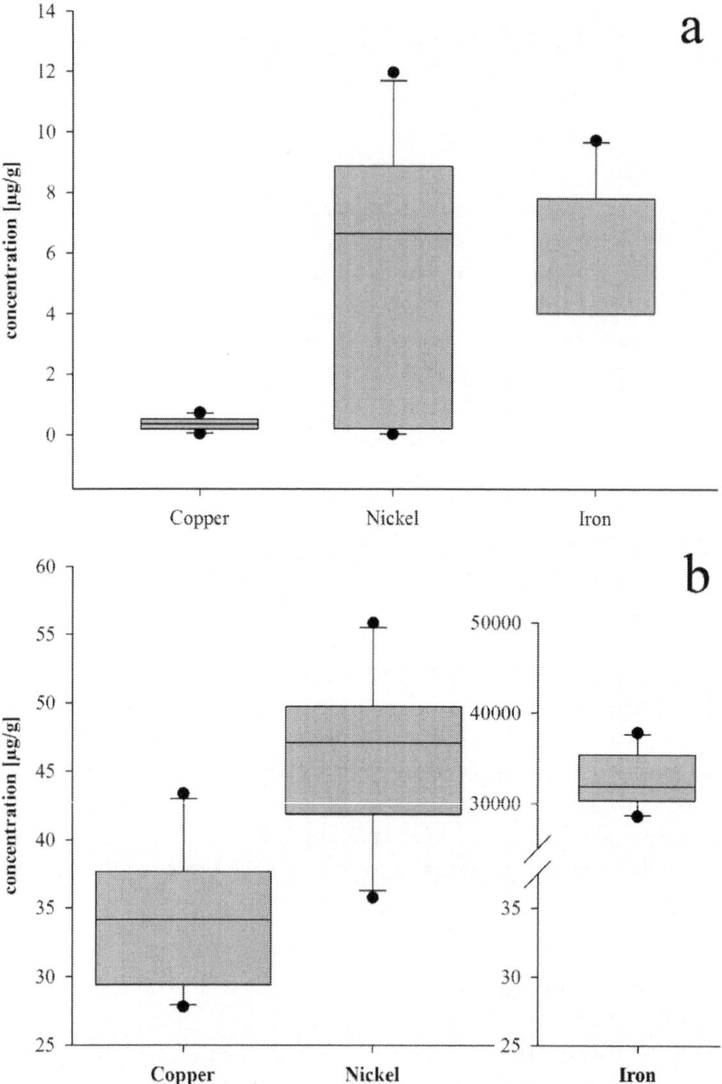

Figure 4. Total average (a) and bioavailable (b) contents of Cu, Fe, and Ni.

Plant performance was measured by biomass which varied greatly between the treatments. On an average, the control plots and the inoculated topsoil subplots had a total dry biomass of about 1400 g. A significant increase in the total dry biomass was seen in the topsoil amended subplots with double inoculation (MBMS; Fig. 5), caused mainly by the

Table 2. Pearson index for bioavailable contents.

		Copper	Iron	Nickel
Copper	Pearson Correlation	1	0.347	0.041
	Sig (2-tailed)		0.326	0.911
	N	10	10	10
Iron	Pearson Correlation	0.347	1	–0.706*
	Sig (2-tailed)	0.326		0.023
	N	10	10	10
Nickel	Pearson Correlation	–0.041	–0.706*	1
	Sig (2-tailed)	0.911	0.023	
	N	10	10	10

*Correlation is significant at the 0,05 level (2-tailed).

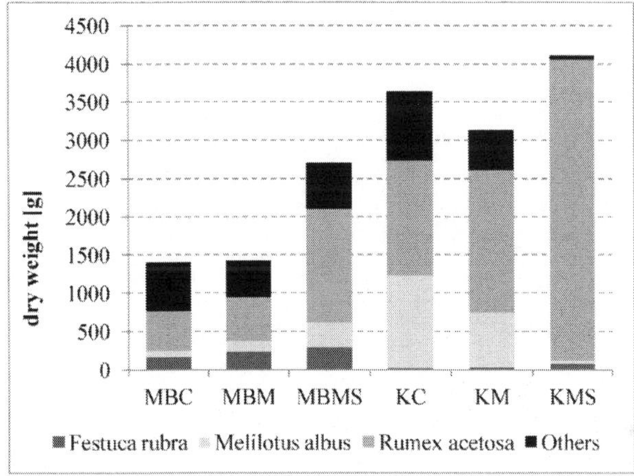

Figure 5. The total dry weight of different plant species.

weed *Rumex acetosa*. In the compost plot, *Melilotus* could grow, and again, *R. acetosa* had invaded. On the subplots inoculated with mycorrhiza and streptomycetes, the competition clearly shifted in favor of invasive species, which accounted for over 95% of the biomass. On the nutrient-enriched soil of the compost field, a significantly lower number of individuals were sufficient to achieve the highest dry biomass of more than 2 kg. All in all, plant growth was significantly improved by amendments of only 5 cm of topsoil or compost.

Metals are often chelated for better absorption. Iron can be released again from the chelator when it is reduced from Fe^{3+} to Fe^{2+} at the root surface. The chelator can diffuse back into the environment and react again with a new metal ion. After uptake, the ion is kept soluble by chelation with other organic compounds like citric acid and is then transported in the xylem. As Fe, Cu and Ni can undergo reversible oxidations, they have important roles in electron transfer and energy transformation and are thus essential elements for both plant and microbe physiology. Above a critical concentration in plant biomass, however, the toxic effect prevails at 20–100 mg kg^{-1} for Cu and 10–100 mg kg^{-1} for Ni. Plants are stressed both by deficiencies and excesses of metals. This is highly dependent on the plant species and their specific abilities to cope with the elements. Even within the same species, considerable variation is found between cultivars or ecotypes. To assess metal contents, three to five tufts of *Festuca* were homogenized and measured. No significant differences in Cu or Fe contents were seen in the above-ground biomass (Fig. 6). This reflects the tight homeostasis exerted for Fe, which is needed for hem-proteins, dehydrogenases and ferrodoxins (Kabata-Pendias and Pendias 2001).

Cu is used in various oxidases, different plastocyanins and ceniloplasmin. If too much Cu is present, intervenal chlorosis, white leaf margins and tips, as well as damaged root tips occur. The average plant contents are between 7.4–15 mg kg^{-1} for grasses (Anke et al. 1972). In the treatments, Cu reached the mean values for grasses which promoted a healthy growth.

The only known plant enzyme that contains Ni is urease. It catalyzes the hydrolysis of urea to form ammonia and carbon dioxide and therefore is essential for plant growth. High amounts of Ni result in leaf chlorosis, gray-green leaves and stunted growth. The average contents for grass are 1.3–2.5 mg kg^{-1} (Szentmihalyi et al. 1980).

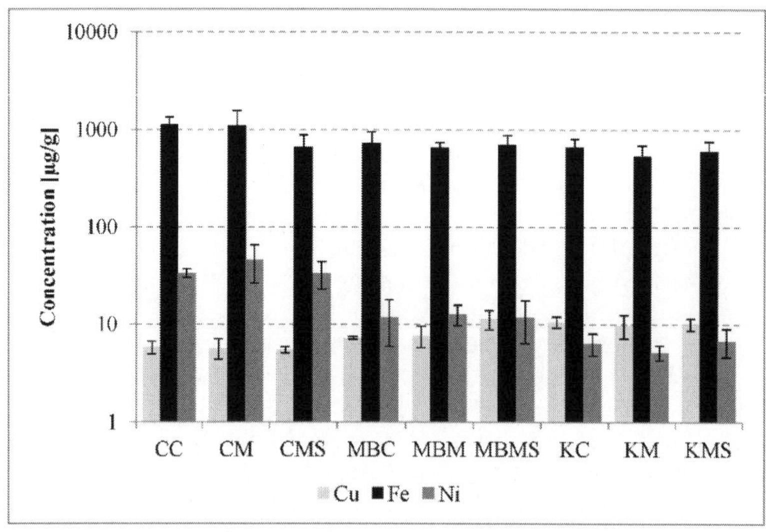

Figure 6. The metal content of the above-ground dry mass of *Festuca rubra*.

Festuca rubra showed increased Ni content in the above-ground biomass. With respect to Ni, increased plant contents were found with amounts about 5 times higher than average. No statistically relevant differences were visible between the roots. However, for Ni a clearly higher mobility was reflected in 5–10 fold higher roots content.

Transfer and Bioconcentration

The translocation factor for root-shoot transport is calculated as:

(1) TF = heavy metal in dry matter shoot (mg kg^{-1})/heavy metal in dry mass root (mg kg^{-1}).

Plants with a translocation factor above 1 for a particular heavy metal are considered suitable for phytoextraction. For *Festuca rubra,* no translocation factors larger than 1 could be found for the heavy metals Cu, Fe, and Ni. While Cu was transported from root to shoot with translocation factors ranging from 0.3–0.8, the highest values reached only on compost amended plots, Fe showed translocation factor values between 0.3–0.9, with the lowest values on topsoil amended plots irrespective of inoculation. For Ni, a trend for higher translocation upon inoculation and double inoculation was observed, with values between 0.4 and 0.9. Thus, this grass cannot be considered a hyperaccumulator, making this combination of *F. rubra* and inoculation with microorganisms a more suitable strategy for phytostabilization of Cu, Fe and Ni.

The transfer coefficients for the individual treatment types and elements are usually calculated from the total metal contents in soil. Since Ni was present in relatively high bioavailable concentrations in the non-amended control substrate, Ni uptake and transfer coefficients were higher on that plot with a decrease following inoculation. The nickel consumption was overall within the limits of plant uptake (Scheffer et al. 2008). In contrast, transfer coefficients were relatively small for Fe and Cu, which does include microbial assisted *Festuca rubra* among phytostabilization approaches.

However, more of interest for the ecotoxicological risk assessment is the bioconcentration factor, which is calculated not from the total, but the bioavailable metal concentration. This bioaccumulation factor is considered low at values lower than 250, while those over 1,000 are considered suitably high for bioextraction strategies. The plant-soil bioconcentration ratio (BCR) for plant dry mass relative to dry soil (soil solids) concentration is calculated:

(2) BCR = concentration in dry plant tissue (μg g^{-1})/concentration in dry soil (μg g^{-1})

The values found at the field site were all low to moderate (Fig. 7). The only significant difference between the treatments is due to the relatively low bioavailable concentrations in the compost amended plots. Taken together, a total extraction of metals from the entire test field area could be seen in the soil treated, which was negligible on the unamended control plot. This is mainly due to the low biomass *Festuca* achieved on the low fertility substrate. On the topsoil subplots, altogether 0.3–0.5 g Cu, 2–5 g Fe and 0.2–0.4 g Ni were extracted per 16 m^2. Per hectare, this would amount to 0.25 kg pure Cu, 2.2 kg Fe and 0.2 kg Ni on an average in one summer, despite the fact that plant concentrations were below threshold values.

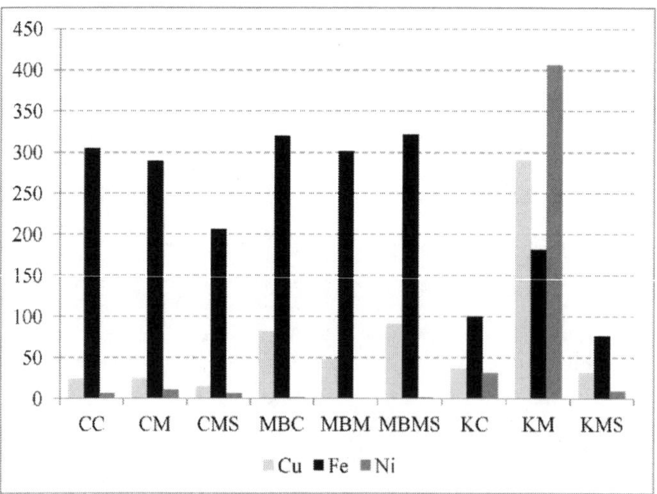

Figure 7. The bioconcentration factors (BCF) calculated with the bioavailable metal content of the substrate.

On the compost subplots, levels were a factor 10 lower. However, here a clear increase in extraction was seen when double inoculation was performed, this increase amounting to a 8 fold higher plant uptake for each of the three tested metals. Hence, microbial inoculation by both, streptomycetes and mycorrhizal (in contrast to mycorrhiza alone) could have a visible effect on plant metal uptake specifically, if additional nutrition and beneficial soil parameters were present due to compost addition.

Conclusion

In situ remediation is an environmentally sustainable solution for treatment of contaminated sites. In our case, a combination of phytoremediation coupled with bioaugmentation based on the addition of a low amount of topsoil or compost together with the inoculation of heavy metal resistant streptomycetes and mycorrhizal fungi, affected soil properties with a shift to more neutral conditions, reduction of conductivity and the enhancement of microbial activity. Especially the compost treatment provided nutrients with the organic matter, which additionally acts as buffer and reduces their mobility. The application of organic matter like biosolids, manure, compost or biochar is known for the reduction of element mobility by binding of mobile ions to functional groups in the organic matter or clay minerals (Walker et al. 2004, Clemente et al. 2005, Chan and Xu 2009).

In cultivation experiments, the microbial community was found to contain mainly spore forming bacteria, which, however, did not necessarily sporulate as seen by spore germination experiments. Spores are an advantage for many soil associated microbes, as they allow surviving unfavourable or stressful conditions in a metabolically inactive form (Nicholson et al. 2000, Nicholson et al. 2002). The reduction of spore formation in the heavy metal contaminated soil was unexpected, because spore formation is a typical stress induced process in endospore producing bacteria. However, spore resistance is affected by metal ion concentrations, which might have led to the reduced number of germinating spores after heat treatment. The exospore forming streptomycetes have been shown to produce a reduced number of spores under metal stress (Schmidt et al. 2005). Thus, specifically the endospore forming bacilli were found to be over-represented in our cultivation experiments by a higher number of fertile, but more heat sensitive spores.

Microbial communities are known to be dramatically affected in composition and function by heavy metals. The observed loss of non-adapted microorganisms from the addition of a comparatively small amount of compost was slower than expected, since one year after soil application, the measured and the predicted communities were very similar. This suggests that applications of small soil quantities, especially of compost, can have a major impact on both, soil properties and soil functions, over a longer period of time.

The impact of small soil quantities was also seen in metal mobility of the observed elements. Especially in the compost treated plots, the amount of mobile Ni and Cu was reduced, making this strategy of combined soil amendments and inoculation a suitable strategy for phytostabilization. The reduction of the bioavailable heavy metal through soil amendments specifically can minimize the wash-out by rainfall, thus lowering the risk of heavy metal ecotoxicity. Further, the combined addition of soil and microorganisms enhanced plant growth increasing the potential of heavy metal removal from the site. The use of plants to stabilize or extract heavy metals from the environment mediated by application of microorganisms is a promising approach to remediate contaminated sites in a non-invasive way.

Acknowledgements

The work was supported financially by the excellence graduate school JSMC and DFG GRK1257 "Alteration and element mobility at microbe-mineral interphases" and further the EU-FP 7 project UMBRELLA. The authors would particularly like to thank to FSU Jena Applied Geology staff especially G. Büchel, D. Merten, U. Buhler, I. Kamp and G. Weinzierl.

We thank the staff of the Bio Pilot Plant at the Hans Knöll Institute for large scale preparation of streptomycetes biomass.

References

Adriano, D.C. 2001. Trace elements in terrestrial environments : biogeochemistry, bioavailability, and risks of metals. New York, Springer.

Amoroso, M.J., D. Schubert, P. Mitscherlich, P. Schumann and E. Kothe. 2000. Evidence for high affinity nickel transporter genes in heavy metal resistant *Streptomyces* spec. J. Basic Microbiol. 40: 295–301.

Anke, M., B. Groppel, H. Lüdke, H. Felkl and J. Kleemann. 1972. Die Spurenelementversorgung der Wiederkäuer in der Deutschen Demokratischen Republik. Arch. Tierernahr. 22: 233–248.

Baath, E. 1989. Effects of Heavy-Metals in Soil on Microbial Processes and Populations (a Review). Water Air Soil Poll. 47: 335–379.

Basil, A.J., J.L. Strap, H.M. Knotek-Smith and D.L. Crawford. 2004. Studies on the microbial populations of the rhizosphere of big sagebrush (*Artemisia tridentata*). J. Ind. Microbiol. 31: 278–288.

BBodSchV. 1999. BBodSchV-Bundes-Bodenschutz- und Altlastenverordnung (BBodSchV). B. d. Justiz, Bundesministerium der Justiz. BGBl. I S. 1554.

Besnard, E., C. Chenu and M. Robert 2001. Influence of organic amendments on copper distribution among particle-size and density fractions in Champagne vineyard soils. Environ. Pollut. 112: 329–337.

Brookes, P.C. and S.P. McGrath. 1984. Effect of metal toxicity on the size of the soil microbial biomass. J. Soil Sci. 35: 341–346.

Büchel, G., H. Bergmann, G. Ebenå and E. Kothe. 2005. Geomicrobiology in remediation of mine waste. Chemie der Erde-Geochemistry 65, Supplement. 1: 1–5.

Carlsson, E. and G. Büchel. 2005. Screening of residual contamination at a former uranium heap leaching site, Thuringia, Germany. Chemie der Erde-Geochemistry 65, Supplement. 1: 75–95.

Chan, K.Y. and Z.H. Xu. 2009. Biochar—Nutrient Properties and their Enhancement. Biochar for Environmental Management: Science and Technology. J. Lehmann and S. Joseph. London, UK, Earthscan. 67–84.

Chander, K., P.C. Brookes and S.A. Harding. 1995. Microbial biomass dynamics following addition of metal-enriched sewage sludges to a sandy loam. Soil Biol. and Biochem. 27: 1409–1421.

Chaudri, A.M., S.P. McGrath and K.E. Giller. 1992. Survival of the indigenous population of *Rhizobium-Leguminosarum* biovar *trifolii* in soil spiked with Cd, Zn, Cu and Ni salts. Soil Biol. Biochem. 24: 625–632.

Clemente, R., D.J. Walker and M.P. Bernal. 2005. Uptake of heavy metals and As by *Brassica juncea* grown in a contaminated soil in Aznalcóllar (Spain): The effect of soil amendments. Environ. Pollut. 138: 46–58.

Crowley, D. 2008. Impacts of Metals and Metalloids on Soil Microbial Diversity and Ecosystem Function. Revista de la ciencia del suelo y nutrición vegetal. 8: 6–11.

Dauphin, L.A., B.D. Moser and M.D. Bowen. 2009. Evaluation of five commercial nucleic acid extraction kits for their ability to inactivate *Bacillus anthracis* spores and comparison of DNA yields from spores and spiked environmental samples. J. Microbiol. Methods. 76: 30–37.

Dimkpa, C., A. Svatos, D. Merten, G. Buchel and E. Kothe. 2008. Hydroxamate siderophores produced by *Streptomyces acidiscabies* E13 bind nickel and promote growth in cowpea (*Vigna unguiculata* L.) under nickel stress. Can. J. Microbiol. 54: 163–172.

Dimkpa, C.O., D. Merten, A. Svatos, G. Buchel and E. Kothe. 2009. Metal-induced oxidative stress impacting plant growth in contaminated soil is alleviated by microbial siderophores. Soil Biol. Biochem. 41: 154–162.

Ellis, M. and F. Weightman. 2003. Cultivation-dependent and -independent approaches for determining bacterial diversity in heavy-metal-contaminated soil. Appl. Environ. Microbiol. 69: 3223–3230.

García-Sánchez, A., A. Alastuey and X. Querol. 1999. Heavy metal adsorption by different minerals: application to the remediation of polluted soils. Sci. Total. Environ. 242: 179–188.

Grawunder, A., M. Lonschinski, D. Merten and G. Buchel. 2009. Distribution and bonding of residual contamination in glacial sediments at the former uranium mining leaching heap of Gessen/Thuringia, Germany. Chemie Der Erde-Geochemistry. 69: 5–19.

Grawunder, A., M. Lonschinski, D. Merten and G. Büchel. 2009. Distribution and bonding of residual contamination in glacial sediments at the former uranium mining leaching heap of Gessen/Thuringia, Germany. Chemie der Erde-Geochemistry 69, Supplement 2: 5–19.

Harter, R.D. and R. Naidu. 1995. Role of Metal-Organic complexation in metal sorption by Soils. Advances in Agronomy. L. S. Donald, Academic Press. 55: 219–263.

Hopwood, D.A. 2006. Soil to genomics: The *Streptomyces* chromosome. Annual Review of Genetics. Palo Alto, Annual Reviews. 40: 1–23.

Isermeyer, H. 1952. Eine einfache Methode zur Bestimmung der Bodenatmung und der Karbonate im Boden. Zeitschrift für Pflanzenernährung, Düngung, Bodenkunde. 56: 26–38.

Iwamoto, T. and M. Nasu. 2001. Current bioremediation practice and perspective. J. Biosci. Bioeng. 92: 1–8.

Johnson, D.B. 2003. Chemical and Microbiological Characteristics of Mineral Spoils and Drainage Waters at Abandoned Coal and Metal Mines. Water Air Soil Pollut. 3: 47–66.

Kabata-Pendias, A. and H. Pendias. 2001. Trace elements in soils and plants. Boca Raton, Fla. [u.a.], CRC Press.

Kandeler, F., C. Kampichler and O. Horak. 1996. Influence of heavy metals on the functional diversity of soil microbial communities. Biol. Fertility Soils. 23: 299–306.

Khamna, S., A. Yokota and S. Lumyong. 2009. Actinomycetes isolated from medicinal plant rhizosphere soils: diversity and screening of antifungal compounds, indole-3-acetic acid and siderophore production. World J. Microbiol. Biot. 25: 649–655.

Lee, I.S., O.K. Kim, Y.-Y. Chang, B. Bae, H.H. Kim and K.H. Baek. 2002. Heavy metal concentrations and enzyme activities in soil from a contaminated Korean shooting range. J. Biosci. Bioeng. 94: 406–411.

Merdy, P., L.T. Gharbi and Y. Lucas. 2009. Pb, Cu and Cr interactions with soil: Sorption experiments and modelling. Colloids and Surfaces A: Physicochem Eng Asp. 347: 192–199.

Nicholson, W., P. Fajardo-Cavazos, R. Rebeil, T. Slieman, P. Riesenman, J. Law and Y. Xue. 2002. Bacterial endospores and their significance in stress resistance. Antonie van Leeuwenhoek. 81: 27–32.

Nicholson, W.L., N. Munakata, G. Horneck, H.J. Melosh and P. Setlow. 2000. Resistance of *Bacillus* endospores to extreme terrestrial and extraterrestrial environments. Microbiol. Mol. Biol. Rev. 64: 548–572.

Norrish, K. 1975. The geochemistry and mineralogy of trace elements. Trace Elements in Soil-Plant-Animal System. D. D. J. Nicholas and A. R. Egan. New York: 55.

Oliveira, A. and M.E. Pampulha. 2006. Effects of long-term heavy metal contamination on soil microbial characteristics. J. Biosci. Bioeng. 102: 157–161.

Pennanen, T., J. Perkiömäki, O. Kiikkilä, P. Vanhala, S. Neuvonen and H. Fritze. 1998. Prolonged, simulated acid rain and heavy metal deposition: separated and combined effects on forest soil microbial community structure. FEMS Microbiol. Ecol. 27: 291–300.

Priha, O., K. Hallamaa, M. Saarela and L. Raaska. 2004. Detection of *Bacillus cereus* group bacteria from cardboard and paper with real-time PCR. J Ind Microbiol. 31: 161–169.

Rogival, D., J. Scheirs and R. Blust. 2007. Transfer and accumulation of metals in a soil–diet–wood mouse food chain along a metal pollution gradient. Environ. Pollut. 145: 516–528.

Salt, D.E., M. Blaylock, N.P.B.A. Kumar, V. Dushenkov, B.D. Ensley, I. Chet and I. Raskin. 1995. Phytoremediation—a Novel Strategy for the Removal of Toxic Metals from the Environment Using Plants. Bio-Technol. 13: 468–474.

Scheffer, F., P. Schachtschabel, H.P. Blume and S. Scheffer. 2008. Lehrbuch der Bodenkunde. Heidelberg [u.a.], Spektrum, Akad. Verl.

Schindler, F., M. Gube and E. Kothe. 2012. Bioremediation and Heavy Metal Uptake: Microbial Approaches at Field Scale. Bio-Geo Interactions in Metal-Contaminated Soils. E. Kothe and A. Varma, Springer Berlin Heidelberg. 31: 365–383.

Schmidt, A., G. Haferburg, M. Sineriz, D. Merten, G. Büchel and E. Kothe. 2005. Heavy metal resistance mechanisms in actinobacteria for survival in AMD contaminated soils. Chemie der Erde-Geochemistry 65, Supplement. 1: 131–144.

Sineriz, M.L., E. Kothe and C.M. Abate. 2009. Cadmium biosorption by *Streptomyces* sp. F4 isolated from former uranium mine. J. Basic Microbiol. 49: S55–S62.

Speir, T. W. 2008. Chapter 13 Relationship between biochemical activity and metal concentration in soil amended with sewage sludge. Developments in Soil Science. A. B. M. A.E. Hartemink and N. Ravendra, Elsevier. 32: 261–279.

Szentmihalyi, S., A. Regius, M. Anke, M. Grun, B. Groppel, D. Lokay and J. Pavel. 1980. The nickel supply of ruminants in the GDR, Hungary and Czechoslovakia dependent on the origin of the basic material for the formation of soil. 3rd Trace Element Symposium on Nickel. M. Anke, H. J. Schneider and C. Bruckner. 229–236.

Valsecchi, G., C. Gigliotti and A. Farini. 1995. Microbial biomass, activity and organic-matter accumulation in soils contaminated with heavy metals. Biol Fertility Soils. 20: 253–259.

Violante, A., V. Cozzolino, L. Perelomov, A.G. Caporale and M. Pigna. 2010. Mobility and bioavailability of heavy metals and metalloids in soil environments. J. Soil Sci. Plant Nutrition. 10: 268–292.

Walker, D.J., R. Clemente and M.P. Bernal. 2004. Contrasting effects of manure and compost on soil pH, heavy metal availability and growth of Chenopodium album L. in a soil contaminated by pyritic mine waste. Chemosphere. 57: 215–224.

Yang, R., J. Tang, X. Chen and S. Hu. 2007. Effects of coexisting plant species on soil microbes and soil enzymes in metal lead contaminated soils. Appl. Soil Ecol. 37: 240–246.

Zeien, H. and G.W. Bruemmer. 1989. Chemische Extraktion zur Bestimmung von Schwermetall Bindungsformen im Böden. Mitt. Dtsch. Bodenkdl. Ges. 59: 505–515.

Zhuang, P., M.B. McBride, H. Xia, N. Li and Z. Li. 2009. Health risk from heavy metals via consumption of food crops in the vicinity of Dabaoshan mine, South China. Sci. Total. Environ. 407: 1551–1561.

CHAPTER 7

Cooperative Activity of Actinobacteria and Plants in Soil Bioremediation Processes

Mariana C. Atjian,[1] Marta A. Polti,[1,2,*] María J. Amoroso[1,3] and Carlos M. Abate[1,2,3]

Introduction

Hexavalent chromium, Cr(VI), is widely used in many industrial processes such as electroplating, wood preservation, etc. The chromium manufacturing industry produces a large quantity of solid and liquid waste containing hexavalent chromium. Cr(VI) compounds are highly water soluble, toxic and carcinogenic in mammals (Jeyasingh and Philip 2005). Cr(VI) as chromate and dichromate is considered one of the principal pollutants in the United States by the Environment Protection Agency (Polti et al. 2009). In contrast, trivalent chromium, Cr(III), is considered nontoxic as it precipitates at higher pH than 5.5 with the formation of insoluble oxides and hydroxides in soil and water systems (Jeyasingh and Philip 2005).

[1]Planta Piloto de Procesos Industriales y Microbiológicos (PROIMI), CONICET, Av. Belgrano y Pasaje Caseros, 4000 Tucumán, Argentina.
[2]Facultad de Ciencias Naturales e Instituto Miguel Lillo, Universidad Nacional de Tucumán, Miguel Lillo 205, Tucumán.
[3]Facultad de Bioquímica, Química y Farmacia, Universidad Nacional de Tucumán, Ayacucho 471, Tucumán.
*Corresponding author: mpolti@proimi.org.ar

The behavior of chromium (Cr) in water and soil and its subsequent bioavailability have been the subject of active investigation because of the existence of chromium in two environmentally important oxidation states (+3 and +6) and the ease with which they can complexes with naturally occurring chemical species in the soil (Mishra et al. 1995). Many technologies are currently used to clean up soils contaminated with heavy metal. The most commonly used ones are soil removal and land filling, stabilization/ solidification, physico-chemical extraction and soil washing. These techniques generally release secondary contamination and are expensive when applied to large areas. Bioremediation is a promising technology. The bioremediation strategy pursue detoxify Cr(VI) in the soil reducing it to Cr(III), immobilizing in the soil matrix. Besides eliminating the toxicity of Cr(VI) by its reduction to Cr(III), the latter forms a particularly insoluble $Cr(OH)_3$ in the pH range of 6–9 (Ksp, 6.7×10^{-31}) severely restricting its ability to migrate to groundwater (Jeyasingh and Philip 2005).

Many microbes were reported to reduce Cr(VI) under aerobic and anaerobic conditions and Cr(VI) reducing microbial population is widespread in soil with potential to aerobic reduction of this metal (Jeyasingh and Philip 2005).

Although most studies have been carried out with Gram negative bacteria (Puzon et al. 2002, Cheung and Gu 2007, Chai et al. 2009, Mohanty and Patra 2011), several Gram positive bacteria, including certain actinobacteria, have also shown to possess Cr(VI) reducing ability (Das and Chandra 1990, Amoroso et al. 2001, Desjardin et al. 2003, Polti et al. 2009).

Actinobacteria represent an important component of the microbial population in soils. Their metabolic diversity and specific growth characteristics make them well suited as agents for bioremediation (Albarracín et al. 2008, Benimeli et al. 2008, Polti et al. 2009, Siñeriz et al. 2009).

The first report on Cr(VI) reduction by actinobacteria was by Das and Chandra (1990). Amoroso et al. (2001) reported on Cr(VI) bioaccumulation by two *Streptomyces* strains, whereas Laxman and More (2002) determined Cr(VI) reduction by *Streptomyces* griseus. However, there are only a few studies on Cr(VI) bioreduction by actinobacteria and their potential for bioremediation processes in soil (Desjardin et al. 2003, Polti et al. 2009).

Polti et al. (2009) showed that a *Streptomyces* strain, previously isolated from sugarcane, was able to reduce 90% of Cr(VI) bioavailability in soil samples supplemented with 50 mg kg^{-1} after seven days of incubation without the addition of any substrate or pretreatment. Also, they showed that the reduction of Cr(VI) by *Streptomyces* sp. MC1 is a useful mechanism for the bioremediation of soils and this could improve biotechnological processes. *Streptomyces* sp. MC1 could not only be applied to semi-liquid or liquid systems, but also to solid systems such as soil.

Traditionally, detection of environmental contamination was carried out using physical and chemical techniques, such as spectroscopy or chromatography, which are powerful tools to detect, quantify and identify several contaminants, at level of parts per million or even parts per billion. However, the metal harmful to living organisms is related to the available fraction (bioavailability) and it is different from the total concentration. These cases are convenient using different extraction methodology and/ or bioindicators and biomonitors, which can change the response with variations of the contaminant concentrations (Csillag et al. 1999, D'Souza 2001, Atlas and Bartha 2002, Markert 2007, Udovic and Lestan 2010). Bioindicators and biomonitors have proven to be excellent tools in many of these cases and could provide information which cannot be derived from technical measurements alone (D'Souza 2001, Markert 2007, Prasad et al. 2010).

The metal accumulation in the rizosphere favors its transference to plants and animals and produces biomagnification (Notten et al. 2005, Rogival et al. 2007). Chromium compounds are extremely toxic to plants and affect their growth and development (Sharma et al. 2003, Shanker et al. 2005). Plants are immobile organisms and, in metal contaminated environments, roots are the primary contact zones with soil pollutants. In order to survive, plants must develop efficient mechanisms by which excessive amounts of heavy metals are taken up and transformed into physiologically tolerable forms. An excess of toxic heavy metal ions induces the formation of free radicals and reactive oxygen species, resulting in oxidative stress, damaging macromolecules (including lipids, proteins and nucleic acids), as well as mitochondrial respiration and carbohydrate metabolism (Labra et al. 2006, Prado et al. 2010). Decrease in root growth is a well documented effect due to heavy metals in trees and crops (Shanker et al. 2005).

Zea mays L. is an important crop species cultivated in a wide range of environmental conditions. Maize is considered to be tolerant to heavy metals, though, Mallick et al. (2010) reported that roots of chromium treated *Zea mays* were stunted and root hair formation was greatly impaired. The leaves had a wilted appearance and chromium was accumulated primarily in the roots with a low translocation rate to the leaves.

In this chapter we can see as the use of a microorganism resistant to Cr(VI), *Streptomyces* sp. MC1, under non sterile conditions, with maize plants as a bioindicator, is a exellent strategy to reduce the bioavailability of chromium in soils.

Selection of Cr(VI) Concentration

Polti et al. (2007) demonstrated the ability of *Streptomyces* sp. MC1 to remove more than 95 percent of Cr(VI) from a liquid medium, at an initial concentration of 50 mg L^{-1}. Therefore, we decided to evaluate which total chromium concentration produces 50 mg kg^{-1} of bioavailable chromium in soil samples. Total bioavailable chromium was determinate in soil samples contaminated with 100, 200 and 400 mg kg^{-1} of Cr(VI) and Cr(III) respectively.

Non-polluted soil samples were taken from an experimental northwest site of Tucumán (Argentina). They were taken from the surface (5–15 cm deep) and kept at 10–15°C in the dark until use. Soil chemical characterization was carried out by Polti et al. (2009): pH 8.1, organic matter 4.16%, nitrogen 1.7 g L^{-1}, organic carbon 24.1 g L^{-1}. Glass pots were filled with 200 g of soil at 20% moisture. Cr(VI) was added as $K_2Cr_2O_7$. A 1,000x stock solution was prepared by dissolving $K_2Cr_2O_7$ in water and filter sterilized prior to use. Similarly, Cr(III) was added as $CrCl_3$, a 1,000x stock solution was prepared by dissolving $CrCl_3$ in water and filter sterilized prior to use. $K_2Cr_2O_7$ or $CrCl_3$ solution was added to obtain a final concentration of 100, 200 or 400 mg kg^{-1} dw of soil for each metal (Polti et al. 2009). The experimental condition labels of soil samples are: SS_0 = soil sample without metal addition; SS_3 = soil sample with Cr(III); and SS_6 = soil sample with Cr(VI).

Total and bioavailable chromium concentrations were determined by atomic absorption spectrometry using a Perkin Elmer Analyst 400 (AAS) in soil and solid samples. The samples were first digested by dissolving soils (1 g) or plants in HNO_3c (2 mL, 10 N). Potentially bioavailable chromium in the soil was measured by a physical method: 100 g of soil was centrifuged at 5050 *g* during 60 min, to reproduce the maximal plant suction (soil water potential: –1500 kPa, conventional wilting point) (Csillag et al. 1999). After centrifugation, the supernatant was recovered, filtered at 0.45 mm and analyzed by AAS for Cr content (A.P.H.A. 1989). Statistical analyses were conducted using the Microcal Origin Working Model Version 6.0. Paired t-test and one way anova variance analysis were used with a probability level of $p < 0.05$.

The bioavailable chromium fraction was between 34 and 38% in SS_6, while, it was 0.04 up to 0.10% in SS_3 (Table 1).

These results agree with those previously found (Kotas and Stasicka 2000, Stewart et al. 2003, Shanker et al. 2005, Mandiwana et al. 2007), where chromium is bioavailable only as Cr(VI). Stewart et al. (2003) studied the sorption patron of Cr(III) and Cr(VI): they found significant variation as a function of soil type and horizon and the contaminant oxidation state. The bioaccessibility of surface-bound Cr(VI) was significantly greater than that of Cr(III). Between 42 and 108% of the total adsorbed Cr(VI) was

Table 1. Bioavailable chromium.

Soil Sample	100[a]	200	400
SS_6	34.02 ± 3.46[b] (34%)	67.16 ± 15.34 (34%)	151.75 ± 9.60 (38%)
SS_3	0.07 ± 0.01 (0.07%)	0.16 ± 0.01 (0.04%)	0.39 ± 0.05 (0.10%)

a: Metal concentration added (mg kg^{-1}).
b: Bioavailable chromium (mg kg^{-1}), mean ± standard deviation.

bioaccessible when it was compared with totally adsorbed Cr(III), which was only 3 and 14% bioaccessible (Stewart et al. 2003).

This behavior is explained by the mechanisms of Cr(III) and Cr(VI) interaction with soils, where Cr is present mostly as insoluble $Cr(OH)_3$.aq or as Cr(III) adsorbed to soil components, which prevents Cr leaching into the groundwater or its uptake by plants. Cr(III) adsorption to humic acids renders it insoluble, immobile and unreactive; this process is most effective within the pH range of 2.7–4.5. On the other hand, in neutral-to-alkaline soils, hexavalent Cr exists mostly in soluble, but also in moderately-to-sparingly soluble chromates. In more acidic soils (pH < 6) $HCrO_4^-$ becomes a dominant form. The chromate ions are the most mobile forms of Cr in soils. They can be taken up by plants and easily leached out into the deeper soil layers causing ground and surface water pollution. Some minor amounts of Cr(VI) are bound in soils, depending on the mineralogical composition and pH of the soil (Cifuentes et al. 1996, Stewart et al. 2003). For further assays the concentration selected was 200 mg kg^{-1} of Cr(VI), because the bioavailable chromium was 60 mg kg^{-1} approximately.

Evaluation of Chromium Removal Ability of *Streptomyces* sp. MC1 in Non Sterile Soil Samples

After this, we performed the evaluation of chromium removal ability of *Streptomyces* sp. MC1 in non sterile soil samples. Cr(VI)-resistant *Streptomyces* sp. MC1, previously isolated and characterized, was used for this work (Polti et al. 2007, 2010, 2011). The strain was maintained on starch-casein agar slants (SC agar) containing (g L^{-1}): starch, 10.0; casein, 1.0; K_2HPO_4, 0.5; agar, 12. The pH was adjusted to 7.0 prior to sterilization.

The precultures for soil sample inoculation were obtained by inoculating *Streptomyces* sp. MC1 in TSB containing (g L^{-1}): tryptone, 15; soy peptone, 3: NaCl, 5; K_2HPO_4, 2.5; glucose, 2.5. Cultures were incubated for three days at 30°C (200 rpm) in Erlenmeyer flasks filled with 50 mL of medium (Polti et al. 2009).

Soil samples were inoculated with precultured *Streptomyces* sp. MC1 at a concentration of 0.5 mg kg^{-1} of soil (wet weight/dry weight). Soil pots were incubated at 30°C, for 14, 28 and 42 days (t–0 = assay total time: 14 days;

t–14 = assay total time: 28 days and t-28 = assay total time: 42 days) and all assays were carried out in triplicate. Experimental condition labels of soil samples: SS_{0+S} = soil sample without metal addition with *Streptomyces* sp. MC1; SS_{3+S} = soil sample with Cr(III) with *Streptomyces* sp. MC1 and SS_{6+S} = soil sample with Cr(VI) with *Streptomyces* sp. MC1.

The bioavailable chromium was 40 mg kg^{-1}, 28 mg kg^{-1} and 18 mg kg^{-1}, in SS_{6+S}, after 14, 28 and 42 days of incubation, respectively (Fig. 1). The maximal removal reached after 42 days was 73% in SS_{6+S}, while in SS_6 it was only 54%. At t–14, *Streptomyces* sp. MC1 removed 35% of bioavailable chromium, reducing by half the required time (from 28 to 14 days), which demonstrated a positive cooperation between *Streptomyces* sp. MC1 and the autochthonous microflora.

There is little information about the quantification of Cr(VI) reduction in soils by autochthonous microflora. Previous works have concluded that addition of organic compounds and high humidity have positive effects on Cr(VI) reduction (Losi et al. 1994, Cifuentes et al. 1996, Turick et al. 1998, Vainshtein et al. 2003); on the other hand, Polti et al. (2009) showed that *Streptomyces* sp. MC1 was able to remove almost 100% of Cr(VI) in different soils, without the addition of nutrients, under normal humidity conditions (20%), with an initial Cr(VI) concentration of 50 mg kg^{-1}.

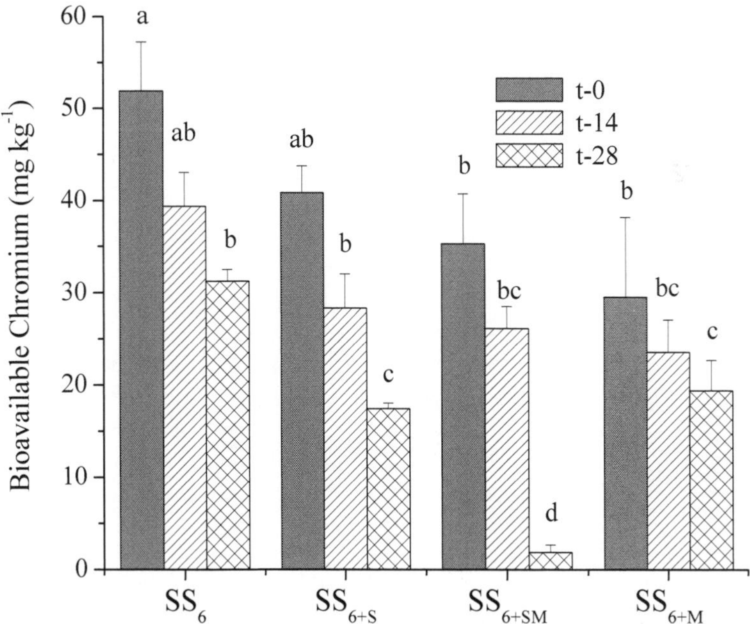

Figure 1. Bioavailable chromium in SS_6, after 14 days (t–0), 28 days (t–14) and 42 days (t-28), at 25–28°C. Different letters indicate significant differences ($p < 0.05$).

Furthermore, *Streptomyces* sp. MC1 is able to tolerate an initial concentration of 200 mg kg^{-1}, and also, it is able to remove the chromium in presence of the native flora, during 42 days of incubation.

Evaluation of Inhibitory Effect of Chromium on *Z. mays* Seedlings

Maize plants were used as model because of its importance in animal and human nutrition. Also, it is known that maize can accumulate heavy metals, included Cr(VI), in roots (Mishra et al. 1995, Mallick et al. 2010).

Seeds of maize, surface sterilized using 2% mercuric chloride, were placed on agar-agar in Petri dishes. Seeds were germinated at room temperature (26–28°C). Seven days old maize seedlings were planted in SS_{6+M} (soil sample with Cr(VI) with *Z. mays*), SS_{3+M} (soil sample with Cr(III) with *Z. mays*) or SS_{0+M} (soil sample without metal addition with *Z. mays*).

Following germination, three seedlings for each treated-group were chosen, transplanted (one using a pot), and grown in a soil sample with chromium (200 mg kg^{-1} dry weight of soil) and without the metal as control, following the design showed above, varying the time of planting:

t–0 = metal contamination on day 0, *Z. mays* planting on day 0 (assay total time: 14 days).

t–14 = metal contamination on day 0, *Z. mays* planting on day 14 (assay total time: 28 days).

t–28 = metal contamination on day 0, *Z. mays* planting on day 28 (assay total time: 42 days).

The pots were kept at room temperature (26–28°C) in a germinator; using 16/8 h light/night cycles; sterile distilled water was added to each pot every alternate day to maintain a moisture level of 20%.

After 14 days, the plants were removed from pots and roots and stems and leaves were measured. Each plant was shaken carefully to remove the bulk soil. The soil that still adhered to the roots was removed by washing with distilled water. The effect of chromium on root, steam and leaf was studied by measuring their length, total and bioavailable chromium, and Cr(VI) content. Maize biomass was evaluated as dry weight, drying at 105°C until constant weight.

Plants developed in presence of Cr(VI) or Cr(III) displayed several macroscopic damages: leaf size and pigmentation decreased, especially with Cr(VI). These damages were more apparent at t–0 and t–14 (data not shown). Similar results had been reported previously (Cervantes et al. 2001, Sharma et al. 2003, Shanker et al. 2005).

Sharma et al. (2003) described several lesions in plants grown with Cr(VI), including interveinal chlorosis in patches in leaves, curled margins, pale appearance, etc.

Also, maize biomass decreased in presence of Cr(VI) and Cr(III) (Fig. 2). Reduction in SS_{6+M} was 86, 71 and 54% at t–0, t–14 and t–28, respectively, compared to biomass obtained in SS_{0+M}. The reduction in SS_{3+M} was 54, 56 and 26 percent at t–0, t–14 and t–28 (Fig. 2).

Root, stem and leaf lengths were also affected (Table 2). With Cr(VI), root, stem and leaf lengths decreased up to 25, 73 and 80%, respectively, regardless of the seed timing. On the other hand, with Cr(III), root and stem lengths decreased up to 13 and 54%, respectively, regardless of the seed method, leaf lengths were affected only in t–0 and t–14, decreasing up to 45 percent, but it was not affected at t–28 (Table 2).

Mallick et al. (2010) evaluated growth of maize plants and development with chromium and concluded that Cr(VI) has negative effects on maize biomass and root length. General response of decreased root growth due to Cr toxicity could be due to inhibition of root cell division/root elongation or the extension of cell cycle in the roots. Under high concentration of both the Cr species combination, the reduction in root growth could be due to the direct contact of seedling roots with Cr in the medium causing a collapse and subsequent inability of the roots to absorb water from the medium (Shanker et al. 2005).

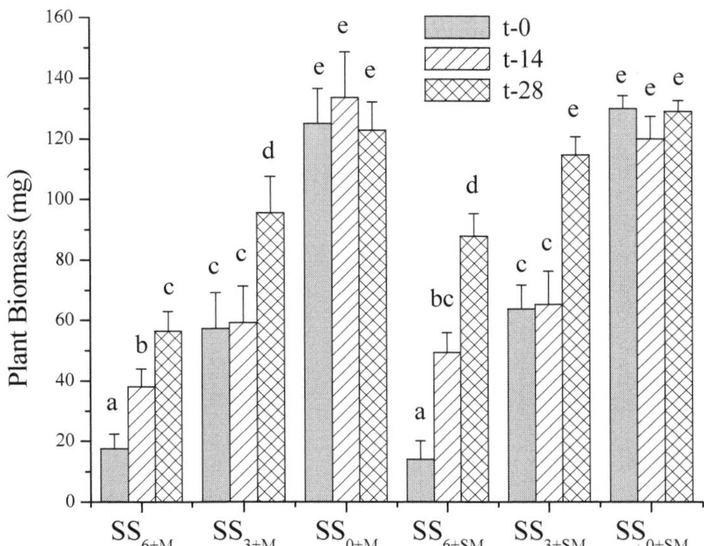

Figure 2. Maize biomass in SS_6 and SS_3 and SS_0, with and without *Streptomyces* sp. MC1, planted at t–0, t–14 and t–28, and harvesting after fourteen days of growth. Different letters indicate significant differences ($p < 0.05$).

Table 2. Maize length after fourteen days of growth in soil samples.

	SS_{0+M}			SS_{6+M}			SS_{3+M}		
	t–0	t–14	t–28	t–0	t–14	t–28	t–0	t–14	t–28
Root	9.8 ± 0.0[a]	9.3 ± 0.1	9.7 ± 0.0	7.0 ± 0.1	7.7 ± 0.3	7.7 ± 0.3	9.4 ± 0.3	8.9 ± 0.0	8.3 ± 0.2
Stem	5.0 ± 0.3	6.7 ± 0.0	7.6 ± 0.1	1.7 ± 0.1	2.0 ± 0.1	2.3 ± 0.1	3.5 ± 0.6	5.0 ± 0.7	8.0 ± 0.2
Leaf	11.1 ± 0.5	16.0 ± 0.5	20.0 ± 0.4	3.1 ± 0.3	5.0 ± 0.6	6.0 ± 0.2	7.2 ± 0.2	13.0 ± 0.8	15.3 ± 0.3

a: Length (cm), mean ± standard deviation.

Shanker et al. (2005) proposed a dual mechanism to explain this reduction in plant height. On the one hand, the reduced root growth and consequent lesser nutrients and water transport to the above parts of the plant and on the other, Cr transport to the aerial part of the plant could have a direct impact on leaf cellular metabolism.

Furthermore, at t–14 and t–28, lesser toxic effects were observed on the seedlings, in comparison with those observed at t–0. The native microflora in the soil could be responsible for this effect. It is well known that non-polluted soil could have indigenous microbial community not only resistant to chromium, but also, with potential to reduce Cr(VI), and consequently, its toxicity (Pal and Paul 2004, Wani et al. 2007).

Chromium accumulation by *Z. mays* plants was also evaluated. Total Cr accumulation observed at t–0 was 110.1 mg g^{-1} and 37.3 mg g^{-1}, in SS_{6+M} and SS_{3+M}, respectively, however at t–14 and t–28, accumulation was up to 4.3 mg g^{-1} (Table 3).

The mechanism of Cr uptake and accumulation by plants has been discussed by several authors (Mishra et al. 1995, Zayed et al. 1998, Cervantes et al. 2001, Vandecasteele et al. 2006, Mandiwana et al. 2007, Gheju et al. 2009, Mallick et al. 2010). Cr(III) and Cr(VI) have independent uptake mechanisms; whereas Cr(VI) uptake depends on metabolic energy, Cr(III) uptake does not. However an active uptake of both Cr species has also been reported (Cervantes et al. 2001). The sulfate transport system is apparently involved in chromate uptake by plants. It was demonstrated that the translocation and accumulation of Cr inside the plant depends on the state of oxidation of the metal. Plants accumulate more Cr when they were grown with Cr(VI) than with Cr(III) (Mishra et al. 1995, Zayed et al. 1998). After the uptake, Cr is transported to the aerial parts by the xylem.

Table 3. Chromium accumulation inside maize plants.

	SS_{6+M}			SS_{3+M}		
	t–0	t–14	t–28	t–0	t–14	t–28
a) Total chromium accumulation						
Stem	10.3 ± 1.0	0.4 ± 0.0	1.8 ± 0.5	4.9 ± 0.3	0.5 ± 0.0	2.1 ± 0.1
Leaf	9.9 ± 0.9	1.1 ± 0.0	1.9 ± 0.0	2.8 ± 0.8	0.9 ± 0.0	0.9 ± 0.0
Whole plant	110.1 ± 10.0	4.3 ± 0.6	3.1 ± 1.6	37.3 ± 3.0	4.1 ± 0.4	1.2 ± 0.1
b) Cr(VI) accumulation						
Root	1.6 ± 0.0[a]	0.2 ± 0.0	0.1 ± 0.0	0.7 ± 0.0	0.2 ± 0.0	0.2 ± 0.0
Stem	1.9 ± 0.0	0.2 ± 0.0	0.2 ± 0.0	0.6 ± 0.0	0.3 ± 0.0	0.3 ± 0.0
Leaf	2.5 ± 0.0	0.2 ± 0.0	0.1 ± 0.0	0.4 ± 0.0	0.2 ± 0.0	0.6 ± 0.0
Whole plant	1.0 ± 0.3	0.3 ± 0.1	1.9 ± 0.1	0.5 ± 0.2	0.2 ± 0.1	1.1 ± 0.0

a: Metal accumulation in seedlings (mg g^{-1}), mean ± standard deviation.

Chromate moves faster in the xylem than Cr(III), presumably because there is an electrostatic interaction with the vessel walls (Cervantes et al. 2001).

On the other hand, Cr(VI) concentration was determined, using the Cr(VI) specific colorimetric reagent 1,5-diphenylcarbazide, dissolved in acetone to a final concentration of 0.05% (A.P.H.A. 1989). Cr(VI) accumulation was up to 2.5 mg g^{-1} at t–0 in SS_6, less than 1 percent of total chromium accumulated (Table 3). Zayed et al. (1998) found similar results, working with several vegetable crops, where plants were grown with either Cr(III) or Cr(VI), but only Cr(III) was found in their tissues. These observations suggest that Cr(VI) is transformed to Cr(III), mainly inside the root cells, but also in the aerial part of the plant (Cervantes et al. 2001).

Moreover, differential accumulation of Cr in plant organs was observed. At t–0, in SS_6, total chromium was accumulated specially in stems and leaves (10.3 and 9.9 mg g^{-1}, respectively), while, at t–14 and t–28, accumulation was not significant (Table 3a). In SS_3, at t–0, higher accumulation was observed in roots and stems (3.9 and 4.9 mg g^{-1}, respectively) (Table 3a).

Mallick et al. (2010) observed maximal chromium accumulation in roots, with low translocation to leaves. In general, roots accumulate 10–100 times more Cr than shoots and other tissues, independently of plant species (Zayed et al. 1998, Cervantes et al. 2001). However, Mishra et al. (1995) found a higher accumulation of Cr in root, when *Z. mays* was exposed to Cr(III), while, under Cr(VI), a higher accumulation was observed in leaves. A high accumulation in plant roots could be because Cr is immobilized in the vacuoles of the root cells, decreasing is toxicity, which may be a naturally toxic response of the plant (Shanker et al. 2005).

Also, Cr(VI) distribution was evaluated. In SS_6, higher Cr(VI) accumulation was observed in leaves, at t–0 (Table 3b). At t–0, Cr(VI) is absorbed by roots and translocated to aerial parts, where it could be reduced to Cr(III), while at t–14 and t–28, Cr(VI) reduction could be carried out by native microflora, which reduces its absorption and translocation by the vegetal tissue. Gheju et al. (2009) found similar results working with 40 mg kg^{-1} of Cr(VI), *Z. mays* roots concentrated Cr(VI), which was slowly translocated within the plant from the roots to the stems, and very slowly translocated further to leaves.

Use of *Z. mays* as Bioindicator

Z. mays was used as bioindicator to confirm the bioremediation success by *Streptomyces* sp. MC1.

Maize plants biomass planted in SS_{6+SM} (soil sample with Cr(VI) with *Streptomyces* sp. MC1, with *Z. mays*) and SS_{3+SM} (soil sample with Cr(III) with *Streptomyces* sp. MC1, with *Z. mays*) was evaluated. At t–0 and t–14

Streptomyces sp. MC1 did not produce significant differences on maize biomass, compared to SS_{6+M} and SS_{3+M}, respectively. However, at t–28 *Streptomyces* sp. MC1 increased *Z. mays* biomass, from 56 mg in SS_{6+M}, to 88mg in SS_{6+SM} (Fig. 2).

Streptomyces sp. MC1 promoted leaf and stem growth at t–28 in SS_{0+SM}, compared to SS_{0+M}. No effect was observed in SS_{3+SM} (Table 4).

In SS_{3+SM} and SS_{0+SM} (soil sample without metal addition with *Streptomyces* sp. MC1, with *Z. mays*), *Streptomyces* sp. MC1 inhibited development of roots compared to SS_{3+M} and SS_{0+M}. However, in SS_{6+SM}, *Streptomyces* sp. MC1 promoted roots and leaves growth, in particular, root lengths increased significantly ($p < 0.05$) from 8 (SS_{6+M}) to 18 cm (SS_{6+SM}) at t–28.

Maize biomass and lengths improved in the presence of *Streptomyces* sp. MC1 in soil samples contaminated with Cr(VI), indicating a decrease of metal toxicity. These results are the first indication about a successful use of *Z. mays* as a bioremediation bioindicator.

In SS_{6+SM} at t–0 the highest total chromium accumulation (70 mg g^{-1}) was detected, but it was 36% less than the accumulation found in SS_{6+M} (Table 5). In SS_{6+SM}, at t–0, total chromium accumulation decreased by 78, 15 and 63% in roots, stems and leaves, respectively, compared to SS_{6+M}. At t–14, accumulation decreased by 17, 43 and 58% in roots, stems and leaves, respectively. At t–28, accumulation decreased by 46% in roots and stems. However, in SS_{3+SM} it was not found significantly different compared to SS_{3+M}. *Streptomyces* sp. MC1 also affected Cr(VI) distribution inside *Z. mays*. Cr(VI) accumulation decreased significantly ($p < 0.05$) in SS_{6+SM} at t–0 in roots (50%). This could indicate a protective effect of *Streptomyces* sp. MC1 on *Z. mays* tissue.

Albarracín et al. (2010) found similar results working with *Z. mays* as bioindicator, but using soil samples contaminated with copper and *Amycolatopsis tucumanensis* as bioremediator, decreasing up to 20% copper accumulation in roots and leaves. Even *Z. mays* could not be considered as phytoremediator (Gheju et al. 2009) as bioavailable chromium was affected by this organism. Diminution of 42% of bioavailable chromium was observed in the absence of *Streptomyces* sp. MC1 (SS_{6+M}), chromium could be adsorbed by *Z. mays* roots, and reduced to Cr(III) by the activity of root exudates (Mishra et al. 1995).

At t–0 and t–14 bioavailable chromium was not significantly different between SS_{6+M} and SS_{6+SM}. However, at t–28 bioavailable chromium decreased up to 96% in SS_{6+SM} (Fig. 1), showing a synergic activity between *Streptomyces* sp. MC1 and *Z. mays* platelets, with enhanced bioremediation efficiency. This is the first report of a cooperative activity between an actinobacteria and *Z. mays* to bioremediate Cr(VI) contaminated soil.

Table 4. Maize length after fourteen days of growth in soil samples in the presence of *Streptomyces* sp. MC1.

	SS_{0+SM}			SS_{6+SM}			SS_{3+SM}		
	t–0	**t–14**	**t–28**	**t–0**	**t–14**	**t–28**	**t–0**	**t–14**	**t–28**
Root	7.8 ± 0.0[a]	4.6 ± 0.8	6.1 ± 0.3	10.7 ± 0.4	12.0 ± 0.1	18.0 ± 0.4	7.5 ± 0.0	5.9 ± 0.2	4.4 ± 0.3
Stem	3.7 ± 0.1	6.0 ± 0.7	9.8 ± 0.3	2.0 ± 0.1	2.5 ± 0.1	2.7 ± 0.1	3.2 ± 0.2	5.0 ± 0.6	7.5 ± 0.2
Leaf	14.3 ± 1.0	18.5 ± 1.0	22.0 ± 0.8	6.0 ± 0.0	4.5 ± 0.5	7.0 ± 0.0	14.5 ± 1.0	11.0 ± 0.9	14.6 ± 1.0

a: Length (cm), mean ± standard deviation

Table 5. Chromium accumulation inside maize plantlets in the presence of *Streptomyces* sp. MC1

	SS_{6+SM}			SS_{3+SM}		
	t–0	t–14	t–28	t–0	t–14	t–28
a) Total chromium accumulation						
Root	1.4 ± 0.1[a]	0.2 ± 0.0	1.6 ± 0.2	2.9 ± 0.4	0.2 ± 0.0	1.8 ± 0.0
Stem	8.8 ± 0.4	0.2 ± 0.0	1.0 ± 0.1	2.2± 0.7	0.2 ± 0.0	1.5 ± 0.1
Leaf	3.7 ± 0.4	0.5 ± 0.0	1.8± 0.1	1.4 ± 0.3	0.2 ± 0.0	0.6 ± 0.2
Whole plant	70.0 ± 6.2	2.5 ± 0.3	2.5 ± 0.2	39.3 ± 2.0	2.4 ± 0.2	2.5 ± 0.1
b) Cr(VI) accumulation						
Root	0.8 ± 0.0[a]	0.2 ± 0.0	0.1 ± 0.0	0.5 ± 0.0	0.1 ± 0.0	0.2 ± 0.0
Stem	0.4 ± 0.0	0.1 ± 0.0	0.2 ± 0.0	0.4 ± 0.0	0.2 ± 0.0	0.3 ± 0.0
Leaf	2.5 ± 0.0	0.2 ± 0.0	0.1 ± 0.0	0.4 ± 0.0	0.1 ± 0.0	0.4± 0.0
Whole plant	0.9 ± 0.0	0.3 ± 0.0	1.4 ± 0.1	0.5 ± 0.0	0.3 ± 0.0	1.4 ± 0.2

a: Metal accumulation in seedlings (mg g^{-1}), mean ± standard deviation.

Acknowledgments

The authors gratefully acknowledge the financial assistance of Consejo de Investigaciones de la Universidad Nacional de Tucumán (CIUNT), Consejo Nacional de Investigaciones Científicas y Técnicas (CONICET) and Agencia Nacional de Promoción Científica y Tecnológica (ANPCyT), Argentina.

References Cited

Albarracín, V.H., A.L. Ávila, M.J. Amoroso and C.M. Abate. 2008. Copper removal ability by *Streptomyces* strains with dissimilar growth patterns and endowed with cupric reductase activity. FEMS Microbiol. Lett. 288: 141–148.

Albarracín, V.H., M.J. Amoroso and C.M. Abate. 2010. Bioaugmentation of copper polluted soil microcosms with *Amycolatopsis tucumanensis* to diminish phytoavailable copper for Zea mays plants. Chemosphere. 79: 131–137.

Amoroso, M.J., G.R. Castro, A. Durán, O. Peraud, G. Oliver and R.T. Hill. 2001. Chromium accumulation by two *Streptomyces* spp. isolated from riverine sediments. J. Ind. Microbiol. Biotechnol. 26: 210–215.

A.P.H.A. 1989. Standard methods for the examination of water and wastewater, 17 ed. American Public Health Association, Washington D.C.

Atlas, R.M. and R. Bartha. 2002. Ecología microbiana y microbiología ambiental. Pearson Educación.

Benimeli, C.S., M.S. Fuentes, C.M. Abate and M.J. Amoroso. 2008. Bioremediation of lindane-contaminated soil by *Streptomyces* sp. M7 and its effects on *Zea mays* growth. Int. Biodeterior. Biodegradation. 61: 233–239.

Cervantes, C., J. Campos-García, S. Devars, F. Gutiérrez-Corona, H. Loza-Tavera, J.C. Torres-Guzmán and R. Moreno-Sánchez. 2001. Interactions of chromium with microorganisms and plants. FEMS Microbiol. Rev. 25: 335–347.

Chai, L., S. Huang, Z. Yang, B. Peng, Y. Huang and Y. Chen. 2009. Cr(VI) remediation by indigenous bacteria in soils contaminated by chromium-containing slag. J. Hazard. Mater. 167: 516–522.

Cheung, K.H. and J.D. Gu. 2007. Mechanism of hexavalent chromium detoxification by microorganisms and bioremediation application potential: a review. Int. Biodeterior. Biodegradation. 59: 8–15.

Cifuentes, F.R., W.C. Lindemann and L.L. Barton. 1996. Chromium sorption and reduction in soil with implications to bioremediation. Soil Science. 161: 233–241.

Csillag, J., G. Partay, A. Lukacs, K. Bujtas and T. Nemeth. 1999. Extraction of soil solution for environmental analysis. Int. J. Environ. Anal. Chem. 74: 305–324.

Das, S. and A. Chandra. 1990. Chromate reduction in *Streptomyces*. Experientia. 46: 731.

Desjardin, V., R. Bayard, P. Lejeune and R. Gourdon. 2003. Utilisation of supernatants of pure cultures of *Streptomyces thermocarboxydus* NH50 to reduce chromium toxicity and mobility in contaminated soils. Water Air Soil Pollut. 3: 153–160.

D'Souza, S.F., 2001. Microbial biosensors. Biosens. Bioelectron. 16: 337–353.

Gheju, M., I. Balcu and M. Ciopec. 2009. Analysis of hexavalent chromium uptake by plants in polluted soils. Ovidius University Annals of Chemistry. 20: 127–131.

Jeyasingh, J. and L. Philip. 2005. Bioremediation of chromium contaminated soil: optimization of operating parameters under laboratory conditions. J. Hazard. Mater. 118: 113–120.

Kotas, J. and Z. Stasicka. 2000. Chromium occurrence in the environment and methods of its speciation. Environ. Pollut. 107: 263–283.

Labra, M., E. Gianazza, R. Waitt, I. Eberini, A. Sozzi, S. Regondi, F. Grassi and E. Agradi. 2006. *Zea mays* L. protein changes in response to potassium dichromate treatments. Chemosphere. 62: 1234–1244.

Laxman, R.S. and S. More. 2002. Reduction of hexavalent chromium by *Streptomyces griseus*. Minerals Engineering. 15: 831–837.

Losi, M.E., C. Amrhein and W.T. Frankenberger. 1994. Factors affecting chemical and biological reduction of hexavalent chromium in soil. Environ. Toxicol. Chem. 13: 1727–1735.

Mallick, S., G. Sinam, R. Kumar Mishra and S. Sinha. 2010. Interactive effects of Cr and Fe treatments on plants growth, nutrition and oxidative status in *Zea mays* L. Ecotoxicol. Environ. Saf. 73: 987–995.

Mandiwana, K.L., N. Panichev, M. Kataeva and S. Siebert. 2007. The solubility of Cr(III) and Cr(VI) compounds in soil and their availability to plants. J. Hazard. Mater. 147: 540–545.

Markert, B., 2007. Definitions and principles for bioindication and biomonitoring of trace metals in the environment. J. Trace Elem. Med. Biol. 21: 77–82.

Mishra, S., V. Singh, S. Srivastava, R. Srivastava, M.M. Srivastava, S. Dass, G.P. Satsangi and S. Prakash. 1995. Studies on uptake of trivalent and hexavalent chromium by maize (*Zea mays*). Food Chem. Toxicol. 33: 393–397.

Mohanty, M. and H.K. Patra. Attenuation of chromium toxicity by bioremediation technology. pp. 1–34. *In:* Whitacre, D.M. (Ed.). 2011. Reviews of Environmental Contamination and Toxicology, vol. 210. Springer. New York, USA.

Notten, M.J.M., A.J.P. Oosthoek, J. Rozema and R. Aerts. 2005. Heavy metal concentrations in a soil-plant-snail food chain along a terrestrial soil pollution gradient. Environ Pollut. 138: 178–190.

Pal, A. and A.K. Paul. 2004. Aerobic chromate reduction by chromium-resistant bacteria isolated from serpentine soil. Microbiol. Res. 159: 347–354.

Polti, M.A., M.J. Amoroso and C.M. Abate. 2007. Chromium(VI) resistance and removal by actinomycete strains isolated from sediments. Chemosphere. 67: 660–667.

Polti, M.A., R.O. García, M.J. Amoroso and C.M. Abate. 2009. Bioremediation of chromium(VI) contaminated soil by *Streptomyces* sp. MC1. J. Basic. Microbiol. 49: 285–292.

Polti, M.A., M.J. Amoroso and C.M. Abate. 2010. Chromate reductase activity in *Streptomyces* sp. MC1. J. Gen. Appl. Microbiol. 56: 11–18.

Polti, M.A., M.J. Amoroso and C.M. Abate. 2011. Intracellular chromium accumulation by *Streptomyces* sp. MC1. Water Air Soil Pollut. 214: 49–57.

Prado, C., L. Rodríguez-Montelongo, J.A. González, E.A. Pagano, M. Hilal and F.E. Prado. 2010. Uptake of chromium by *Salvinia minima*: effect on plant growth, leaf respiration and carbohydrate metabolism. J. Hazard. Mater. 177: 546–553.

Prasad, M.N.V., H. Freitas, S. Fraenzle, S. Wuenschmann and B. Markert. 2010. Knowledge explosion in phytotechnologies for environmental solutions. Environ. Pollut. 158: 18–23.

Puzon, G.J., J.N. Petersen, A.G. Roberts, D.M. Kramer and L. Xun. 2002. A bacterial flavin reductase system reduces chromate to a soluble chromium(III)-NAD*þ* complex. Biochem. Biophys. Res. Commun. 294: 76–81.

Rogival, D., J. Scheirs and R. Blust. 2007. Transfer and accumulation of metals in a soildiet-wood mouse food chain along a metal pollution gradient. Environ. Pollut. 145: 516–528.

Shanker, A.K., C. Cervantes, H. Loza-Tavera and S. Avudainayagam. 2005. Chromium toxicity in plants. Environ. Int. 31: 739–753.

Sharma, D.C., C.P. Sharma and R.D. Tripathi. 2003. Phytotoxic lesions of chromium in maize. Chemosphere. 51: 63–68.

Siñeriz, M.L., E. Kothe and C.M. Abate. 2009. Cadmium biosorption by *Streptomyces* sp. F4 isolated from former uranium mine. J. Basic. Microbiol. 49: S55–S62.

Stewart, M., P. Jardine, C. Brandt, M. Barnett, S. Fendorf, L. McKay, T. Mehlhorn and K. Paul. 2003. Effects of contaminant concentration, aging, and soil properties on the bioaccessibility of Cr(III) and Cr(VI) in soil. Journal of Soil Contamination. 12: 1–21.

Turick, C.E., C. Graves and W.A. Apel. 1998. Bioremediation potential of Cr (VI)-contaminated soil using indigenous microorganisms. Bioremediat. J. 2: 1–6.

Udovic, M. and D. Lestan. 2010. *Eisenia fetida* avoidance behavior as a tool for assessing the efficiency of remediation of Pb, Zn and Cd polluted soil. Environ. Pollut. 158: 2766–2772.

Vainshtein, M., P. Kuschk, J. Mattusch, A. Vatsourina and A. Wiessner. 2003. Model experiments on the microbial removal of chromium from contaminated groundwater. Water Res. 37: 1401–1405.

Vandecasteele, B., C. Buysse and F. Tack. 2006. Metal uptake in maize, willows and poplars on impoldered and freshwater tidal marshes in the Scheldt estuary. Soil Use Manage. 22: 52–61.

Wani, R., K. Kodam, K. Gawai and P. Dhakephalkar. 2007. Chromate reduction by *Burkholderia cepacia* MCMB-821, isolated from the pristine habitat of alkaline crater lake. Appl. Microbiol. Biotechnol. 75: 627–632.

Zayed, A., C.M. Lytle, J.H. Qian and N. Terry. 1998. Chromium accumulation, translocation and chemical speciation in vegetable crops. Planta. 206: 293–299.

CHAPTER 8

Cadmium Bioremediation by a Resistant *Streptomyces* Strain

Siñeriz Louis Manuel,[1,3,*] Kothe Erika[3] and Abate Carlos Mauricio[1,2,4]

Introduction

Cadmium is a metallic element that is rarely found in pure form. It is present in various types of rocks, soils and water, as well as coal and oil. Cadmium is a highly toxic heavy metal that is mainly used for Ni-Cd batteries and pigments in plastics and glassware, stabilizers for PVC, protective plating on steel and alloys contributing to environmental pollution (Wilson 1988, Scoullos et al. 2001). It is also found in sewage sludge from industrial areas and in agricultural phosphate fertilizers at low concentrations.

Cadmium in soils derives both from natural and anthropogenic sources. It has no biological function with one exception. The average cadmium concentration in non-volcanic soil ranges from 0.01 to 1 mg/kg, but in

[1]Planta Piloto de Procesos Industriales y Microbiológicos (PROIMI), CONICET, Av. Belgrano y Pasaje Caseros, 4000 Tucumán, Argentina.
[2]Facultad de Ciencias Naturales e Instituto Miguel Lillo, Universidad Nacional de Tucumán, Miguel Lillo 205, Tucumán.
[3]Friedrich-Schiller-Universität, Institut für Mikrobiologie, Mikrobielle Phytothologie, Neugasse 25, Jena, Germany.
[4]Facultad de Bioquímica, Química y Farmacia, Universidad Nacional de Tucumán, Ayacucho 471, Tucumán.
*Corresponding author: msineriz@yahoo.com

volcanic soil, levels of up to 4.5 mg/kg have been found [Korte 1983]. Polluted soil near an electroplate factory in Nanjing, China, has been found to contain 87.6 mg kg^{-1} Cd (Sheng and Xia 2006) and higher levels can be found in the vicinity of metal processing factories and mining areas.

The Environmental Protection Agency of the United States (EPA) has identified cadmium as a carcinogen B1, probable human carcinogen and is usually listed among the 10 priority pollutants in this country, exhibiting both acute and chronic toxicity.

Cadmium is a very mobile metal in soil and can be transferred and accumulated in the roots, leaves and stems of plants (Mench et al. 1989). A solution to cadmium contamination would be the cleaning of polluted soils by conventional treatments used in industry: removal of the soil with deposition in land fill, or, e.g., physico-chemical treatments. The adsorption properties of mineral and organic amendments have been exploited (Bailey et al. 1999), but heavy metal leakage can occur in the soil and with the water path, especially in humid climatic conditions (Antoniadis and Alloway 2001).

In the last decade, microbial bioremediation processes emerged as a viable alternative to traditional methods. Different microorganisms like bacteria and fungi, including yeasts, have a potential ability of metal biosorption, but very little information is available on actinobacteria used for this purpose (Abbas and Edwards 1989, Kefala et al. 1999, Jézéquel et al. 2005, Polti et al. 2007, Jézéquel and Lebeau 2008, Albarracín et al. 2008). The actinobacteria are a group of Gram positive bacteria with a high G+C content, they are metabolically versatile with a highly active secondary metabolism and are found mainly in soil, constituting a substantial group of the microflora.

Actinobacteria produce many extracellular enzymes in soil. By decomposing complex mixtures of organic polymers with degrading exoenzymes and extracellular peroxidases (Wang et al. 1989, Mason et al. 2001), they are also important to the recycling of nutrients associated with recalcitrant polymers (McCarthy and Williams 1992). An advantage over Gram-negative soil bacteria is their ability to spread through relatively dry soil via hyphal growth and to survive in adverse conditions by formation of durable spores (Kieser et al. 2000).

In this chapter it is meant to show a highly cadmium resistant actinobacteria strain, *Streptomyces* sp. F4, with a capacity to bioremediate cadmium in liquid medium and soil microcosms. Also it is important to show the study of resistance mechanisms and cadmium uptake in soil by this microorganism, making it potentially useful on a larger scale in areas polluted with this metal.

Isolation and Selection of Cd Resistant Actinobacteria Strains

In order to find actinobacterias for bioremediation purposes, 46 strains were isolated from two contaminated sites and one uncontaminated site: eight from soils from Wismut, a former uranium mine in Germany polluted with several heavy metals; four from the Rio Hondo Dam, Tucumán, Argentina, polluted by effluents from sugar mills, paper and mining industry. 34 strains were isolated from the marine sediments in Ushuaia, Argentina, considered a non-polluted area. The microorganisms were isolated on Starch-Casein Agar (SCA) and Artificial Soil Agar (ASA). All isolates showed typical actinobacteria morphology with both substrate mycelium and pigmented, aerial, branched hyphae with spores. Ushuaia was included in this study to find out if actinobacteria with resistance mechanisms against cadmium from pristine sites are present. The isolates were maintained by regular transfers to Minimal Medium (MM) agar slants tubes.

Primary qualitative screening showed that 25 of the 46 colonies, isolated from both polluted and non-polluted sites, were Cd-resistant and selected for the semi quantitative screening: 18 from Ushuaia, four from Wismut and three from Rio Hondo Dam (Fig. 1). Qualitative assays were carried out on plates containing MM agar. Rectangular wells were made in the center of the plates by aseptically removing agar strips. The resulting ditch was filled with 500 µl of a 10 mg L^{-1} $CdCl_2$ solution. Spore suspensions of the strains to be tested were streaked onto the plates, completely covering the surface, except for the wells; plates were then incubated for 7 days at 30°C. Microbial growth close to the well was used as the qualitative parameter for metal resistance according to Amoroso et al. (1998).

For semi-quantitative assays of cadmium resistance, 50 µL of $CdCl_2$ of different concentrations (1–100 mg L^{-1}) were used to fill the wells of MM agar Petri dishes previously inoculated with 100 µL of spore suspensions (1×10^9 CFU mL^{-1}) of the strain to be tested. The growth inhibition diameter (D) was measured after incubation at 30°C during 4 days and strains were classified into resistant ($D < 7$ mm) and non-resistant ($D > 7$ mm). Five of the 25 actinobacteria were able to grow at 100 mg L^{-1} (USH1.12, USH1.9, USH3.15, F4 and R22), six at 50 mg L^{-1} (USH1.6, USH3.22, USH3.24, USH3.25, F3.8 and R10), and five at 25 mg L^{-1} $CdCI_2$ (USH3.10, USH3.11, USH3.2, F3.6 and R25) (Fig. 1).

Strains resistant to 100 mg L^{-1} of Cd were tested for Cd toxicity on MM plates containing defined amounts of this metal (0–200 mg L^{-1}). USH1.12 and R22 were resistant to 2 mg L^{-1}, USH 3.15 to 6 mg L^{-1}, USH 1.9 to 10 mg L^{-1} and F4 up to 150 mg L^{-1} Cd. The control strains *Streptomyces lividans* TK24 and *S. coelicolor* A2 (3) were able to grow at 2 and 0 mg L^{-1} Cd, respectively. In liquid MM, none of the strains assayed, including *S. lividans* TK24 and

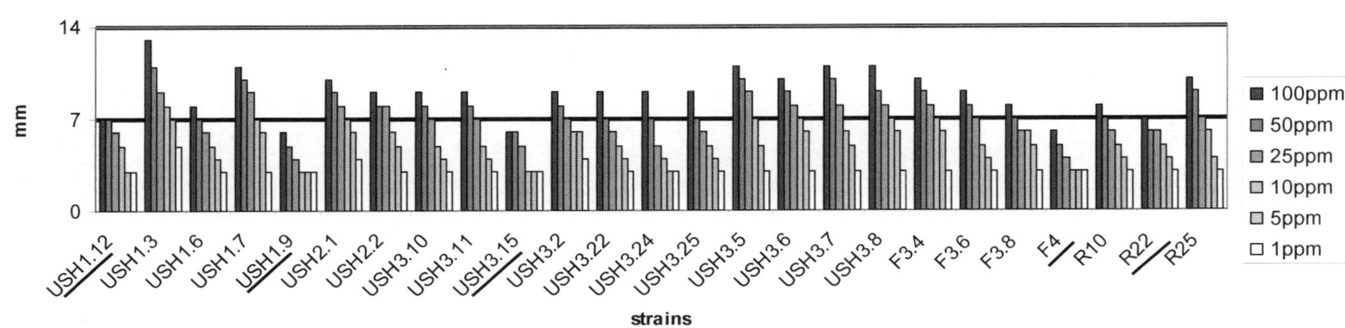

Figure 1. Semi-quantitative resistance of actinobacteria at 1, 5, 10, 25, 50 and 100 mgL^{-1} Cd measured as growth inhibition diameter (in mm). Actinobacteria were isolated from Ushuaia, Argentina (USH), from sediments at a former uranium mine, Germany (F), and from the Rio Hondo dam, Argentina (R). The horizontal line indicates an arbitrary limit to consider cadmium-resistant (below) and non-resistant (above) strains.

S. coelicolor A2 (3), were not able to grow at 2 mg L^{-1} Cd except for *Streptomyces* sp. F4, which was able to grow at 15 mg L^{-1} Cd after two weeks, but not at 20 mg L^{-1}. Since F4 was the most Cd-resistant strain it was chosen for subsequent assays.

Taxonomic Description and Physiological Characterization of *Streptomyces* sp. F4

It was characterized morphologically, having *Retinaculiaperti* spore chains with a white to gray spore mass, which produced no diffusible pigment, and light yellow substrate mycelium (Amoroso et al. 2000). Colours were analyzed using the RAL color code (RAL K5 2006). On media with tyrosine the substrate color was traffic yellow (RAL 1023) and without tyrosine zinc yellow (RAL 1018). The spores were pearl white (RAL 1013).

Physiological parameters were determined for further characterization. *Streptomyces* sp. F4 could tolerate up to 7.5% NaCl and grew well at 100 µg ml^{-1} lysozyme. Good growth was also observed at pH between 6 and 10. Glucose, xylose, arabinose and rhamnose allowed good growth, whereas sucrose, raffinose and cellulose were less favorable as carbon sources. Excellent growth was obtained with inositol, manitol and fructose.

Biochemical assaying was performed with API Zym, API Coryne and API 20E in duplicate according to the manufacturer (API, BioMérieux, France). *Streptomyces* sp. F4 was able to carry out the following activities: acid phosphatase, alkaline phosphatase, naphthol-AS-BI-phosphohydrolase, β-galactosidase, α-glucosidase, catalase, gelatin hydrolysis, and oxidation of glucose, mannose, inositol, sorbitol and arabinose.

The following activities were negative: esterase, esterase lipase, lipase, leucine arylamidase, valine arylamidase, cystine arylamidase, trypsin, α-chymotrypsin, α-galactosidase, α-glucuronidase, β-glucuronidase, β-glucosidase, N-acetyl-β-glucosaminidase, α-mannosidase and α-fucosidase, reduction of nitrate, pyrazinemidase, pyrrolidonyl arylamidase, urease, arginine dihydrolase, lysine decarboxylase, ornithine decarboxylase, citrate utilization, H_2S production, urease, tryptophane deaminase, indole production and acetoin production. Fermentation or oxidation of rhamnose, saccharose, melibiose and amygdalin was not observed. Neither was reduction of nitrates to nitrites and nitrogen gas.

Based on the 16S rDNA sequence, *Streptomyces* sp. F4 (GenBank accession number DQ141201) shows 99% homology with *Streptomyces tendae* (Siñeriz et al. 2009).

Growth Inhibition by Cadmium

Spore suspensions were used as inoculum. In MM medium without Cd, F4 reached the exponential phase on the 4th day (1.492 g L^{-1}) and biomass decreased between day 6 and 8 (1.378 g L^{-1} and 1.150 g L^{-1}, respectively), probably because of bacterial lysis. In medium with 8 mg L^{-1} Cd, less growth was observed with a maximum of 0.304 g L^{-1} on day 8. Growth on day 4 with Cd was 20% (five times) less as compared to growth without the metal (Fig. 2).

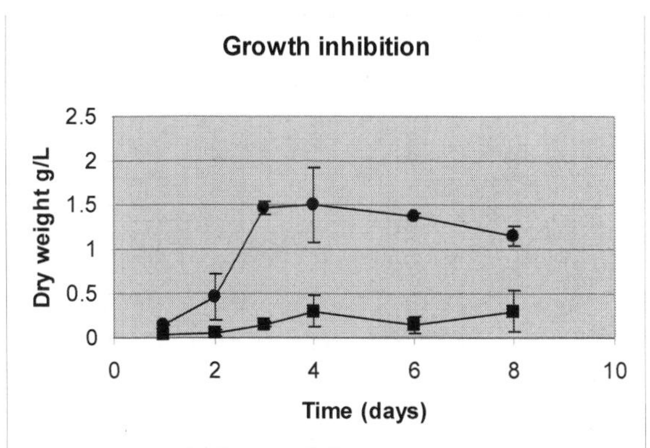

Figure 2. *Streptomyces* sp. F4 growth in MM with 8 mg L^{-1} (■) and without Cd (●) measured as biomass (dry weight).

Cadmium Removal from Liquid Medium

Bacterial biomass decreased from 1.66 g L^{-1} (0 mg L^{-1} Cd) to 1.08 g L^{-1} with 8 mg L^{-1} Cd. Surprisingly, at 8 mg L^{-1} growth was higher (1.08 g L^{-1}) than at 6 mg L^{-1} (0.84 g L^{-1}), but the difference is within the standard deviation (data not shown). At the same time, Cd absorption increased from 1.03 mg (2 mg L^{-1}) to 1.78 mg (8 mg L^{-1}). Maximum biosorption was observed at 2 mg L^{-1} (51.6%). At 4 mg L^{-1} of Cd(II), this parameter was 20.8%, at 6 mg L^{-1} 17.9% and at 8 mg L^{-1} 22.2% (data not shown). Specific biosorption by *Streptomyces* sp. F4 increased with increasing Cd concentration in the medium, with a maximum value of 2.1 mg Cd g $cells^{-1}$ (8 mg L^{-1}). Cadmium concentrations in supernatant and biomass were determined by atomic absorption spectrometry using an ICP-AES (Inductively Coupled Plasma-Atomic Emission Spectrometry).

Haq et al. (1999) found that one *Enterobacter cloacae* and two *Klebsiella* strains were able to remove 100 mg L^{-1} Cd within 24 h from LB medium, corresponding to 85–87% Cd. However, LB medium is richer than the MM used in our study, thus Cd bioavailability is less because of the metal complexation in the medium. In another work, El-Helow et al. (2000) found that a cadmium-resistant strain of *Bacillus thuringiensis* was able to grow in optimized Estuarine salt broth (ESB) medium supplemented with 0.25 mM $CdCl_2$ (281 mg L^{-1} Cd), removing 79% of the metal ions within 24 h with a specific biosorption capacity of 21.57 mg Cd g^{-1} dry weight. On the other hand, similar studies were conducted using copper solutions; in this case Albarracín et al. (2005) found specific biosorption capacities of up to 25 mg Cu g $cells^{-1}$ in copper-resistant *Streptomyces* strains.

In order to determine the localization of biosorbed Cd, subcellular fractionation was performed. The highest amount of Cd was found in the cell wall, 41.2%. Only 7.4% was found in the exopolysaccharide layer, while the cytosol contained 39.4%. The ribosome and membrane fraction contained 12% of the Cd (Fig. 3).

Albarracín et al. (2008) made a subcellular fractionation of a copper-resistant actinobacteria, *Amycolatopsis* sp. AB0, after 7 days incubation in MM supplemented with 32 mg L^{-1} Cu(II). They observed that 40% of copper was in an exopolysaccharide, while 60% was in the cell. The copper present in the cell was distributed into 85.65% in the cytosol, 10.9% in the cell wall and 3.45% in the ribosome and membrane fraction.

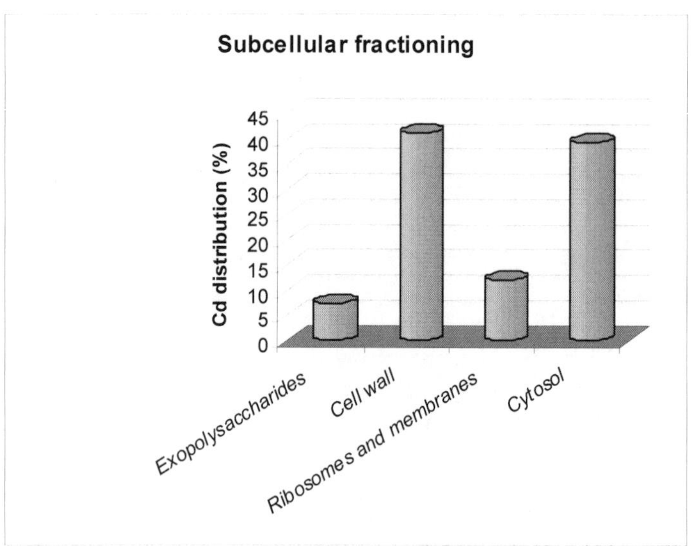

Figure 3. Cadmium distribution in the subcellular fractions of *Streptomyces* sp. F4 grown with 10 mg L^{-1} Cd after 5 d.

Electron Microscopy (TEM and SEM)

Cd localization was verified by electron microscopy. Thin sections of Cd-stressed (8 mg L^{-1}) cells of *Streptomyces* sp. F4 showed a cytoplasm with dark granulate appearance when compared to untreated cells. These dark grains were not observed in *Streptomyces* sp. F4 grown without the metal (Fig. 4). No significant differences were observed in the cell wall, even though 41.2% of the metal was detected in this fraction.

Electron microscopy was performed at the Laboratorio de Microscopía Electrónica del Noroeste, Tucumán, Argentina. Pellets of 1 week old cultures, grown in MM with 8 mg L^{-1} Cd, were rinsed twice with sterile distilled water. Samples were prepared according to standard TEM procedures (Albarracín 2007) and analyzed with a Zeiss EM109.

Cunningham and Lundie (1993) detected localized sites of cadmium precipitation on the surfaces of non-starved *Clostridium thermoaceticum* cells. In another work, El-Helow et al. (2000) found dark stains on the outer cell wall surface of *Bacillus thuringiensis* grown with Cd. Roane et al. (2001) observed Cd sequestration by the EPS layer as a dark precipitate surrounding *Arthrobacter* D9 and *Pseudomonas* I1A cells, while in *Pseudomonas* H1 and *Bacillus* H9, an intracellular accumulation of cadmium was observed.

With a Scaning Electron Microscope we observed *Streptomyces* sp. F4 grown in MM with 10 mg L^{-1} Cd for 1 week, and a control without the metal. In the control we observed hyphae of similar size and thickness. There were no dead or deformed cells (Fig. 5). In the presence of Cd, we observed hyphae of varying sizes, some thicker and others thinner.

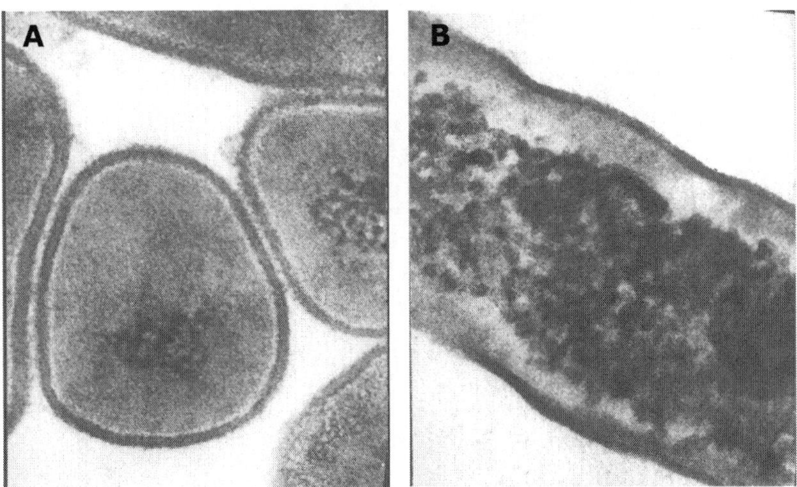

Figure 4. Transmission electron micrographs of *Streptomyces* sp. F4 grown without (A) and with 8 mg L^{-1} Cd (B) after one week. Magnification 140, 600x.

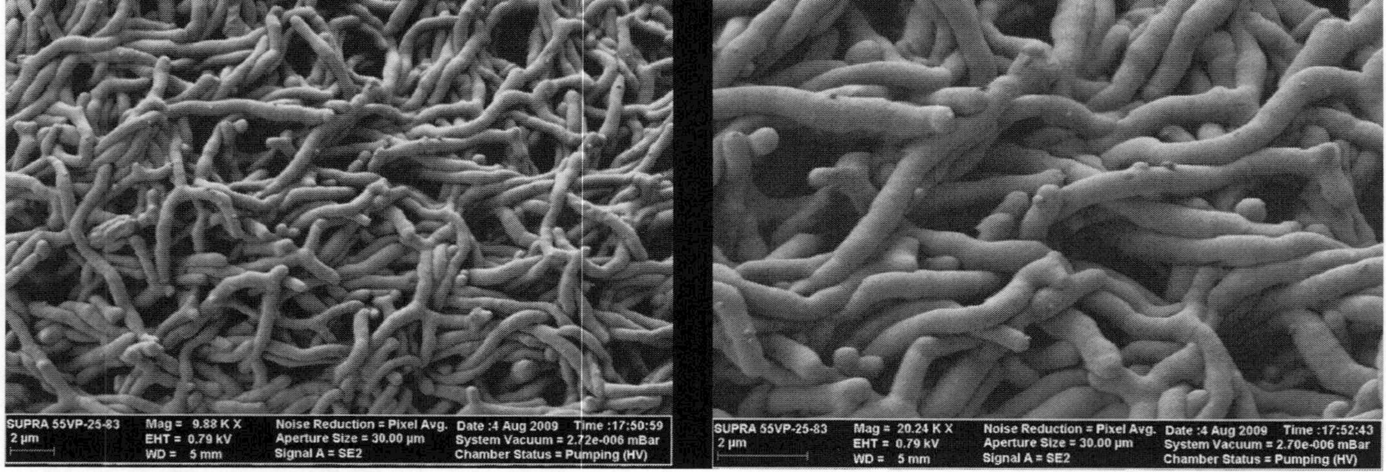

Figure 5. Scanning electron micrographs of *Streptomyces* sp. F4 grown on MM without Cd for 1 week. Magnification: 10.000 and 20.000x.

Many dead cells were visualized, and others deformed (Fig. 6). Although *Streptomyces* sp. F4 has a high resistance to Cd, the metal produced many changes and cell damage.

Lu et al. (2006) found a heavy metal-resistant bacterium, *Enterobacter* sp. J1, which after growing in liquid medium added with Cd, had a cell surface distorted and damaged as compared to the control without metal. Nithya et al. (2011) isolated heavy metal resistant strains from marine sediments and a strain of *Bacillus pumilus* after growing in the presence of Cd did not show changes in morphology, whereas in the presence of other metals such as Se, Hg, Pb and As exhibit differences, like change of shape, or excessive production of exopolysaccharide.

Cd uptake and complexation by *Streptomyces* sp. F4 could be due to two different mechanisms. One possible mechanism is binding of Cd by adsorption to the cell wall, which was confirmed by subcellular fractionation (Fig. 3). The other mechanism could involve the presence of intracellular protective proteins like metallothioneins, which bind the metal that enters the cell (Fig. 4).

Bioremediation assays on soils

To see if *Streptomyces* sp. F4 was able to immobilize Cd in soil, (so that it is not bioavailable), microcosm tests were conducted. Bioremediation assays were made in soil samples (sandy and with low organic matter) from the west of Tucuman province. Samples were inoculated in 200 g of soil, in glass jars, adjusted to 20% humidity, suplemented with 1% glucose and 30% glass beads, a chemically inert substrate used to aerate the ground, so the bacteria could grow. Samples were inoculated with mycelium of *Streptomyces* sp. F4, grown in rich medium TSB (Triptic Soy Broth). Samples were incubated at 30°C and the CFU were calculated by counting in Petri dishes. The Cd bioavailable fraction was extracted from the soil by centrifugating at a speed representing the conventional wilting point of plants, i.e., the conventional upper limit of the potential exerted by plant roots. Then, total Cd was measured with an ICP-AES.

Growth of *Streptomyces* sp. F4 in soil with 100 mg L^{-1} Cd for six weeks, with and without the metal, was compared. The CFU was calculated. Growth was similar in soil with Cd compared with soil without Cd; even at two, three and six weeks of incubation. Growth of the strain was higher in the soil with 100 mg L^{-1} Cd, demonstrating the great *Streptomyces* sp. F4 capacity to tolerate Cd.

Bioavailable Cd values, calculated from the centrifugation method, allowed to determine that Cd at the initial time was 39.27 mg L^{-1}, increased slightly after a week (43.08 mg L^{-1}) and then decreased to 15.9 mg L^{-1} after eight weeks (data not shown).

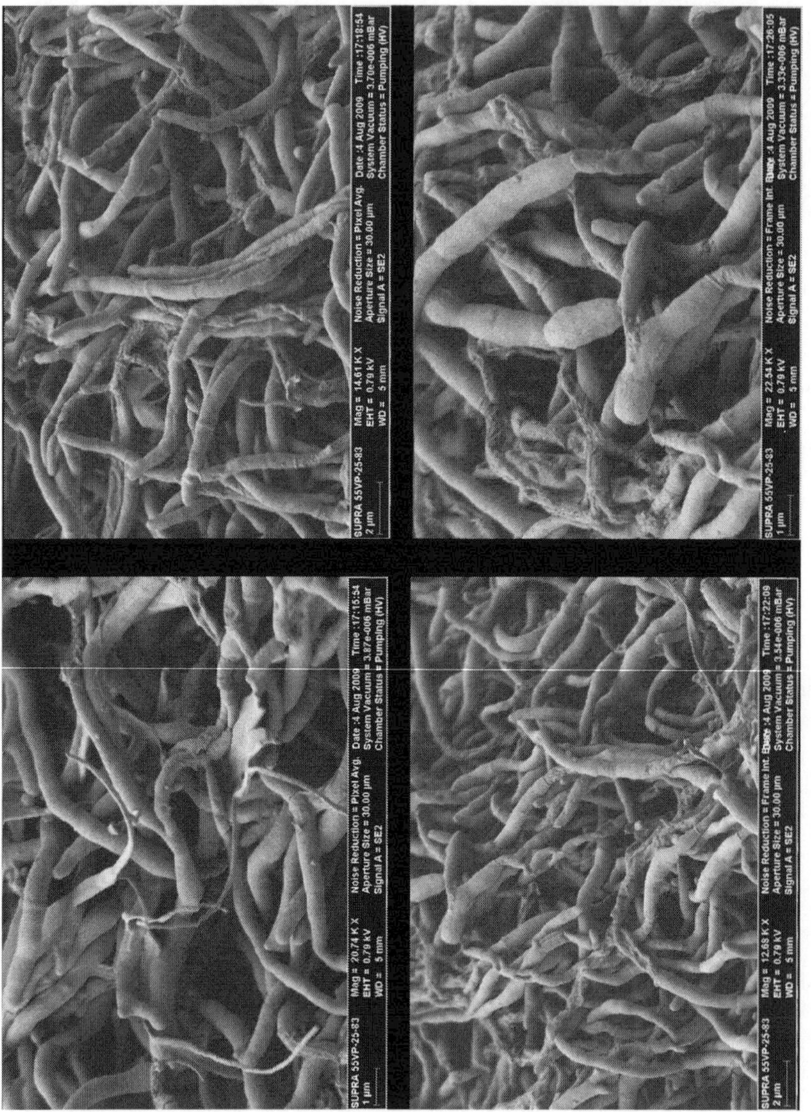

Figure 6. Scanning electron micrographs of *Streptomyces* sp. F4 grown on MM with 10 mg L^{-1} Cd during 1 week. Magnification: 12.500–22.500x.

In other assays in non-sterile and sterile soil, we observed that in sterile soils inoculated with *Streptomyces* sp. F4, bioavailable Cd decreased from 50 to 22 mg L^{-1} approximately (Fig. 7). In the non inoculated control the final bioavailable Cd was 39 mg L^{-1} approx. It was also observed that in non-sterile soil, the indigenous bacteria contributed to decrease the bioavailable Cd, compared to sterile soil.

Jézéquel and Lebeau (2008) did a similar approach, inoculating contaminated soil with 25 mg kg^{-1} Cd with a *Bacillus* sp. and a *Streptomyces* sp., to reduce the potentially available Cd for plants. After 3 weeks, the Cd bioavailable was reduced sustantialy and the bacteria survived and colonized the soil.

The assays realized with *Streptomyces* sp. F4 demonstrate the ability of this strain to colonize the soil, and to uptake the Cd, so it is not bioavailable to other organisms in the food chain.

All these results are important because Cd resistance by actinobacteria is a subject that deserves to be continuously studied because there are few publications related to the use of these microorganisms in Cd bioremediation in soils.

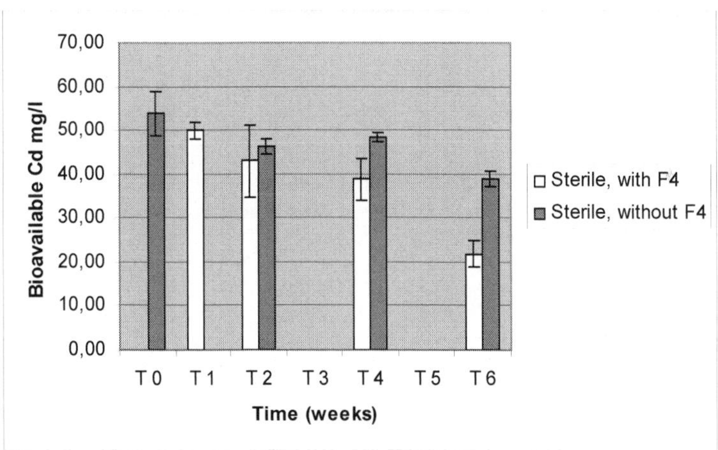

Figure 7. Bioavailable Cd on sterile soil added with 100 mg L^{-1} Cd, inoculated with *Streptomyces* sp. F4 (white bars) and non inoculated (grey bars).

Acknowledgments

This work was supported by Consejo de Investigaciones de la Universidad Nacional de Tucumán (CIUNT), Agencia Nacional de Promoción Científica y Tecnológica (ANPCyT) and Consejo Nacional de Investigaciones Científicas y Técnicas (CONICET). I would like to thank Dr. María Julia Amoroso for giving me some of the strains, and for corrections.

References Cited

Abbas, A. and C. Edwards. 1989. Effects of metals on a range of *Streptomyces* species. Appl. Environ. Microbiol. 55: 2030–2035.

Albarracín, V.H., M.J. Amoroso and C.M. Abate. 2005. Isolation and characterization of indigenous copper-resistant actinomycete strains. Chemie der Erde. 65: 145–156.

Albarracín V.H. 2007. Estudios de los aspectos fisiológicos, bioquímicos y moleculares de la resistencia al cobre en cepas de actinomycetes. PhD Thesis. Fac. de Bioquímica, Química y Farmacia. UNT. Tucumán, Argentina.

Albarracín, V.H., B. Winik, E. Kothe, M.J. Amoroso and C.M. Abate. 2008. Copper bioaccumulation by the actinobacterium *Amycolatopsis* sp. AB0. J. Basic Microbiol. 48: 323–330.

Amoroso, M.J., G.R. Castro, F.J. Carlino, N.C. Romero, R.T. Hill and G. Oliver. 1998. Screening of heavy metal-tolerant actinomycetes isolated from the Salí River. J. Gen. Appl. Microbiol. 44: 129–132.

Amoroso, M.J., D. Schubert, P. Mitscherlich, P. Schumann and E. Kothe. 2000. Evidence for high affinity nickel transporter genes in heavy metal resistant *Streptomyces* spec. J. Basic Microbiol. 40: 295–301.

Antoniadis, V. and B.J. Alloway. 2001. Availability of Cd, Ni and Zn to ryegrass in sewage sludge-treated soils at different temperatures. Water Air Soil Pollut. 132: 201–214.

Bailey, S.E., T.J. Olin, R.M. Bricka and D.D. Adrian. 1999. A review of potentially low-cost sorbents for heavy metals. Water Res. 33: 2469–2479.

Cunningham, D.P. and L.L. Lundie. 1993. Precipitation of cadmium by *Clostridium thermoaceticum*. Appl. Env. Microbiol. 59: 7–14.

El-Helow, E.R., S.A. Sabry and R.M. Amer. 2000. Cadmium biosorption by a cadmium resistant strain of *Bacillus thuringiensis*: regulation and optimization of cell surface affinity for metal cations. BioMetals. 13: 273–280.

Haq, R., S.K. Zaidi and A.R. Shakoori. 1999. Cadmium resistant *Enterobacter cloacae* and *Klebsiella* sp. isolated from industrial effluents and their possible role in cadmium detoxification. World J. Microbiol. Biotechnol. 15: 283–290.

Jézéquel, K., J. Perrin and T. Lebeau. 2005. Bioaugmentation with a *Bacillus* sp. to reduce the phytoavailable Cd of an agricultural soil: comparison of free and immobilized microbial inocula. Chemosphere. 59: 1323–1331.

Jézéquel, K. and T. Lebeau. 2008. Soil bioaugmentation by free and immobilized bacteria to reduce potentially phytoavailable cadmium. Bioresour. Technol. 99: 690–698.

Kefala, M.I., A.I. Zouboulis and K.A. Matis. 1999. Biosorption of cadmium ions by actinomycetes and separation by flotation. Environ. Pollut. 104: 2: 283–293.

Kieser, T., M.J. Bibb, M.J. Buttner, K.F. Chater and D.A. Hopwood (Ed.). 2000. Practical *Streptomyces* Genetics. John Innes Foundation, Norwich.

Korte, F. 1983. Ecotoxicology of cadmium: general overview. Ecotoxicol. environ. Saf. 7: 3–8. *In*: Cadmium-environmental aspects. Environmental Health Criteria, World Health Organization. Geneva. 135, 1992.

Lu, W.B, J.J. Shi, C.H. Wang and J.S. Chang. 2006. Biosorption of lead, copper and cadmium by an indigenous isolate *Enterobacter* sp. J1 possessing high heavy-metal resistance. J. Hazard. Mater. 134: 80–86.

Mason, M.G., A.S. Ball, B.J. Reeder, G. Silkstone, P. Nicholls and M.T. Wilson. 2001. Extracellular heme peroxidases in actinomycetes: a case of mistaken identity. Appl. Env. Microbiol. 67: 4512–4519.

McCarthy A.J. and S.T. Williams. 1992. Actinomycetes as agents of biodegradation in the environment—a review. Gene. 115: 189–192.

Mench, M., J. Tancogne, A. Gomez and C. Juste. 1989. Cadmium bioavailability to *Nicotiana tabacum* L., *Nicotiana rustica* L., and *Zea mays* L. grown in soil amended with cadmium nitrate. Biol. Fertil. Soils. 8: 48–53.

Nithya, C., B. Gnanalakshmi and S.K. Pandian. 2011. Pattern of cadmium accumulation and essential cations during growth of cadmium-tolerant fungi. Mar Environ. Res. 71: 283–294.

Polti, M.A., M.J. Amoroso and C.M. Abate. 2007. Chromium(VI) resistance and removal by actinomycete strains isolated from sediments. Chemosphere. 67: 660–667.

RAL K5 Colour selector. 2006. Deutsches Institut für Gütesicherung und Kennzeichnung E.V.

Roane, T.M., K.L. Josephson and I.L. Pepper. 2001. Dual-bioaugmentation strategy to enhance remediation of cocontaminated soil. Appl. Env. Microbiol. 67: 3208–3215.

Scoullos, M., G. Vonkeman, I. Thornton and Z. Makuch. 2001. Mercury, cadmium, lead: Handbook for sustainable heavy metals policy and regulation. Kluwer Academic Publishers, Dordrecht. *In*: Nordic Council of Ministers. Cadmium Review, 1, 4, January 2003.

Sheng, X.F. and J.J. Xia. 2006. Improvement of rape (*Brassica napus*) plant growth and cadmium uptake by cadmium-resistant bacteria. Chemosphere. 64: 6: 1036–1042.

Siñeriz, M.L., E. Kothe and C.M. Abate. 2009. Cadmium biosorption by *Streptomyces* sp. F4 isolated from former uranium mine. J. Basic Microbiol. 49: 55–62.

Wang, Z., D.L. Crawford, A.L. Pometto and F. Rafii. 1989. Survival and effects of wild-type, mutant, and recombinant *Streptomyces* in a soil ecosystem. Can. J. Microbiol. 35: 535–543.

Wilson, D.N. 1988. Cadmium-market trends and influences. *In*: Cadmium 87. Proceedings of the 6th International Cadmium Conference, London, Cadmium Association 9–16.

CHAPTER 9

Streptomyces from Soils Contaminated with Boron Compounds

Norma Beatriz Moraga,[1,2] María Julia Amoroso[3,4] and Verónica Beatriz Rajal[1,2,5,*]

Introduction

Boron (B) is a metalloid studied widely mainly because of its pharmacological, industrial and agrochemical properties. Its importance, both as a nutrient for the growth of some plants (Warington 1923) and an essential element in plant structure and cell walls, has long being known (O'Neill et al. 2004). Evidence suggests that some animals and unicellular eukaryotes including humans (Nielsen 2000) also require boron, but the concentrations needed vary widely according to the species. Cyanide bacteria (Mateo et al. 1986) and *Bacillus boroniphilus* sp. nov. (Ahmed et al. 2007a) are the only bacteria in which boron has been determined to be essential for growth. On the other hand, there are many well known metabolites containing boron synthesized

[1]Instituto de Investigaciones para la Industria Química (INIQUI), CONICET-UNSa, Avda. Bolivia 5150, 4400 Salta, Argentina.
[2]Facultad de Ingeniería, Universidad Nacional de Salta, Avda. Bolivia 5150, 4400 Salta, Argentina.
[3] Planta de Procesos Industriales y Microbiológicos (PROIMI), CONICET, Avda. Belgrano y Pje. Caseros, 4000 Tucumán, Argentina.
[4]Facultad de Bioquímica, Química y Farmacia, Universidad Nacional de Tucumán, 4000 Tucumán, Argentina.
[5]Fogarty International Center, University of California in Davis, US
*Corresponding author: vbrajal@gmail.com

by bacteria, like aplasmomycins (Nakamura et al. 1977), tartorlon A and B (Irschik et al. 1995), boromycin (Kohno et al. 1996) and borophycin (Arai et al. 2004).

Although some bacteria are known to require boron for their growth or to be capable of accumulate it, no studies have been published or registered about the capacity of the genus *Streptomyces* and *Lentzea* to tolerate high boron concentrations and their possibility to be used in the process of bioremediation of boron from contaminated environments.

The aim of this chapter is to discuss about seven boron-tolerant *Streptomyces* and one *Lentzea* isolated from soils naturally and anthropogenically contaminated with high concentrations of boron compounds in Salta, northwest of Argentina, where there is aggressive mining exploitation and processing. To the best of our knowledge there is no specific published information about *Streptomyces* or *Lentzea* genus tolerant to high boron concentrations.

Boron and Boron Compounds

Chemistry

Boron is a non-metal whose electronic configuration $2s^2 2p^1$ shows an electron deficiency and orbital availability, this being the main reason to be classified as a strong Lewis acid extremely avid for electrons. This fact will condition its participation in many chemical reactions with the formation of numerous compounds.

The chemical behavior of this element is quite singular as it shows just a few similarities with its group neighbors: aluminum and the remaining elements. But, as expected from its diagonal properties, it presents interesting relations with Silica, one of them being the chemistry of oxidized systems, which despite being less varied, present important structural similarities. The basic structural unit of boron oxidized compounds is the BO_3 triangle, whereas in the case of SiO there is a tetrahedron. Both structures have the capacity to share oxygen generating rings, chains, layers, and other complex structures in common. It is worth remarking that tetrahedral BO_4 structures appear in the frame of various natural compounds.

After the chemistry of carbon hydrides, that of boron is the richest and most complex. This also presents analogies to Si similar compounds, although the latter are less variable and abundant. The chemistry of the plain halides of both elements is also quite similar. Regarding its biological behavior, the ease with which boric acid forms alkyl or aryl esters of the type $B(OR)_3$ (Barán 1989) is remarkable. Boric acid can also form chelates in successive reaction stages with polyhydrolized species.

These compounds in which boron has tetrahedral coordination, made up of five member rings, are very stable since these rings are tension free. The acidity of OH groups bonded to boron exceeds that of the free acid.

Major uses and sources

Over the last 130 years, borate uses have grown remarkably, given the great applications this element has in modern lifestyles. There are over 200 boron bearing minerals, but boron industrial exploitation is feasible only from some of them, which are found in very few places in the world, such as Salta in Argentina, Death Valley in California (Flores 2004) and Kirka in Turkey (Helvaci and Ortí 2004). Borates constitute the main natural source of the boron element. They are boric acid salts or esters containing the radical B_2O_3. Borates form polymerized groups in chains, layers, and multiple isolated groups because the trivalent boron ion is joined to three oxygen atoms in its most stable form. Its properties, main applications in everyday life and percentage in each mineral are described in Table 1.

Table 1. Boron minerals basic features and main uses.

Mineral Empirical Formula	%B_2O_3	% B	pH	Solubility in water at 20°C (g/100 g sol sat)	Main uses
Borax (Tincal) $Na_2B_4O_7.10\ H_2O$	36.5	11.35	9.3	2.40	Abrasive, antifreeze, refractories, paper production, pharmacology, pesticides, paints, metal protectors from atmospheric effects
Kernite $Na_2B_4O_7.4\ H_2O$	51.0	15.83	W/D	W/D	Alloys, electronics (thermistors), artificial fiber (tennis rackets)
Ulexite $NaCa\ B_5O_9.8\ H_2O$	43.0	13.35	8.8	0.43	Boric acid, ceramic industry, fertilizers
Colemanite $Ca_2B_6O_{11}.5\ H_2O$	50.8	15.79	8.7	0.16	Borax, boric acid, plastic reinforcing and thermal insulating blankets
Hydroboracite $CaMg\ B_6O_{11}.6\ H_2O$	50.5	15.70	8.9	0.20	Fertilizer industry and other compounds for agriculture, special ceramic products, melting product for metallurgic furnaces for titanium ore reduction
Inyoite $Ca_2B_6O_{11}.13\ H_2O$	37.6	11.69	W/D	W/D	Fertilizers, "fries" and varnish, glass fiber

W/D: Without Description

For over 200 years boric acid has had external medicinal uses (mainly as antiseptic) and internal (as sedative, spasmolytic and diuretic), but since the 80s, criticism about its use, inefficacy and toxicity in prolonged treatments has increased. In fact, it has been gradually replaced by other more therapeutically active prescriptions. Knowledge about this compound suggests that it should be used more as an auxiliary reactant than as a main active component (Kiegel 1972).

Borax, $Na_2B_4O_7.10\ H_2O$, one of the simplest salts derived from boric acid has also had widespread homeopathic, antiseptic and bactericide applications. But since its therapeutic value has been questioned, its use has been reduced to cosmetics, additives and bathing salts.

Boric acid salts from alkaloids and numerous nitrogen bases have also had medicinal uses, for example borates, namely, Colin borate for liver disorders and tuberculosis, procaine borate as local anesthetic, hexamentilentriamine borate as renal antiseptic, neomycin borate for ophthalmic use and dodecilguanidine borate as fungicide. Compounds of boric acid with lactic acid or triethanolamine have also been used as contraceptives (Barán 1989).

Organic borates are compounds that have gained relevance due to their proven wide scope germicide and fungicide action. They have at least a hydrocarbon substitute above the boron atom and, in the case of triaryl boranes, they are stabilized with amines acting as Lewis bases. Another compound in this family also relevant for its sedative properties, is phenyl boric acid (Kiegel 1972). Some boron-containing compounds termed borinic esters have also shown antibacterial activity (Benkovic 2005).

Among the many functions that boron has, the mineral and bone metabolism is by far the most important one. It has been demonstrated that supplementary dietary boron may balance bone structural and metabolic disturbances produced by the lack of other important elements like Calcium, Vitamin D or Magnesium. Due to its capacity to balance bone metabolism, boron may be an important element in prevention of osteoporosis and treatment for it (Crespo 2001).

Boron has two stable natural isotopes, ^{10}B (20 percent abundant) and ^{11}B (80 percent abundant). The former has an extremely highly efficacious capture section to absorb thermal neutrons (Barth et al. 2007). The nuclear reaction performed is:

$$^{10}B + n \longrightarrow {}^{7}Li + \alpha\ (E = 2.4\ MeV)$$

This reaction is especially attractive for tumor treatments given as both ^{10}B and thermal neutrons are harmless for tissues due to the small size of both the produced particles, which limits their radium action and the highly efficacious capture section for thermal neutrons (ca. 3840 barn, a thousand times superior to those of C, N and O present in all organic matter) (Barán 1989). Even if this reaction liberates high energy radiation which may interfere with cell substance destroying it (Steinberg et al. 1964), this is its only negative point since the possibility of locating neutron beams and its compliance with all the prerequisites indispensable for the use of compounds in human oncological therapies, makes it quite promising (Barth 2003). Over the last fifty years there have been attempts to incorporate boron in different bioactive molecules (Morin 1994), above all in the neutron capture therapy in brain tumors (Sjöberg et al. 1997).

Boron and Biology

Toxicity

It is well known that above certain concentrations, boron is toxic for microorganisms, animals and human cells. Because of this toxic effect, it is used in the treatment of vulvovaginal diseases caused by *Candidas* and *Saccharomyces* (Swate and Leed 1974), as food preservative (Nielsen 2004) and as cockroach repellent insecticide (Habes et al. 2006).

The last review of the World Health Organization about the quality of drinking water (WHO 2009) reported that assays on laboratory animals (rats, mice and dogs) for short and long term exposures to boric acid and borax in food or drinking water affected the male reproductive tract, its main consistent target of toxicity (Truhaut et al. 1964, Weir and Fisher 1972, Green et al. 1973, Lee et al. 1978, NTP 1987). Even in human beings, reproductive functions are affected over certain values (Çöl and Çöl 2003).

The lowest reported lethal doses of boric acid (Stokinger et al. 1981) are 640 mg Kg^{-1} body weight (oral), 8,600 mg Kg^{-1} body weight (dermal), and 29 mg Kg^{-1} body weight (intravenous injection), and death has occurred at total doses of between 5 and 20 g of boric acid for adults and <5 g for children (WHO 2009).

Microbiology

There are only a few microorganism species mentioned in the specialized literature that can grow in high boron concentration in natural environments. However, there are many that are stimulated or not affected by the presence of boron compounds.

Some one-celled eukaryotes require boron to grow, the range of concentration needed according to the species being quite wide. So far, just in cyanobacteria (Mateo et al. 1986) and in *Bacillus boroniphilus* sp. nov. (Ahmed et al. 2007a) B as boric acid has been determined as an essential element for their growth.

In *Azobacter*, boron stimulates fixation mechanisms of N (Anderson et al. 1961) and in marine bacteria such as *Vibrio harveyi*, it intervenes in mediation processes of *Quorum Sensing* (Chen et al. 2002). Also in the *Arthrobacter nicotinovorans* species the catabolism of phenyl boric acid has been described (AFB) (Negrete-Raymond et al. 2003), in which B is eliminated mainly as orthoboric acid [$B(OH)_3$]. Even so, none of these studies shows that bacterial growth is actually limited if this non metal is absent.

Some bacteria have been isolated from soils naturally contaminated with boron minerals in Hisarcik, Turkey, such as *Bacillus boroniphilus* sp. nov. (Ahmed et al. 2007a) that requires boron to grow and can tolerate more than 450 mM boric acid, *Gracilibacillus boraciitolerans* sp. nov. (Ahmed et al. 2007b) that can tolerate up to 450 mM boric acid, *Chimaereicella boritolerans* sp. nov. (Ahmed et al. 2007c) that can tolerate up to 300 mM boric acid, *Lysinibacillus boronitolerans* (Ahmed et al. 2007d) and *Lysinibacillus parviboronicapiens* (Miwa et al. 2009a) that can tolerate up to 150 and 50 mMB as boric acid respectively. Miwa et al. (2008) have also published the existence of V*ariovorax boronicumulans* sp. nov, a Gram-negative bacterium able to accumulate boron inside. This mechanism, the same as plants, enables the removal of contaminating boron from the soil through its accumulation inside the cells. These authors have also isolated three actinobacteria from contaminated soils and active sludge (Miwa and Fujiwara 2009b), two of them belonging to the genus *Rhodococcus* and the other to the genus *Microbacterium.*

Even though actinobacteria make up the most abundant bacterial group present in all soils (representing 90 percentage of micro flora) and show an important biodegrading activity through the secretion of a great number of extra cellular enzymes enabling the metabolism of recalcitrant compounds (Keiser et al. 2000), no information is available so far about the action of the genus *Streptomyces* or *Lentzea* over B compounds. Soil actinobacteria develop different strategies to support their population in cycles ranging from fast proliferation to slow sporulation and growth in periods of stress or nutrient scarcity. These features and the fact that they are infrequent pathogens have endowed them with a great potential to perform bioremediation processes (Ravel et al. 1998).

Transport into the cell

In the species *Arabidopsis thaliana* two boric acid channels have been identified: NIP5; 1 and BOR1 (boric acid/borate exporter). Both proteins are necessary for plant growth in B limiting conditions. Moreover, boron homologues BOR1 are required in homeostasis processes in mammals' cells and they are also involved in tolerance to toxic concentrations in plants and yeasts (Takano et al. 2008).

At physiological pH (5–6) and in the absence of bimolecular interaction, B exists mostly as [$B(OH)_3$] species without charge. This is a weak acid according to Lewis' definition of ([$B(OH)_3$] + H_2O = [$B(OH)_4^-$] + H^+; pKa is 9.24) (Takano et al. 2008).

Protein NIP5; 1 besides easing the acid transport into the cells depending on its concentration gradient, enables the passive diffusion of the non charged acid [$B(OH)_3$] through the plasmatic membrane. On the other hand, BOR1 exports the acid turned into a borate [$B(OH)_4^-$] in the plant cytosol in which the pH is higher than in the apoplasm (7.5) (Frommer and Von Wiren 2002).

Considerations about partition coefficients of ether-water, molecular weight and the number of hydrogen bonds formed by B, have led to the conclusion that the theoretical lipid permeability of boric acid is $8x10^{-6}$ cm s^{-1} (Raven 1980). This relatively high value has become the basis of the theory of the transport mechanism of boric acid passive diffusion through the lipid bilayer.

In *Saccharomyces cerevisiae* mutations of the transporter of urea Dur3 and Pps1p (member of the MIP family with activity in glycerol transport), showed a decrease in B accumulation inside the cells and an increase in tolerance up to 90 mM B (toxic levels) (Nozawa et al. 2006).

Recent studies have established that boric acid is subject to membrane transport through a channel besides passive diffusion, at least as it goes into the root cells (Takano et al. 2008).

Miwa and Fujiwara (2009b) suggested that a regulatory mechanism for intracellular boron may exist in bacteria, although the intracellular level of boron can vary among the species. They also suggested that an active transport mechanism of boron might also play an important role in bacteria to acquire boron under deficient conditions, but no homologs to AtBor1 have been found yet in bacterial genome databases.

Biotechnology

A great number of metabolites exist which are synthesized by bacteria containing Boron such as aplasmomycines (Nakamura et al. 1977), tartrolones A and B (Irschik et al. 1995), boromycines (Kohno et al. 1996) and borophycines (Arai et al. 2004). Cyanobacteria *Nostoc spongiaeforme* var. *tenue* and *Streptomyces griseus* produce other antibiotics containing B (Dembitsky et al. 2002). Negrete and Raymond (2003) have described phenyl boric acid catabolism (BPA) in *Arthrobacter nicotinovorans species*, from which boron is mainly eliminated as ortho boric acid [$B(OH)_3$].

The species *Streptomyces antibioticus* isolated from soils in the Ivory Coast contained boromycin (Hütter et al. 1967). This is a naturally synthesized boron macrolide antibiotic which inhibits the growth of Gram-positive bacteria and protozoans of the genera plasmodiae and babesiae (Prelog et al. 1973). This compound (Fig. 1), used for preventive treatment against coccocidyosis in domestic fowl (Miller and Burg 1975), has no effect on either Gram-negative bacteria or on fungi. Its action is based on its negative effects on the cell cytoplasmic membrane bringing about the loss of potassium ions. Kohno et al. (1996) found that boromycin strongly inhibits HIV-1 both in clinical isolates and the strain cultured *in vitro*. Besides it inhibits protein RNA and DNA synthesis in all *Bacillus subtilis* cells (Pache and Zahner1969).

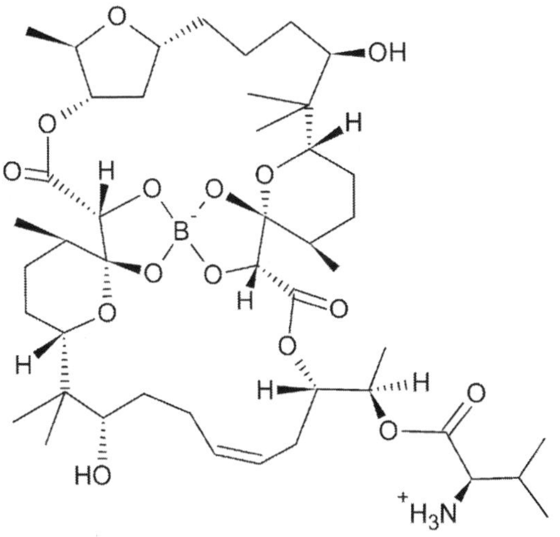

Figure 1. Boromycin structure.

Aplasmomycin, another antibiotic containing B (Fig. 2), was found in *Streptomyces griseous*, a strain isolated from marine sediments in Sagami Bay, Japan (Okami et al. 1976). It is known to inhibit the growth of Gram-positive bacteria including mycobacteria *in vitro*. That strain also produces two minor components, aplasmomycin B and C (Sato et al. 1978), the latter also being produced by actinobacteria isolated from marine sandy sediments in California (Stout et al. 1991). Antibacterial activity of form C of this antibiotic is inferior to variety B.

The antibiotic borophycin (Fig. 3) has been isolated from greenish-blue algae *Nostoc linckia* (Arai et al. 2004). It is made up of two equal halves with a total structure resembling that of other antibiotics containing B.

Tartrolon B boric acid is an ester of Tartrolon A3, whereas A1 and A2 are stereoisomers of Tartrolon A3. Every one of them can become any of the other two. Tartrolon C was also isolated from a *Streptomyces* species (Lewer

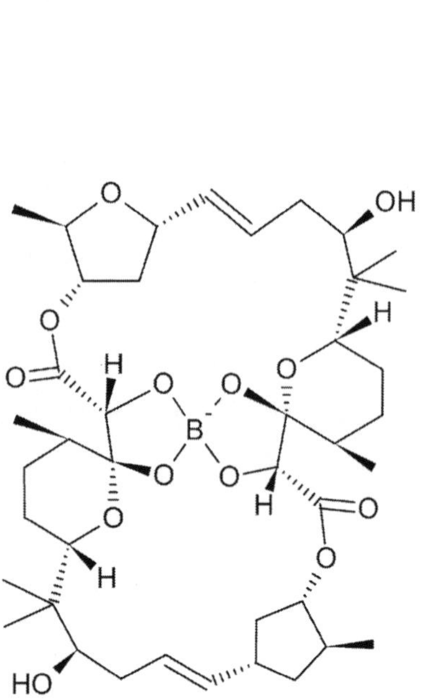

Figure 2. Aplasmomycin structure.

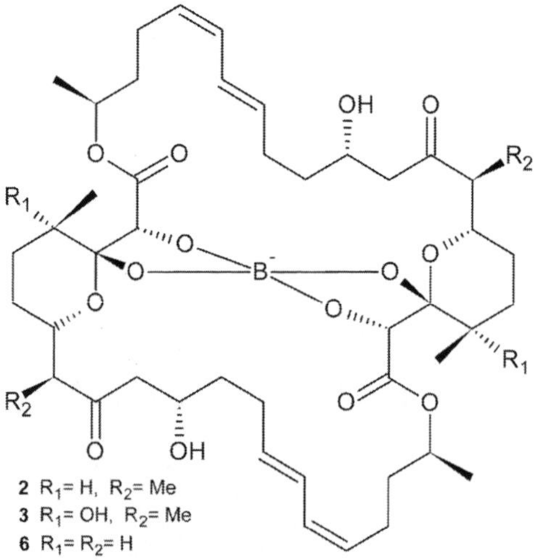

2 R_1= H, R_2= Me
3 R_1= OH, R_2= Me
6 R_1= R_2= H

Figure 3. Borophycin structure.

et al. 2003). Tartrolones become active against Gram-positive bacteria, acting over some specific enzymes and over the whole membrane. Gram-negative bacteria, fungi and yeasts are insensitive but mammals' cells are strongly inhibited, especially by B structure as shown in Fig. 4. This structure inhibits the synthesis of many cell macromolecules in *Staphylococcus aureus,* but has no effect on RNA and DNA polymerases isolated from *Escherichia coli.*

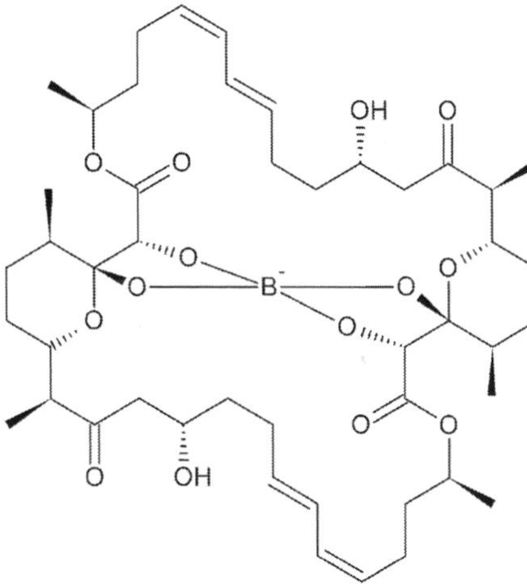

Figure 4. Tartrolon structure.

Boron in the Environment

Environmental fate

Boron compound are originally part of the soils in different areas. The rain, the wind and other natural events facilitate the spread of this element into the air and water bodies. Human activities (exploitation and industrialization) also contribute to the extension of contamination (anthropogenic contamination).

Soil

Mean boron concentration on the earth crust is low, hardly over 10 mg L^{-1}. Soils and rocks hold concentrations of about 450 mg L^{-1}, mainly distributed as sodium, calcium, and magnesium salts whereas in continental waters and the oceans concentrations are about 0.1 mg L^{-1} and 4.6 mg L^{-1}, respectively (SEGEMAR 2002).

Native boron is found in most soils in humid regions as borosilicate with variable amounts of Fe, Al, Mg, Mn, Ca and Na, axinite $Ca_2(Fe,Mn)Al_2[BO_3(OH)Si_4O_{12}]$ and thurmaline (Na,Ca) (Mg,Al) $(Al,Fe,Mn)_6\,(BO_3)_3\,(Si_6O_{18})\,(OH,F)_4$ being two good examples. Boron concentration depends on the type of soil, the amount of organic matter which contains boron, and the amount of rainfall which can remove boron from the soil. Besides, most B available in the soil is provided by the organic fraction, which liberates

B on decomposing it by means of microorganisms which free it from the organic compounds in the soil (Flores 2004).

Air

As borates exhibit low volatility, boron is not expected to be present in the atmosphere as vapour in any significant degree (Sprague 1972).

Volcanic activity, volatilization of boric acid from the sea, mining operations, glass and ceramic manufacturing, the application of agricultural chemicals, and coal-fired power plants may cause atmospheric emissions of borates and boric acid in particulates (< 1–45 μm in size) (WHO 2009).

Water

Boron in earth is present mainly in oceans, where the average concentration is 4.5 mg L^{-1} (Weast et al. 1985). Groundwater boron concentrations range widely throughout the world, from <0.3 to >100 mg L^{-1} (WHO 2009).

Concentrations of boron in surface waters are less than about 0.5 mg L^{-1} (Coughlin 1998, Neal et al. 1998, Wyness et al. 2003), except in highly mineralized areas, as naturally carbonated groundwater, where boron concentration is naturally higher (WHO 2009). Boron is present in groundwater primarily as a result of natural leaching from rocks and soils containing boron compounds (WHO 2009).

As in natural waters, boron exists primarily as undissociated boric acid with some borate ions, waterborne boron may be adsorbed by soils and sediments. These adsorption–desorption processes are expected to be the only significant mechanism influencing the fate of boron in water (Rai et al. 1986). Water pH affects the extent of boron adsorption directly in soils and sediments, the major adsorption range being between pH 7.5 to 9.0 (Waggott 1969, Keren and Mezuman 1981, Keren et al. 1981).

Food

As boron is present in a number of food items, daily boron intake can be higher than normally expected. Meats and grain are poor sources of boron, although vegetables, legumes, fruits, pulses and nuts are the richest ones (United Kingdom Expert Group on Vitamins and Minerals 2002).

Daily intake estimations of boron have been made for various age and sex groups on the basis of food products. The estimated median, mean and 95th-percentile daily intakes of boron were 0.75, 0.93, and 2.19 mg d^{-1}, respectively, for all groups and 0.79, 0.98 and 2.33 mg d^{-1}, respectively, for adults aged 17 and above (WHO 2009).

Situation in Salta

The Puna region, located in the central Andes, has the largest borate deposits in Argentina. This region extends about 100,000 Km^2 in the north western territory at an average altitude of 3,700 m above sea level on the western area of Jujuy, Salta and the northern boundary of Catamarca. It is a desert region with alluvial soils, salines, and volcanic strata.

There are several natural boron mineral deposits in the Salta province (Fig. 5). The main tertiary ones are Tincalayu and Sijes, the only borax deposits in the world other than in Death Valley in California (Flores 2004) and Kirka in Turkey (Helvaci and Ortí 2004). The largest deposits lie in Salar del Hombre Muerto, Cauchari and Diablillos, Salta (Alonso 2007). Tincalayu reserves are estimated at 5.0 MT with a 12 percentage law in B_2O_3 (12 percent of commercial value in pure boric oxide). Salta is the first world producer of hydroboracite (116,099 T in 2010, mineral plentifully found in the salt quarries of Sierra of Sijes), the first Latin American borate producer and the third world producer of borates (Secretaría de Minería de Salta 2011).

Despite the natural presence of boron compounds there is also anthropogenic contamination in Salta mainly due to borate extraction and transportation to the processing area (lack of suitable infrastructure in Puna forces enterprises to set up their processing plants some 400 km away from the salt mines). Also the industrialization/refining process to obtain the

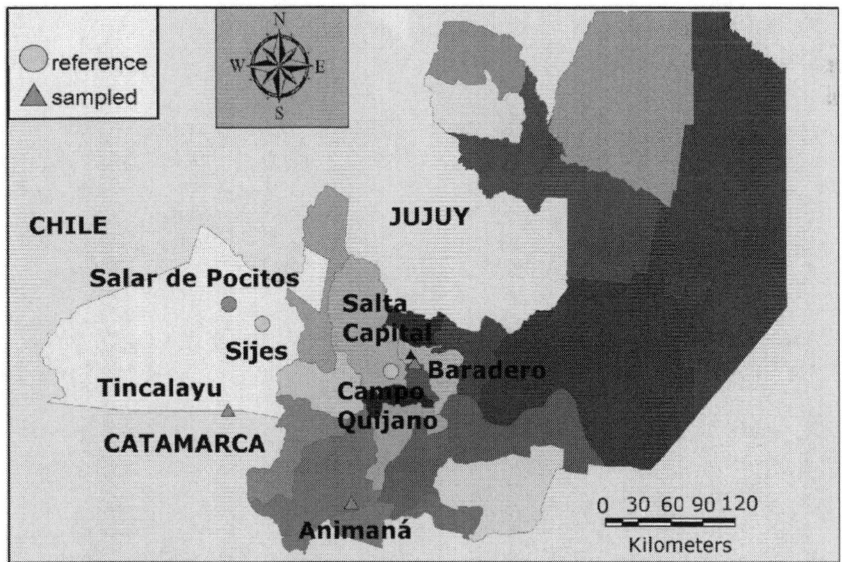

Figure 5. Boron mineral deposits and anthropogenically contaminated locations in the province of Salta, Argentina.

required borate derivative contribute to that, namely lixiviation, separation of fluids from solids and crystallization cooling of useful remnants, concentration and purifying operation of residues or remnants.

There are many treatments to reduce the environmental contamination produced by the industrialization of boron products, such as direct waste disposal *in situ,* recycling of process streams, waste reutilization and the re-engineering of operations and processes. Thus, to diminish dust contamination and treat effluents in order to avoid harmful effects from current and future exploitations, different methods are applied: ionic treatment, removal with organic solvents, metal hydroxide adsorption, alkaline precipitation and processes with membranes among others (Flores 2004).

Environmental contamination due to boron can be observed in many areas and contaminated soils are hardly recoverable. Despite the application of contamination mitigating methods and the existing environmental regulations setting control values for boron concentrations in air, water and soil in farming and cattle breeding areas, there are many polluted sites.

Baradero, a densely inhabited area of about 46,620 m^2, where a company worked until 1993 producing borax and boric acid, has an important environmental burden (Fig. 6). The place has the highest boron concentration

Figure 6. Boron anthropogenically contaminated area, Baradero and sampling scheme.

values detected in the air, soil and water in the area. Between the years 1997 and 2000, studies conducted there by the Salta Mining Secretariat and Municipal Government (PNUMA 2004), showed that the soil had a high content of soluble boron and salinity, thus showing that the land was not suitable for human, farming and domestic animal breeding activities.

Samples taken from surface waters in Arenales River, contained boron at concentrations ranging between 4 and 26 mg L^{-1} in areas rich in boron-containing soils. In other areas of the same river, concentrations were below 0.3 mg L^{-1}. These values were to be expected because of two factors: 1) Baradero is about 8 Km from the river and boron concentrations in water are largely dependent on the leaching from the surrounding geology and, to a lesser extent, wastewater discharges from domestic washing agents (borates are components of most domestic washing agents) (ISO 1990); and 2) boron is not removed by conventional wastewater and drinking-water treatment methods (WHO 1998).

Microorganisms Isolation and Characterization

Area of study and sampling scheme

As mentioned earlier, Salta is a great producer of borates and the exploitation and industrialization spread contamination. Three locations in this province were selected as research sites. Baradero, located in the aforementioned urban area, which presents anthropogenic contamination with boron compounds, Tincalayu and Animaná, which are naturally high in boron content (see sampled sites in Fig. 5). The samples from Tincalayu were obtained randomly from three different sectors (called A, B, and C) from an exploitation mine. The third site, Animaná, was also sampled randomly from three different sectors of the river side (called D, E and F). Considering that the depth where most microorganisms should be concentrated is between a 0–15 cm deep surface layer, 10 g of soil were taken from each sector in the three places sampled.

Systematic sampling of Baradero soil (see sampling scheme in Fig. 6) was conducted on a rectangular grid (Mason 1992), following the recommendations for homogeneous areas below five hectares (Valencia and Hernández 2002). The total area of 46,620 m^2 was divided into nine 70×74 m sectors of 5,180 m^2 and each of them was subdivided into five equidistant points to take the 10 g samples. Thus, 50 g of soil was obtained per sector in only three (S1, S2, S3) of the nine sectors (totaling 150 g of composite sample) according to Instituto Nacional de Ecología de México (2005) for this type of analysis.

Physicochemical characterization of the soil

The physicochemical characterization of the soils from Baradero (sectors S1, S2, and S3), Tincalayu (A, B, and C), and Animaná (D, E, and F) was carried out. The samples were analyzed for water content, organic matter content, pH, B_2O_3 and total boron following the method recommended in Análise de Corretivos, Fertilizantes e Inoculantes (1988). The results obtained are shown in Table 2.

The sample obtained from sector S3 in Baradero soil differed from the other two as it was more humid and sandy and with incipient vegetation, so this sector was used as the low boron contamination control for Baradero soil samples. The concentrations of boron and its derivatives were measured to detect the highest and lowest boron concentration values to which the microorganisms were exposed in the environment to be used later as a reference in tolerance studies.

Table 2. Physicochemical characterization of soils from Baradero (sectors S1, S2, and S3), Tincalayu (A, B, and C) and Animaná (sectors D, E, and F).

Parameters	Baradero			Tincalayu			Animaná		
	S1	S2	S3	A	B	C	D	E	F
Organic matter content (%)	2.36	3.44	2.78	3.58	3.22	3.71	0.25	1.18	1.06
Humidity at 110°C (%)	5.49	5.83	6.06	12.53	2.2	7.61	0.20	0.68	0.78
pH in KCl	8.25	8.80	7.03	9.13	9.03	9.00	7.90	7.85	7.83
$[B_2O_3]$ (g/100 g soil)	1.08	2.72	0.56	10.38	0.84	5.24	2.21	2.26	2.37
B_{total} (g/100 g soil)	0.34	0.84	0.17	3.20	0.26	1.62	0.69	0.70	0.74

Isolation and preservation of strains

The samples from each soil sector were quartered; 10 g were taken and added to 90 mL of 1 percent sodium hexametaphosphate sterile solution, used as extracting agent. It was vortexed ten minutes and then let stand for 30 min for the solids to settle. Three successive tenfold dilutions from the previous solution were performed in sterile water. They were cultured in a general medium, Nutrient Agar (NA) (Britania), and in Starch Casein Agar (SCA) (1 percentage starch, 0.1 percentage hydrolyzed casein, 0.05 percentage K_2HPO_4, and 1.5 percentage agar, pH 7± 0.5), a specific medium for flora present in soils. Inoculation was performed by surface in Petri dishes that were incubated at 30°C for 7 days.

All the colonies obtained were isolated in SCA because distinctive features between them were observed, facilitating their identification and purity control. Pure cultures were stored at 4°C in SCA tubes until used.

Characterization and identification of the microorganisms

A total of 127 strains were initially isolated from the contaminated soils. Fifty of them (yeast and fungi were not included) were subjected to further studies (see below) to assess their ability to grow under different conditions. From all the strains isolated, only eight were in the end identified as actinobacteria. Thus only the results corresponding to them are presented in detail in this chapter.

Ability to grow at different temperatures and with NaCl

The 50 strains initially selected were studied in their ability to grow in nutrient broth (NB) at different temperatures and saline concentrations. For that, 0.1 percentage v/v of fresh cultures were inoculated to NB and cultured at 4 (for 4 d), 15 (48 hr), and 30°C (48 hr). Simultaneously, they were inoculated in NB with 0.5 and 1 percentage w/v NaCl, respectively, pH adjusted to 7 ± 0.2 prior to sterilization (and verified after sterilization), and cultured at 30°C.

No growth was detected in any culture before 48 hrs of incubation at 4°C. Conversely, all the strains grew in 24 hrs at 30°C, even with 1 percentage w/v NaCl (Table 3) and only one (strain 132) required the presence of NaCl to grow. According to Zahran (1997) and Marguesin and Schiner's (2001) definitions, all the strains can be classified as halotolerant.

Table 3. Culture of the eight isolated strains in nutrient broth with and without (wo NaCl) NaCl at different temperatures.

Strain	4°C wo NaCl 4 d	15°C wo NaCl 48 h	30°C wo NaCl 48 h	30°C 0.5% w/v NaCl 24 h	30°C 1% w/v NaCl 24 h
002	–	–	+	+	+
048	+	+	+	+	+
050	+	+	+	+	+
053	+	+	+	+	+
128	–	+	+	+	+
130	+	+	+	+	+
132	–	–	–	+	+
133	+	+	+	+	+

+ Positive; – Negative

Macroscopic and microscopic observations

Fresh cultures of the isolates (NB, 24 hr) were Gram stained in order to perform a preliminary classification. *Salmonella* sp. and *Staphylococcus aureus* were used as positive controls for Gram-negative and Gram-positive bacteria, respectively.

From these analyses, thirty three strains out of the fifty were identified as Gram-negative bacilli. The remaining 17 Gram positive isolates were later studied to characterize them according to physiological and chemical features. Eight of the 17 Gram positive isolates seemed to belong to the actinobacteria group, based upon macro and microscopic morphology, mycelium color, spore formation, colony consistency and their distinctive "earthy" odor because of the geosmin present (Polti et al. 2007). All of these strains produced in Starch Casein Aagar extensively branching, primary mycelium transformed during their life cycle into aerial mycelium bearing and into typical spores (Table 4).

Table 4. Macroscopic characteristics of the isolated strains when grown in Starch Casein Agar at 30°C.

Strain	Mycelium color	Spore color	Colony appearance
002	Red	White	Rugose
048	Yellow	Gray	Rugose
050	White	Gray	Rugose
053	Yellow	White	Smooth warty
128	Red	White	Rugose
130	White	Gray	Rugose
132	White	Gray	Rugose
133	Red	Grey	Rugose

Physiological and chemical characterization

The identification of all the Gram negative isolates (33 in total) taking into account physiological and chemical features was carried out using the commercial kit API® 20 NE (BioMérieux, France) following the manufacturer's protocol. The other 17 isolated strains were also tested using this kit, just to elucidate some information related to their capacity of assimilation and degradation of some compounds present in this kit. This standardized system combines eight conventional assays: nitrate reduction (NO_3) into nitrites, indol formation from L-tryptophan (TRP), D-glucose fermentation (GLF), esculin (ESC) and gelatin (GEL) hydrolysis, presence of enzymes such as arginine dihydrolase (ADH), urease (URE) and β-galactosidase (PNPG, 4-nitrofenil-βD-galactopiranoside hydrolysis) and 12 assimilation tests of different carbohydrates: D-glucose (GLU), L-arabinose (ARA), D-mannose (MNE), D-manitol (MAN), N-acetyl-glucosamine (NAG), D-maltose (MAL), potassium gluconate (GNT), capric acid (CAP), adipic acid (ADI), malic acid (MLT), trisodium citrate (CIT) and phenyl acetic acid (PAC). Only the results for eight strains are presented in Tables 5 and 6.

Table 5. Conventional tests with API® 20 NE (BioMérieux) (NO_3-: nitrate reduction; TRP: indol formation from L-triptophane; GLF: D-glucose fermentation; ADH: presence of arginine dihydrolase enzyme; URE: presence of urease enzyme; ESC: esculine hydrolysis ferric citrate; GEL: gelatine hydrolysis; PNPG: 4-nitrofenil-βD-galactopiranoside hydrolysis).

Strain	NO_3	TRP	GLF	ADH	URE	ESC	GEL	PNPG
002	–	–	–	–	–	(+)	–	+
048	–	–	–	–	+	+	–	+
050	–	–	–	–	–	+	+	–
053	–	–	–	–	–	+	+	+
128	–	–	–	–	–	+	–	(+)
130	–	–	–	–	–	+	+	(+)
132	–	–	–	–	–	+	+	–
133	–	–	–	–	(+)	+	–	+

+ Positive; – Negative; (+) Weakly positive

Table 6. Assimilation tests of different carbohydrates with API® 20 NE (BioMérieux) (GLU: D-glucose; ARA: L-arabinose; MNE: D-mannose; MAN: manitol; NAG: N-acetil-glucosamine; MAL: D-Maltose; GNT: potassium gluconate; CAP: capric acid; ADI: adipic acid; MLT: malic acid; CIT: trisodium citrate; PAC: phenyl acetic acid).

Strain	GLU	ARA	MNE	MAN	NAG	MAL	GNT	CAP	ADI	MLT	CIT	PAC
002	+	+	+	–	–	(+)	+	–	(+)	(+)	–	–
048	+	(+)	+	+	(+)	+	+	–	–	(+)	+	(+)
050	+	+	+	+	+	+	(+)	–	–	+	+	(+)
053	+	+	+	+	+	+	+	–	–	+	+	+
128	(+)	–	–	–	–	–	+	–	–	–	–	–
130	+	–	+	–	–	–	–	–	–	–	+	–
132	+	–	(+)	(+)	–	+	+	–	(+)	+	+	–
133	+	(+)	+	+	+	–	–	–	–	+	+	–

+ Positive; – Negative; (+) Weakly positive

Molecular identification

DNA from the eight strains suspected to be *Streptomyce*s was isolated, amplified and sequenced in order to determine their taxonomic position. The sequences were compared with closely related clones from GenBank and located in clusters according to the classifications proposed by Williams et al. (1989) for the *Streptomyces* group in the Bergey's Manual of Determinative Bacteriology (1994). In the proposed classification, the definitions of cluster-groups and clusters were based on phylogenetic similarity coefficients.

All the strains isolated from boron contaminated soils were confirmed as belonging to the actinobacteria group.

The generated phylogenetic tree (Fig. 7) was divided into five major clusters on the basis of their evolutionary distances calculated through the neighbour-joining method. Most of the species located in cluster I were incorporated in the largest cluster-group A and widely distributed in different clusters (19, 23, 12, 18, 15, 16, etc.) according to the proposed classification. This first cluster was further divided into two sub-clusters: in sub-cluster Ia, the isolated strain 048 (HQ538723) was closely associated with *S. achromogenes* (99.4 percent) while strain 133 (HQ538729) was associated with *S. griseosporeus* (98.5 percent) and in sub-cluster Ib, the isolated strain 053 (HQ538724) was closely associated with *S. albogriseolus* (100 percent) and the isolated strain 002 (HQ538731) was closely associated with *S. iakyrus* (100 percent). The species located in cluster II were incorporated in the cluster-group G, and located only in cluster 68. In cluster II, the isolated strain 050 (HQ538725) was closely associated with *S. fradiae* (100 percent). The species located in cluster III were incorporated in cluster-group F and only in cluster 61. In cluster III, the isolated strain 130 (HQ538727) was closely associated with *S. polychromogenes* (100 percent). Species located in cluster IV are incorporated in cluster-group A and located in clusters 18 and 19. In cluster IV, the isolated strain 128 (HQ538726) was closely associated with *S. lincolnensis* (99.8 percent). Finally, in cluster V, the isolated strain 132 (HQ538730) was associated with the species of the genus *Lentzea* (97.6 –97.8 percent).

Growth in the Presence of Boron

Only a few microorganisms have been reported previously as having boron tolerance. Among Gram-positive bacteria *Lysinibacillus parviboronicapiens,* low boron-containing bacteria can tolerate up to 6 percentage w/v NaCl, but only 50 mM boron (Miwa et al. 2009a). *Lysinibacillus boronitolerans* can

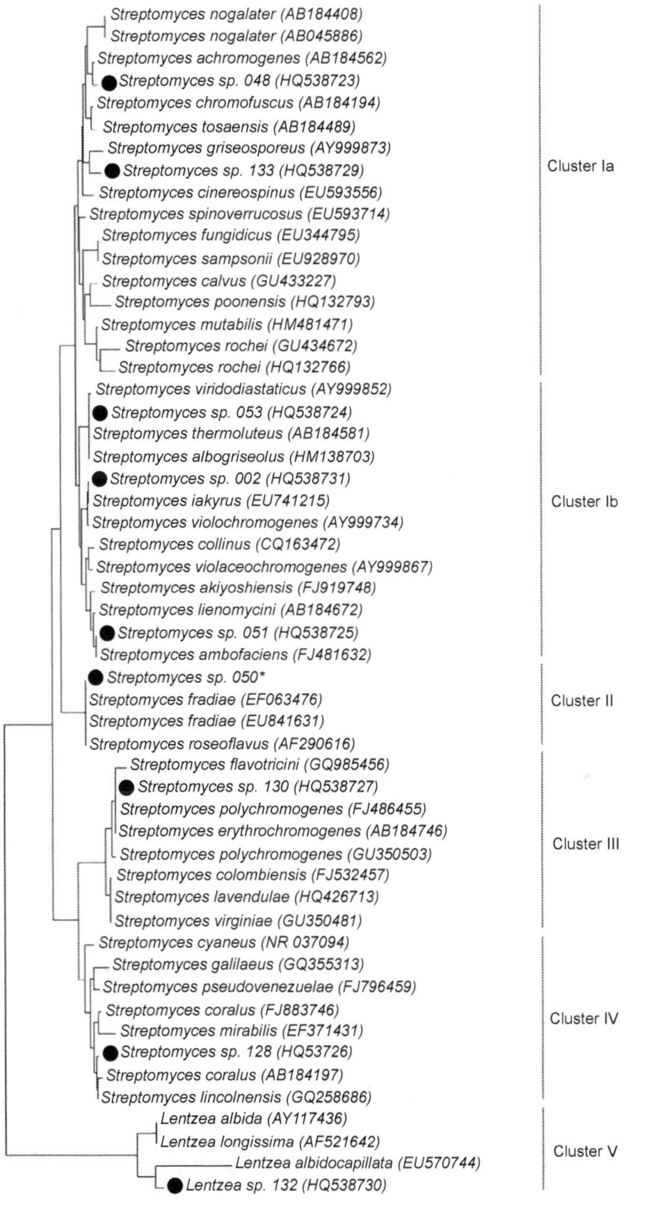

Figure 7. The phylogenetic tree of actinobacteria species inferred by the Neighboring-Joining method (Saitou and Nei 1987), using Kimura's evolutionary distance (Kimura 1980) and based on the comparison of a nearly complete 16S rDNA sequence of 49 organisms. Accession numbers of 16S rDNA sequences are given in parentheses. The label • shows the strains isolated in this study.

tolerate up to 150 mM boron (Ahmed et al. 2007d), while *Bacillus boroniphilus* sp. nov. (Ahmed et al. 2007a) and *Gracilibacillus boraciitolerans* sp. nov. (Ahmed et al. 2007b) can tolerate over 450 mM of boric acid. *G. boraciitolerans* actually requires boron to grow. Some actinobacteria were isolated and the amount of boron as boric acid they could accumulate was determined. For *Microbacterium foliorum* intracellular boron concentration was less than 0.4 nmol g^{-1} DW and less than 0.7 nmol g^{-1} DW for both *Rhodococcus* sp. and *Rhodococcus erythropolis* (Miwa and Fujiwara 2009b).

Gram-negative bacteria are also able to tolerate high boron concentration such as *Chimaereicella boritolerans* sp. nov. up to 300 mM B (Ahmed et al. 2007c) and *Variovorax boronicumulans* sp. nov. can accumulate intracellular boron (Miwa et al. 2008). This mechanism, the same as that in some plants, allows contaminating boron to be removed and accumulated inside their cells. Recently, a study from Ahmed and Fujiwara (2010) based on B-uptake, showed that there is an efflux/exclusion mechanism in the B-tolerant *B. boroniphilus* that helps the microorganisms keeps the intracellular soluble B concentration low when it is exposed to high external B concentrations.

In this case only the eight isolated actinobacteria strains were studied in their ability to grow in the presence of boron. The strains were 048 and 050 from sector S2 (highest boron contamination in Baradero), 002 from S1, 053 from S3, 128 and 130 from A (highest boron contamination in Tincalayu), and 132 and 133 from sectors E and F respectively, in Animaná.

Boric acid in concentrations similar to those in the study area soils, was used to prove B tolerance. This acid was selected because of its high B percentage (17.49 percent) and low molecular weight which enables B to enter the cell more easily. Besides, it is one of the most water soluble B compounds (5.49 percentage at 20ºC) and transport channels have been identified in the cells (Takano et al. 2008).

Qualitative screening

A qualitative screening assay was carried out in Petri dishes containing SCA. Rectangular troughs were cut-off in the center of plates and filled with 600 µL of sterile boric acid solutions of five different concentrations: 80, 150, 340, 390, and 440 mM. Isolates were inoculated by streaking 3 cm perpendicularly to the troughs and incubated at 30°C for 7 d (Amoroso et al. 1998). Microbial growth was used as the qualitative parameter of boron tolerance. Three categories depending on the length of growth were defined: limited (up to 1 cm), partial (up to 2 cm) and total (up to 3 cm).

All the studied microorganisms were able to grow in all the boron concentrations tested, although not always closer to the trough. The results were neither clear nor consistent. Growth was uncertain since the true boron concentration at different points in the agar gel was unknown due to diffusion of boron into the agar possibly because of non-homogeneous initial striking.

Semi-quantitative screening

Petri dishes containing SCA were inoculated with 200 μL of a spore suspension of the strains (concentrations ranging between 1×10^5 and 8×10^6 CFU mL^{-1}). After that 5 mm diameter portions were removed forming wells that were filled up with 50 μL of seven H_3BO_3 solution concentrations (80, 150, 210, 270, 340, 390, and 440 mM). Sterile water was used as control. After a 24 hr incubation at 30°C the growth inhibition diameter was measured. The strains were considered tolerant when the diameter of inhibition was less than 7 mm, and non-tolerant when it was higher, according to an arbitrary criterion (Amoroso et al. 1998).

All the eight strains were able to grow at 80 mM, while only 50 and 13 percent of them grew at 210 mM and 440 mM, respectively. According to the inhibition criteria adopted, all the strains were B-tolerant up to 270 mM (Fig. 8). Interestingly, strain 048 from Baradero (sector S2) was able to grow without any inhibition over the entire range of boric acid concentrations used, while strain 128 from Tincalayu (sector A) showed tolerance even to 440 mM of B. The level of B-tolerance is similar to that reported by Ahmed et al. (2007a, 2007b) for other Gram-positive bacteria. Higher concentrations of boric acid were not used due to solubility limitations.

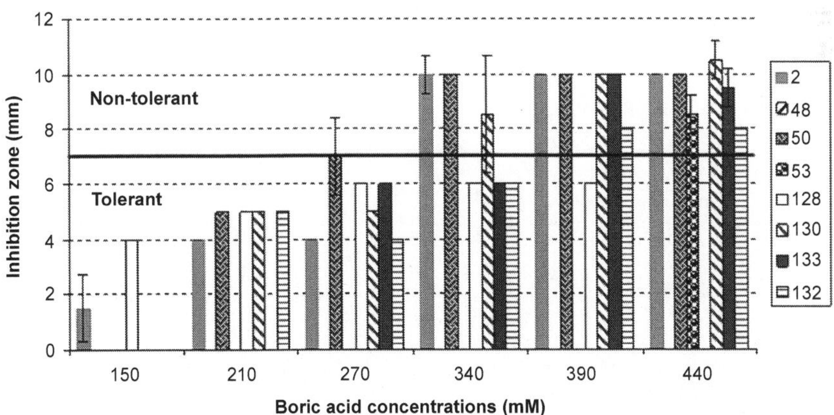

Figure 8. Semi-quantitative assay of boron tolerance measured as the diameter of the inhibition zone. The horizontal line indicates the arbitrary limit used to define boron tolerant (below) and non-tolerant (above) strains.

Growth in liquid medium

The eight strains were cultured in minimal liquid medium (MM) (composition in g L^{-1}: glucose, 10.0; L-asparagine, 0.5; K_2HPO_4, 0.5; $MgSO_4.7H_2O$, 0.20; $FeSO_4.7H_2O$, 0.01, pH 7 ± 0.5) with the addition of boric acid (H_3BO_3) until the final concentrations desired - 20, 40, and 80 mM—were achieved; pH was adjusted to neutrality. One mL of spore suspension was inoculated in each of eight 100 mL Erlenmeyers, containing 20 mL minimal medium with and without boric acid. These were incubated for five to seven days (according to the strain) at 30°C, 250 rpm in an orbital shaker in order to assess growth in boric acid presence.

The sampling was performed at different times by removing a complete Erlenmeyer from the set. The biomass was recovered by filtration through sterile 47 mm diameter and 0.45μm pore-sized Sartorius membranes of cellulose nitrate and determined by dry weight (24 hr at 105°C).

Although the results of growth in Minimal Medium containing boric acid varied with the strain, none of them could grow at 80 mM. Increasing boric concentration affected the growth of S. *lincolnensis* (strain 128, Fig. 9.b) negatively as shown in Fig. 10, *S. fradiae* (strain 050) and *S. griseosporeus* (strain 133) whereas the effect observed for *S. Iakyrus* (strain 002) was the production of an exopolysaccharide.

On the other hand, the increase in boric acid untill 40 mM did not affect *Streptomyces achromogenes* (strain 048, Fig. 9a) (see Fig. 10), *S. polychromogenes* (strain 130) and *S. albogriseolus* (strain 053) significantly.

For upper concentrations (> 80 mM), like those used for the quantitative assays, the growth was strongly inhibited. A possible reason for this is that bacterial cell are in intimate contact with toxic solutions, thus allowing the entry of boric acid into the cell, which did not happen in agar medium because of diffusion problems.

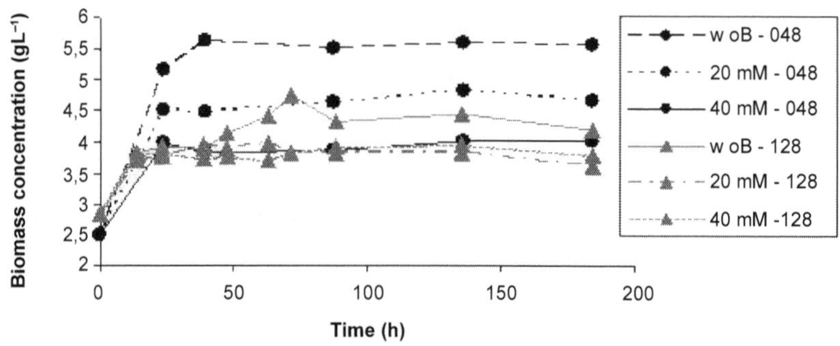

Figure 9. *Streptomyces* growth curves in Minimal Medium with and without (wo) boric acid.

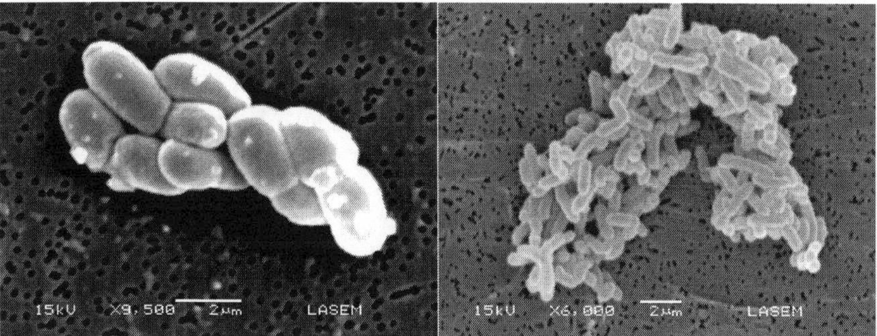

Figure 10. Electronic microgrophies of *Streptomyces* strains a) *achromogenes* (048) and b) *lincolnensis* (128).

Nevertheless, the level of B-tolerance is about 400 times higher than that reported by Miwa and Fujiwara (2009b) for other actinobacteria.

The results obtained so far with the isolated strains show them to be promising for their application in diminishing contamination and eventually in the production of industrially interesting metabolites.

Final Remarks

Actinobacteria is a physiologically diverse bacterial group that has the soil as natural habitat. They are known for their ability to produce antibiotics and metabolites and remove heavy metals. Actinomycetes dominated the community in non-contaminated as well as in contaminated soils (Sandaa et al. 2001).

From the eight isolates obtained in this work, strain 132 from Animaná was identified as *Lentzea sp.*, while the other seven strains were *Streptomyces* sp. (with high similarity to *iakyrus, achromogenes, fradie, ambofaciens, albogriseolus, lincolnensis, polychromogenes,* according to 16S rDNA sequences). Some of the strains can be associated with contaminated soils, e.g., *Lentzea* with heavy-metal contaminated soils (Haferburg 2007). Others were reported for specific abilities to degrade some compounds, such as *S. iakyrus* and parathion (De Pasquale et al. 2008) and *S. albogriseolus* and latex (Gallert 2000), to produce antibiotics such as *S. achromogenes* and streptozotocin, rubradirin (Maharjan et al. 2003) and tomaymycin (Li et al. 2009), *S. ambofaciens* and spiramycin (Karray et al. 2010), *S. lincolnensis* and lyncomicine (Koběrská et al. 2008) and to produce enzymes, such as *S. fradie* with keratolytic activity (Noval and Nickerson 1958). However to the best of our knowledge there are no reports in the literature about *Streptomyces* or *Lentzea* capacity to grow in soils or media with high concentrations of boron compounds.

The assays for boron tolerance and for growth in the presence of boron show these bacteria as promising potential bioremediation agents since these microorganisms are infrequent pathogens, are metabolically versatile and show primary biodegradative activity. They are able to secrete a range of extracellular enzymes that allow them to metabolize recalcitrant molecules. They also have different strategies like rapid proliferation and sporulation to maintain the population by prolonged slow growth and scavenging.

Finally, strains 048, 128, and 130 look promising to deal with environmental contamination with boron. Although Miwa and Fujiwara (2009b) exposed some microorganisms (including three actinobacteria) to concentrations of 0.1 mM of boric acid in liquid media, the strains isolated from this research look even more promising since they are able to grow in the presence of 440 mM boric acid in agar and also in liquid media in the presence of 40 mM of boric acid, therefore they are able to tolerate 400 times higher concentrations than the reported ones.

Whether the bacteria are using the boron for the production of some metabolite like an exopolysaccharide or an antibiotic, or just capturing and accumulating it internally will need to be elucidated through further studies. Also further studies will be conducted to determine the fate of boron during bacterial growth and to examine if these microorganisms can be used to reverse contamination effects on the environment.

References Cited

Ahmed, I., A. Yolota and T. Fujiwara. 2007a. A novel highly boron tolerant bacterium, *Bacillus boroniphilus* sp. nov., isolated from soil, that requires boron for its growth. Extremophiles. 11: 217–224.

Ahmed, I., A. Yolota and T. Fujiwara. 2007b. *Gracibacillus boraciitolerans* sp. nov., a highly boron-tolerant and moderately halotolerant bacterium isolated from soil. Int. J. Syst. Evol. Microbiol. 57: 796–802.

Ahmed, I., A. Yolota and T. Fujiwara. 2007c. *Chimaereicella boritolerans* sp. nov., a boron-tolerant and alkaliphilic bacterium of the family *Flavobacteriaceae* isolated from soil. International Int. J. Syst. Evol. Microbiol. 57: 986–992.

Ahmed, I., A. Yolota, A. Yamazoe and T. Fujiwara. 2007d. Proposal of *Lysinibacillus boronitolerans* gen. nov. sp. nov., and transfer of *Bacillus fusiformis* to *Lysinibacillus fusiformis* comb. nov. and *Bacillus sphaericus* to *Lysinibacillus sphaericus* comb. nov. Int. J. Syst. Evol. Microbiol. 57: 1117–1125.

Ahmed, I. and T. Fujiwara. 2010. Mechanism of boron tolerance in soil bacteria. Can. J. Microbiol. 56: 22–26.

Alonso, R.N. 2007. *Minería de Salta,* Crisol Ediciones, Salta.

Amoroso, M.J., G.R. Castro, F.J. Carlino, N.C. Romero, R.T. Hill and G. Oliver. 1998. Screening of heavy metal-tolerant actinomycetes isolated from Sali River. J. Gen. Appl. Microbiol. 44: 129–132.

Análise de Corretivos, Fertilizantes e Inoculantes. Métodos Oficiais. Laboratorio Nacional de Referência Vegetal (LANARV), 1998. Ministério de Agricultura Secretaria Nacional de Defensa Agropecuária. pp. 49–79.

Anderson, G.R. and J.V. Jordan. 1961. A non-essential growth factor for *Azotobacter choococcum*. Soil Science. 92: 113–116.

Arai, M., Y. Koizumi, H. Sato, T. Kawabe, M. Sugamuna, H. Kobayashi, H. Tomoda and S. Omura. 2004. Boromycin abrogates bleomycin-induced G2 checkpoint. J. Antibiot. 57: 662–668.

Barán, E. 1989. La Nueva Farmacopea Inorgánica X. Compuestos de Boro. Acta Farm. Bonaerense. 8(3): 199–206.

Barth, R.F., W. Yang and J.A. Coderre. 2003. Rat brain tumor models to assess the efficacy of boron neutron capture therapy: a critical evaluation. J. Neurooncol. 62: 61–74.

Barth, R.F., J.A. Coderre, M.G.H. Vicente, T.E. Blue and S.I. Miyatake. 2007. Boron Neutron Capture Therapy of Brain Tumors Current Status and Future Prospects. High-Grade Gliomas Current Clinical Oncology, V. 431–459.

Benkovic, S.J., S.J. Baker, M.R.K. Alley, Y. Woo, Y. Zhang, T. Akama, W. Mao, J. Baboval, P.T. Ravi, Rajagopalan, M. Wall, L. Kahng, A. Tavassoli and L. Shapiro. 2005. Identification of Borinic Esters as Inhibitors of Bacterial Cell Growth and Bacterial Methyltransferases, CcrM and MenH. J. Med. Chem. 48: 7468–7476.

Chen, X., S. Schauder, N. Potier, A.V. Dorsselaer, I. Pelczer, B.L. Bassler and F.M. Hughson. 2002. Structural identification of a bacterial quorum-sensing signal containing boron. Nature. 415: 545–549.

Çöl, M. and C. Çöl. 2003. Enviromental boron contamination in waters of Hisarcik area in the Kutahya Province of Turkey. Food Chem. Toxicol. 41: 1417–1420.

Coughlin, J.R. 1998. Sources of human exposure: overview of water supplies as sources of boron. Biol. Trace Elem Res. 66: 87–100.

Crespo, E. 2001. El boro, elemento nutricional esencial en la funcionalidad ósea. Revista Española de Cirugía Osteoarticular. 36: 88– 95.

De Pasquale, C., R. Fodale, L. Lo Piccolo, E. Palazzolo, P. Quatrini and G. Alonzo. 2008. Degradazione di pesticidi organofosforici in matrici ambientali. XXVI Convegno Nazionale della Societa Italiana di Chimica Agraria. Palermo, Italy.

Dembitsky, V.M., R. Smoum, A.A. Al-Quntar, H.A. Ali, I. Pergament and M. Srebnik. 2002. Natural occurrence of boron-containing compounds in plants, algae and microorganisms. Plant Sci. 163: 931–942.

Flores, H.R. 2004. El Beneficio de los Boratos. Historia, Minerales, Yacimientos, Usos, Tratamiento, Refinación, Propiedades, Contaminación, Análisis Químico, Crisol Ediciones, INBEMI, Salta.

Frommer, W.B. and N. Von Wiren. 2002. Plant biology: ping-pong with boron. Nature. 420: 282–283.

Gallert, C. 2000. Degradation of latex and of natural rubber by *Streptomyces* strain La 7. Syst. Appl. Microbiol. 23: 433–41.

Green, G.H., M.D. Lott and H.J. Weeth. 1973. Effects of boron-water on rats. Proceedings, Western Section, American Society of Animal Science. 24: 254–258.

Habes, D., S. Morakchi, N. Aribi, J.P. Farine and N. Soltani. 2006. Boric acid toxicity to the German cockroach, Blattella germanica: Alterations in midgut structure, and the acetylcholinesterase and glutathione S-transferasa activity. Pestic Biochem. Physiol. 84: 17–24.

Haferburg, G. 2007. Studies on heavy metal resistance of bacterial isolates from a former uranium mining area. PhD Thesis, Friedrich-Schiller-Universität Jena, Germany.

Helvaci, C. and F. Ortí. 2004. Zoning in the kirka borate deposit, western turkey: primary evaporitic fractionation or diagenetic modifications? The Canadian Mineralogist. 42: 1179–1204.

Hütter, R., W. Keller-Schien, F. Knüsel, V. Prelog, G.C. Rodgers jr., P. Suter, G. Vogel, W. Voser and H. Zähner. 1967. Stoffwechselprodukte von Mikroorganismen. 57. Mitteilung. Boromycin. Hel Chim. Acta. 50: 1533–1539.

Instituto Nacional de Ecología, 2005. Muestreo y caracterización de un sitio. http://www.ine.gib.mx/ueajei/publicaciones/libros/459/cap3.html. México.

Irschik, H., D. Schummer, K. Gerth, G. Höfle and H. Reichenbach. 1995. The tartrolons, new boron-containing antibiotics from a *myxobacterium, Sorangium cellulosum*. J. Antibiot. 48: 26–30.

ISO, 1990. Water quality-Determination of borate-Spectrometric method using azomethine-H. Geneva, International Organization for Standardization (ISO 9390:1990).

Karray, F., E. Darbon, N.C. Nguyen, J. Gagnat and J.L. Pernodet. 2010. Regulation of the Biosynthesis of the Macrolide Antibiotic Spiramycin in *Streptomyces ambofaciens*. J. Bacteriol. 192: 5813–5821.

Keiser, T., M.J. Viv, M.J. Buttner, K.F. Chater and D.A. Hopwood. Preparation and analysis of genomic plasmid DNA. pp. 161–210. *In*: T. Keiser, M.J. Viv, M.J. Buttner, K.F. Chater, D.A. Hopwood (Eds.). 2000. Practical *Streptomyces* Genetics. The John Innes Foundation, Norwich, England.

Keren, R. and U. Mezuman. 1981. Boron adsorption by clay minerals using a phenomenological equation. Clays Clay Miner. 29: 198–204.

Keren, R., R.G. Gast and B. Bar-Yosef. 1981. pH-dependent boron adsorption by Namontmorillonite. Soil Sci. Soc. Am. J. 45: 45–48.

Kiegel, W. 1972. Pharmazie. 27: 1–14.

Kimura, M. 1980. A simple method for estimating evolutionary rates of base substitutions through studies of nucleotide sequences. J. Mol. Evol. 16: 111–120.

Koběrská, M., J. Kopecký, J. Olšovská, M. Jelínková, D. Ulanova, P. Man, M. Flieger and J. Janata. 2008. Sequence analysis and heterologous expression of the lincomycin biosynthetic cluster of the type strain *Streptomyces lincolnensis* ATCC 25466. Folia Microbiol. 53: 395–401.

Kohno, J., T. Kawahata, T. Otake, M. Morimoto, H. Mori, N. Ueba, M. Nishio, A. Kinumaki, S. Komatsubara and K. Kawashima. 1996. Boromycin, an anti-HIV antibiotic. Biosci. Biothechnol. Biochem. 60: 1036–1037.

Lee, I.P., R.J. Sherins and R.L. Dixon. 1978. Evidence for induction of germinal aplasia in male rats by environmental exposure to boron. Toxicol. Appl. Pharmacol. 45: 577–590.

Lewer, P., E.L. Chapin, P.R. Graupner, J.R. Gilbert and C. Peacock. 2003. Tartrolone C: a novel insecticidal macrodiolide produced by *Streptomyces* sp. CP1130. J. Nat. Prod. 66: 143–145.

Li, W., S.C. Chou, A. Khullar and B. Gerratana. 2009. Cloning and Characterization of the Biosynthetic Gene Cluster for Tomaymycin, an SJG-136 Monomeric Analog. Appl. Environ. Microbiol. 75: 2958–2963.

Marguesin, R. and F. Schiner. 2001. Potential of halotolerant and halophilic microorganisms for biotechnology. Extremophiles. 5: 73– 83.

Maharjan, J., K. Liou, H.C. Lee, C.G. Kim, J.J. Lee, J.C. Yoo and J.K. Sohng. 2003. Functional identification of rub52 gene involved in the biosynthesis of rubradirin. Biotechnol. Lett. 25: 909–15.

Mason, B. 1992. Preparation of soil sampling protocols; sampling techniques and strategies. US EPA, EPA, 600. R-92. pp. 128–169.

Mateo, P., I. Bonilla, E. Fernández-Valiente and E. Sanchez-Maeso. 1986. Essentiality of boron for dinitrogen fixation in *Anabaena* sp. PCC 7119. Plant Physiol. 94: 1154–1560.

Miller, M.B. and R.W.Burg. 1975. Boromycin as a coccidiostat. US Patent 3864479.

Miwa, H., I. Ahmed, J. Yoon, A. Yokota and T. Fujiwara. 2008. *Variovorax boronicumulans* sp. nov., a boron-accumulating bacterium isolated from soil. Int. J. Syst. Evol. Microbiol. 58: 286–289.

Miwa, H., I. Ahmed, A. Yokota and T. Fujiwara. 2009a. *Lysinibacillus parviboronicapiens* sp. nov., a low boron-containing bacterium isolated from soil. Int. J. Syst. Evol. Microbiol. 59: 1427–1432.

Miwa, H. and T. Fujiwara. 2009b. Isolation and identification of boron-accumulating bacteria from contaminated soils and active sludge. Soil Science and Plant Nutrition. 55: 643–646.

Morin, C. 1994. The chemistry of boron analogues of biomolecules. Tetrahedron. 50: 12521–12569.

Nakamura, H., Y. Iitaka, T. Kitahara, T. Okazaki and Y. Okami. 1977. Structure of aplasmomycin. J. Antibiot. 30: 714–719.

Neal, C., K.K. Fox, M. Harrow and M. Neal. 1998. Boron in the major UK rivers entering the North Sea. Sci. Total Environ. 210–211: 41–52.

Negrete-Raymond, A., B. Weder and L. Wackett. 2003. Catabolism of Arylboronic Acids by *Arthrobacter nicotinovorans* Strain PBA. Appl. Environ. Microbiol. 69: 4263–4267.

Nielsen, F.H., 2000. The emergence of boron as nutritionally important throughout the life cycle. Nutrition. 16: 512–514.

Nielsen, F.H. Boron. *In*: E. Marian, M. Anke, M. Stoeppler (Eds.). 2004. pp. 1251–1260. Elements and their compounds in the environment, 2nd Edn. Wiley-VCH Verlag GmbH and Co KGaA Weinheim, Germany.

Noval, J.J. and W.J. Nickerson. 1958. Decomposition of native keratin by *Streptomyces fradie.* J. Bacteriol. 77: 251–263.

Nozawa, A., J. Takano, M. Kobayashi, N. Von Wirén and T. Fujiwara. 2006. Roles of BOR1, DUR3 and FPS1 in boron transport and tolerance of Saccharomyces cereviciae. FEMS Microbiol. Lett. 262: 216–222.

NTP. 1987. Toxicology and carcinogenesis studies of boric acid (CAS no. 10043-35-3) in B6C3F1 mice (food studies). Research Triangle Park, NC, United States Department of Health and Human Services, Public Health Service, National Institutes of Health, National Toxicology Program (NTP Technical Report Series No. 324).

Okami, Y., H. Okazaki, T. Kitahara and H. Umezawa. 1976. Studies on marine microorganisms. 5. New antibiotic, aplasmomycin, produced by a Streptomycete isolated from shallow sea mud. J. Antibiot. 29: 1019–1025.

O'Neill, M.A., T. Ishii, P. Albersheim and A.G. Darvill. 2004. Rhamnogalacturonan II: structure and function of a borate cross-linked cell wall pectic polysacccharide. Annu. Rev. Plant Biol. 55: 109–139.

Pache, W. and H. Zahner. 1969. Metabolic products of microorganisms. 77. Studies on mechanism of action of boromycin. Arch. Microbiol. 67: 156–165.

PNUMA: Programa de las Naciones Unidas para el Medio Ambiente Oficina Regional para América Latina y el Caribe, 2004. Boletín del 29 de Septiembre de 2004.

Polti, M.A, M.J. Amoroso and C.M. Abate. 2007. Chromium (VI) resistance and removal by actinomycete strins isoltated from sediments. Chemosphere. 67: 660–667.

Prelog, V., H. Zaehner and H. Bickel. 1973. Antibiotic A 28829. US Patent 3,769,418.

Rai, D. et al. 1986. Chemical attenuation rates, coefficients, and constants in leachate migration. Vol. 1: A critical review. Report to Electric Power Research Institute, Palo Alto, CA. USA.

Ravel, J., M.J. Amoroso, R.R. Colwell and R.T. Hill. 1998. Mercury resistant actinomycetes form Chesapeake Bay. FEMS Microbiol. Lett. 162: 177–184.

Raven. 1980. Short and long-distance transport of boric acid in plants. New Phytol. 84: 231–249.

Saitou, N. and M. Nei. 1987. The nieghbor-joining method, a new methos for reconstructing phylogenetic trees. J. Mol. Evol. 4: 406–425.

Sandaa, R.A., V. Thorsvik and φ. Enger. 2001. Influence of long-term heavy-metal contamination on microbial communities in soil. Soil Biol. Biochem. 33: 287–295.

Sato, K., T. Okazaki, K. Imaeda and Y. Okami. 1978. New antibiotics, aplasmomycins-B and aplasmomycins-C. J. Antibiot. 31: 632–635.

Secretaría de Minería de Salta, Ministerio de Desarrollo Económico, 2011. Crisol Ediciones, Salta.

SEGEMAR, Servicio Geológico Minero Argentino, 2002. *Publicación Técnica—UNSAM N° 8,* Salta.

Sjöberg, S., J. Carlsson, H. Ghaneolhosseini, L. Gedda, T. Hartman, J. Malmquist, C. Naeslund, P. Olsson and W. Tjarks. 1997. Chemistry and biology of some low molecular weight boron compounds for boron neutron capture therapy. J. Neurooncol. 33: 41–52.

Sprague, R.W. 1972. The ecological significance of boron. Anaheim, CA, United States Borax Research Corporation, 58 pp.

Steinberg, J.H. and A.L. Mcclosekey. 1964. Progress in Boron Chemistry. Vol.1. Eds. Mc Millan, New York.

Stokinger, H.E. Boron. pp. 2978–3005. *In*: G.D. Clayton, F.E. Clayton [eds.]. 1981. Patty's industrial hygiene and toxicology. Vol. 2B. Toxicology, 3rd ed. John Wiley & Sons. New York.

Stout, T.J., J. Clardy, I.C. Pathirana and W. Fenical. 1991. Aplasmomycin-C–structural studies of a marine antibiotic. Tetrahedron. 47: 3511–3520.

Swate, T.E. and J.C. Leed. 1974. Boric acid treatment of Vulvovaginal candidiasis. Obstetrics & Gynecology 43: 893–895.

Takano, J., K. Miwa and T. Fujiwara. 2008. Boron transport mechanisms: collaboration of channels and transporters. Trends Plant Sci. 13: 451–456.

Truhaut R., N. Phu-Lich and F. Loisillier. 1964. Effects of the repeated ingestion of small doses of boron derivatives on the reproductive functions of the rat. Comptes Rendus de l'Académie des Sciences. 258: 5099–5102.

United Kingdom Expert Group on Vitamins and Minerals (2002) Revised review of boron. London, Food Standards Agency (EVM/99/23/P.RevisedAug2002; http://www.food.gov.uk/multimedia/pdfs/boron.pdf).

Valencia, I.C.E. and B.A. Hernández. 2002. Muestreo de suelos preparación de muestras y guía de campo. 1ra Edición, UNAM, México.

Waggott, A. 1969. An investigation of the potential problem of increasing boron concentrations in rivers and water courses. Water Res. 3: 749–765.

Warington, K. 1923. The effect of boric acid and borax on the broad bean and certain other plants. Ann. Bot. 37: 629–672.

Weast, R.C., M.J. Astle, W.H. Beyer. 1985. Eds. CRC handbook of chemistry and physics, 69th ed. Boca Raton, FL, CRC Press, Inc., pp. B-77, B-129

Weir, R.J. and R.S. Fisher. 1972. Toxicologic studies on borax and boric acid. Toxicol. Appl. Pharmacol. 23: 351–364.

Williams, S.T., M. Goodfellow and G. Alderson. Genus *Streptomyces* Waksman and Henrici 1943, 339AL. *In*: S.T. Williams, M.E. Sharpe, J.G. Holt Eds. 1989. Bergey's Manual of Systematic Bacteriology, Vol. 4, PP. 2452–2492. Williams and Wilkins, Baltimore, USA.

WHO, 1998. Boron, International Programme on Chemical Safety (Environmental Health Criteria 204; http://www.inchem.org/documents/ehc/ehc/ehc204.htm), WHO, Geneva.

WHO, World Health Organization, 2009. Boron in drinking-water. Background document for development of WHO *Guidelines for Drinking-water Quality*.

Wyness, A.J., R.H. Parkman and C. Neal. 2003. A summary of boron surface water quality data throughout the European Union. Sci. Total Environ. 314–316: 255–269.

Zahran, H.H. 1997. Diversity, adaptation and activity of the bacterial flora in saline environments. Biology and Fertility of Soils. 25: 211–223.

CHAPTER 10

Biodegradation of Pesticides by Actinobacteria and their Possible Application in Biobed Systems

Gabriela Briceño,[1,2,3,*] Leticia Pizzul[3,4] and María Cristina Diez[1,3]

Introduction

According to the Food and Agricultural Foundation of the United Nations (FAO), food production will have to increase by 70% to feed a population of nine billion people by 2050. To reach this goal, agriculture production will have to increase substantially, and as a consequence, measures to avoid a negative impact of the increased use of pesticides in the environment will be required. Even after all precautions are taken, pollution can still result

[1]Department of Chemical Engineering, Universidad de La Frontera, P.O. Box 54-D, Temuco, Chile.
[2]Department of Chemical Science, Universidad de La Frontera, P.O. Box 54-D, Temuco, Chile.
[3]Scientific and Technological Environmental Biotechnology Center, Scientific and Technological Bioresources Nucleus (BIOREN) and Department of Chemical Engineering and Natural Resources, Universidad de La Frontera, PO Box 54-D, Temuco, Chile.
[4]Uppsala BioCenter, Department of Microbiology, Swedish University of Agricultural Sciences, P.O. Box 7025, 750 07 Uppsala, Sweden.
*Corresponding author: gbriceno@ufro.cl

from accidental spills or be inherited from past industrial activities. Pollution sites represent a significant risk to human health and other living forms because of their toxicity and persisting presence in the environment; such sites therefore need to be cleaned up.

The use of microorganisms or their enzymes to avoid or mitigate soil and water contamination—bioremediation—is an attractive, environment friendly and cost-effective option (Juwarkar et al. 2010). Microorganisms have the ability to interact both chemically and physically with substances, which can lead to structural changes or complete degradation of the target molecule (Raymond et al. 2001, Wiren-Lehr et al. 2002). Actinobacteria, formerly classified as actinomycetes, constitute a significant fraction of the microbial population in soils, with populations that commonly exceed 1 million microorganisms per gram (Goodfellow and Williams 1983, Seong et al. 2001). These microorganisms have been recognized as having great potential for the degradation of numerous organic compounds, including pesticides (De Schrijver and De Mot 1999, Sette et al. 2005, Castillo et al. 2006, Fuentes et al. 2010). However, successful laboratory studies concerning bioremediation are not always reproducible during *in situ* decontamination. Added free-living cells are subjected to several biological and abiotic stresses that decrease their chances of survival. One strategy to overcome this problem involves the use of immobilized cells, i.e., cells adsorbed or entrapped in a carrier that restricts their mobility within a defined space and thereby preserves their catalytic activity (Mrozik and Piotrowska-Seget 2010).

Whereas bioremediation is a feasible technique for the clean-up of pesticide-contaminated soils from, for example, industrial sites, pollution should ideally be prevented in the first place. The use of biobeds originated in Sweden as a response to the need for simple and effective methods to minimize point-source contamination from using pesticides. A biobed is a simple and inexpensive structure on farms that is intended to collect and degrade spills of pesticides biologically (Castillo et al. 2008). The use of biobeds has since expanded to other countries in Europe and recently to Latin America.

This chapter gives a general overview of the role of actinobacteria in the degradation of pesticides and their use in bioremediation, both as free-living and immobilized cells. Their potential as inoculants to increase pesticide degradation in biobeds is discussed in particular.

Pesticide Contamination and Biological Treatment: General Aspects

Pesticides in the environment

Pesticide usage has increased dramatically worldwide over the past two decades, coinciding with changes in farming practices and increasingly intensive agriculture. This widespread use of pesticides for agricultural and non-agricultural purposes has resulted in the presence of pesticide residues in various environmental matrices. Pesticide contamination of surface waters has been well documented (Table 1) and constitutes a major issue that gives rise to concerns at local, regional, national and global scales (El-Nahhal et al. 1997, Planas et al. 1997, Fatoki and Awofolu 2005, Westbom et al. 2008). Pesticides such as atrazine, isoproturon, diuron and mecoprop are the main herbicides found in groundwater in the UK, whereas atrazine, alachlor and metolachlor have been detected in groundwater in other European, American and African countries. Another problem is the persistence of certain pesticides in the soil and their toxicity to other organisms (Chen et al. 2001) even at low concentrations (Crecchio et al. 2001).

In the field, less than 0.1% of the applied pesticide reaches the target pests. The other fraction can be bound to the soil, undergo abiotic or biological transformation or be subjected to diffuse losses, such as volatilization, surface transport during heavy rainfall conditions, and leaching through the soil profile, to finally reach groundwater and superficial waters (Pimentel 1995). How long and how much of the pesticide remains in the soil depends on how strongly it is bound to soil components and how readily it is degraded. It also depends on the environmental conditions at the time of application, on the potential degradation and on the physical and chemical characteristics of the soil system, such as the moisture content, the organic matter and clay contents, the pH and the sorption/desorption capacity (Arias-Estévez et al. 2008).

The occurrence of pesticide residues in surface waters, groundwater and large volumes of soil can be due to the inadequate management of pesticides (Castillo et al. 2008). Approximately 40 to 90% of surface-water contamination is attributed to direct losses, i.e., spillages that result from the filling operation; leaks in the spray equipment; spray leftovers and technical rest volumes in the tank, pump and booms; rinsing water from cleaning the internal tank to avoid carryover effects (damage and residues) onto

Table 1. Residues of pesticide found in groundwater and superficial water in some countries.

Country	Detected pesticides	Levels	Water	Reference
Portugal	Lindane, atrazine, imazine,dimethoate, metribuzin, endosulfan, prometryn, metolachlor.	0.22–17 μg L^{-1}	Groundwater	Barcelo (1991)
	Quinalphos (28.8%), Paraquat (27.9%).	*		Teixeira et al. (2004)
Netherlands	Atrazine, simazine, dieldrin, propazine and lindane (+ -HCH).	100–200 ng L^{-1}	Groundwater	Maanen et al. (2001)
	A broad list of pesticides (27).	**27% ***11%	Groundwater	Schipper et al. (2008)
Greece	Lindane (γ-BHC), chlorpyrifos, propachlor.	0.005–0.01 μg L^{-1}	Groundwater	Karasali et al. (2002)
Canada	Alachlor, metalachlor, atrazine, metribuzin, cianazina.	0.17–0.34 μg L^{-1}	Superficial	Goss et al. (1998)
England	Lindane (α+β-HCH), heptachlor, aldrin, γ-chlordane, endosufan, dieldrin, endrin, 2,4′-DDT, etc.	5.5–160 ng L^{-1}	Superficial	Fatoki and Awofolu (2005)
South Africa	Lindane (α+β-HCH), heptachlor, aldrin, γ- chlordane, endosufan, dieldrin, endrin, 2,4′-DDT, etc.	6–80 ng L^{-1}	Superficial	Fatoki and Awofolu (2005)
Brazil	Alachlor, atrazine, chlorothalonil, endosulfan, simazine, metribuzin, monocrotofos, malathion, chlorpyrifos, metribuzin, etc.	0.001–0.174 μg L^{-1}	Superficial, River, Lakes	Laabs et al. (2002)
Spain	Atrazine, desethylatrazine, simazine, desethylsimazine, metolachlor, desethylterbuthylazine, terbuthylazine, metalaxyl.	Up to 0.63 μg L^{-1} Up to 2.46 μg L^{-1}	Superficial Groundwater	Hildebrandt et al. (2008)
Hungary	Acetochlor, atrazine, carbofuran, diazinon, fenoxycarb, metribuzin, phorate, prometryn, terbutryn, and trifluralin.	**59%	Superficial	Maloschik et al. (2007)
USA	Data from 10 years of study (25 pesticides).	-	Groundwater Streams	Gilliom (2007)

*Represent positive cases of intoxication; **Percentage of the samples containing pesticide; ***Percentage of samples exceeding 0.1 μg L^{-1}. Information obtained from Diez (2010).

the following crop and water from external cleaning of spray equipment, among others. The activities mentioned constitute typical point sources of contamination by pesticides.

Bioremediation of contaminated soils

The development of technologies that guarantee the elimination of pesticide contamination in a safe, efficient and inexpensive manner is required. In the past, primarily physical and chemical methods have been used to remove contaminants from polluted soils. Today, biological methods, including bioremediation, are considered to be relatively cost-effective and environment friendly.

By definition, bioremediation is the use of living organisms, primarily microorganisms, to degrade environmental contaminants into less toxic substances. Bioremediation uses naturally occurring bacteria and fungi or plants to degrade or detoxify substances hazardous to human health and/or the environment (Vidali 2001). Bioremediation processes can be divided into *natural attenuation*, in which contaminants are degraded by native microorganisms present in the soil without amendments; *biostimulation*, in which nutrients and oxygen are applied to the system to improve microbial growth and accelerate biodegradation by the best adapted native degrading microorganisms and *bioaugmentation*, which enhances the biodegradative capacities in contaminated sites by inoculation of single strains or consortia of microorganisms with the desired catalytic capabilities to degrade the target contaminants (Iwamoto and Nasu 2001, Diez 2010, Juwarkar et al. 2010). Bioremediation is still an immature technology and has limitations; for example, local environmental conditions might not be appropriate for the microorganisms to display their degrading potential, and sometimes metabolites with higher toxicities than the parent molecules are generated. In such cases, bioaugmentation with specific microorganisms with a known degradation pathway could represent a solution to the problem (Fantroussi and Agathos 2005). Bioaugmentation could be applied in soils with a low or non-detectable number of contaminant-degrading microorganisms, in soils that contain compounds that require multiple decontamination processes or for small scale sites where the costs of non-biological methods exceed the cost of bioremediation (Mrozik and Piotrowska-Seget 2010). Several factors determine the applicability of bioaugmentation, including temperature, moisture, pH, organic matter content, aeration, nutrient content and soil type (abiotic factors). In contrast, the most important biotic factors include competition between indigenous and exogenous microorganisms for limited carbon sources as well as antagonistic interaction and predation (Mrozik and Piotrowska-Seget 2010).

The role of microorganisms in the dissipation of pesticides, especially in the soil, has long been recognized. The initial strain selection has usually been based on single soil criterion and degradation ability, with no consideration for the potential ability of the strain to proliferate and be active in the target sites (Thompson et al. 2005). Environmental parameters, together with the phenotypic characteristics of the strain and procedures for its introduction to the contaminated soil, determine the activity, persistence and performance of the introduced strains (Vogel and Walter 2001, Thompson et al. 2005, Hosowaka et al. 2009). In general, the use of microorganisms from an ecological site similar to the polluted one is considered the best choice (Fantroussi and Agathos 2005, Hosowaka et al. 2009). Moreover, a suitable strain or consortium should grow fast, withstand high concentrations of contaminants and have the ability to survive in a wide range of environmental conditions (Mrozik and Piotrowska-Seget 2010). Unfortunately, successful laboratory studies concerning bioremediation do not necessarily lead to reproducible *in situ* decontamination. A strategy for improving the effectiveness of bioaugmentation is the immobilization of microorganisms prior to inoculation. Immobilization is defined by IUPAC as the "technique used for the physical or chemical fixation of cells, organelles, enzymes, or other proteins (e.g., monoclonal antibodies) onto a solid support, into a solid matrix or retained by a membrane, in order to increase their stability and make possible their repeated or continued use." In this manner, microorganisms are protected from adverse environmental conditions (high salinity, unfavorable soil pH, extremes in temperature, insufficient or excessive soil moisture, heavy-metal toxicity and the presence of biocides) and the competition with the indigenous microflora is reduced. Moreover, the use of immobilized microorganisms may result in a more rapid degradation of organic compounds in comparison with free-living cells and could also increase the biological stability of cells for extended periods of time (Fantroussi and Agathos 2005, Vancov et al. 2005, Mrozik and Piotrowska-Seget 2010).

Methods such as adsorption, covalent binding, entrapment, encapsulation and crosslinking are used for the immobilization of cells. Immobilization by adsorption is the simplest method and involves reversible surface interactions between cells and a support material with adsorption properties. Microorganisms are mixed together with the support material for a given period of incubation, the immobilized cells are collected and the unbound biological components are removed by extensive washing. Immobilization through covalent binding involves the formation of a covalent bond between the cell and support material. Normally the bond is formed between functional groups present on the surface of the support material and functional groups on the surface of the cell. Immobilization by entrapment differs from adsorption and covalent binding in that the cells are free in solution but restricted in movement by the lattice structure of a polymer-gel matrix. The support material, in this case, acts as a barrier and can be advantageous because it protects the immobilized cells from microbial and environmental contamination. Encapsulation of microorganisms can be achieved by enveloping the biological components within the various forms of semipermeable membranes. Encapsulation is similar to entrapment in that the microorganisms are free in solution but restricted in space. Finally, immobilization through cross linking is support-free and involves connecting (binding) cells to each other to form a large, three-dimensional complex structure; it can be achieved either by chemical or physical methods. Cross linking is rarely used as the only means of immobilization and often used to enhance the other described methods of immobilization (Cassidy et al. 1996). The choice of method varies according to the type of microorganism, the immobilizing matrix and the operational system (Saudagar et al. 2008).

Both natural and synthetic materials are used for the purpose of immobilization. The first category includes dextran, agar, agarose, alginate, chitosan polyacrylamides, k-carrageenan, and agricultural residues, whereas the synthetic materials include poly (carbamoyl) sulfonate, polyacrylamide and polyvinyl alcohol. However, concerns have arisen about the use of synthetic polymers because they are not biodegradable (Plangklang and Reungsang 2009). One of the most suitable methods for cell immobilization is entrapment in calcium alginate because this technique is simple and cheap, nontoxic for the microorganisms and protects them against the toxicity of xenobiotic compounds (Anisha and Prema 2008, Bazot and Lebeau 2008).

Biobeds as a tool to prevent pesticide pollution

Biobeds are biological systems were originated in Sweden in response to the need for a simple and effective system to minimize environmental contamination from pesticide manipulation, especially when filling the spraying equipment, a typical point source of contamination (Tortensson and Castillo 1997, Castillo et al. 2008). In Sweden and other countries in Europe, there are numerous examples of biobeds being used on farms and they have been shown to be efficient at reducing pesticide water-body contamination (Vischetti et al. 2007, Castillo et al. 2008). Biobeds are based on the adsorption and potential degradation of organic biomixtures (Fig. 1) composed of top soil, peat, and straw (25: 25: 50 vol %) that fill a deep pit (60 cm) in the ground and a grass layer that covers the surface (Tortensson and Castillo 1997, Castillo et al. 2008). Biobeds are a low-cost alternative for the treatment of pesticide waste and washings, providing a matrix to absorb the pesticides and facilitate their biodegradation. Straw stimulates the growth of

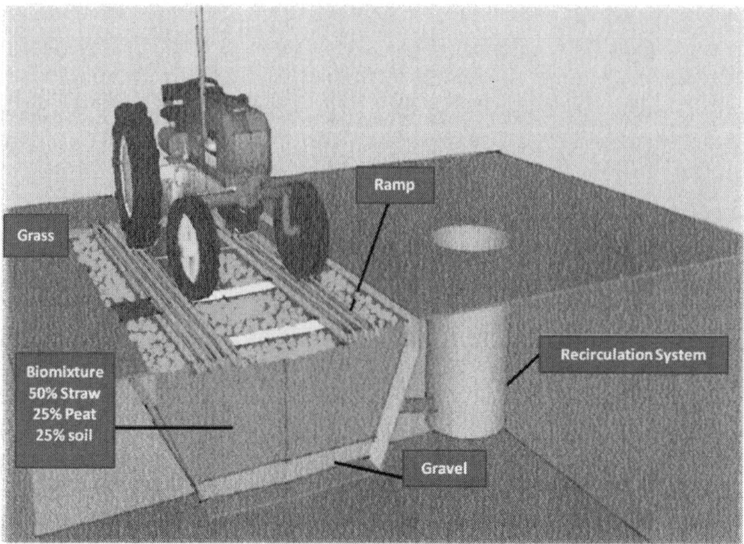

Figure 1. Diagram of a biobed

Color image of this figure appears in the color plate section at the end of the book.

ligninolytic microorganisms and the production of extracellular ligninolytic enzymes, such as peroxidases and phenoloxidases. The peat contributes to sorption capacity, moisture control and abiotic degradation of pesticides and decreases the pH of the biomixture, which is favorable primarily for fungi and their pesticide-degrading enzymes (Torstensson and Castillo 1997, Castillo et al. 2001). The soil enhances the sorption capacity in the biobed, and it is also an important source of pesticide-degrading bacteria, including actinobacteria that can act synergistically with the fungi.

Recently, the use of biobeds has expanded to other countries in Europe and Latin America, and their implementation has led to modifications of the original biobed. For example, modification of the depth was adopted in England to increase the retention time of the pesticide in the bed. In Italy, this technology is known as a biomass bed and utilizes biomixtures as filters through which pesticide-contaminated water is circulated and decontaminated. Because peat is not readily found in Italy, organic material, such as urban and garden compost, peach stones and citrus peel, are being tested as replacements. In Latin America, countries such as Perú, Guatemala, Ecuador and Chile have been or are implementing the system, in many cases building smaller biobeds adapted to smaller-sized farms. A thorough description of the original biobed and its modifications, along with a discussion of the factors that affect their efficiency, is found in the review by Castillo et al. (2008).

Several studies have demonstrated that biobeds can effectively retain and degrade a wide range of pesticides, either alone or in mixtures (Torstensson and Castillo 1997, Castillo et al. 2001, Fogg et al. 2003, Fogg et al. 2004, Vischetti et al. 2004, Castillo and Torstensson 2007, Vischetti et al. 2008). Niels et al. (2006) evaluated the degradation and leaching of 21 pesticides. They determined that no traces of 10 out of 21 applied pesticides were detected in the percolate. Fogget al. (2003) evaluated the ability of biobeds to degrade pesticide mixtures (isoproturon and chlorothalonil) and the concentration effect. They found that, with the exception of isoproturon at concentrations greater than 11 mg kg^{-1}, degradation was more rapid in the biomix than in topsoil. The degradation of either isoproturon or chlorothalonil was unaffected by the presence of the other pesticide.

Bioremediation of pesticide-contaminated soils by actinobacteria

General characteristics of actinobacteria

Actinobacteria, formerly classified as actinomycetes, are Gram-positive bacteria with a high G+C content in their DNA. They can be found in a wide range of habitats and are particularly abundant in soil, where they are usually present in numbers of 10^4–10^6 colony-forming units per gram of soil (Goodfellow and Williams 1983, Seong et al. 2001, Shrivastava et al. 2008). Actinobacteria present a morphology that varies among the different genera, from cocci and pleomorphic rods to branched filaments that break down into spherical cells or aerial mycelium with long chains of spores. The spores represent a semi-dormant stage in the life cycle of the bacteria, and the spores can survive in soil for long periods of time in conditions with low nutrients and low water availability (McCarthy and Williams 1992). In general, the optimal conditions for their growth are temperatures of 25–30°C and pH among 5–9 with an optimum close to 7.

In soils, actinobacteria play an important role in the recycling of organic carbon and are able to degrade complex polymers (Goodfellow and Williams 1983). Many strains have the ability to solubilize lignin and degrade lignin-related compounds by producing cellulose-and hemicellulose-degrading enzymes and extracellular peroxidases (Ball et al. 1989, Pasti et al. 1990, Mason et al. 2001). They also occur in other rich environments in organic matter, such as composts, in both the mesophilic and thermophilic phases (Steger 2006) and sewage sludge, where the mycolic acid-containing actinobacteria are associated with the extensive and undesirable formation of stable foams and scum (Goodfellow et al. 1996, Seong et al. 1999). Some actinobacteria are important as human pathogens. However, most of the interest in this group of microorganisms lies in their ability to produce several structurally diverse bioactive compounds of pharmaceutical and agricultural importance. Although the order Actinomycetales encompasses more than 80 different genera, two-thirds of the microbial antibiotics known today are produced, primarily by *Streptomyces* species (Bèrdy 2005, Monciardini et al. 2002). Actinobacteria also produce secondary metabolites that exhibit bioactivities other than antibiotics, such as enzyme inhibitors, immunosuppressive, phytotoxins and pesticides (Park et al. 2002, Bèrdy 2005, Imada 2005).

In recent years, actinobacteria have received attention as candidates for applications in the bioremediation of contaminated soils (Fuentes et al. 2010, 2011) because of their ability to produce extracellular enzymes that are able to degrade a wide range of complex organic compounds and spores that are resistant to desiccation. In addition, the frequently occurring filamentous growth favors the colonization of soil particles and minimizes the need for mixing (Ensign 1992, Pogell 1995). Many genera of actinobacteria have been shown to be tolerant to and remove both inorganic pollutants, such as mercury and copper (Ravel et al. 1998, Albarracín et al. 2008) and organic pollutants, such as polyaromatic hydrocarbons (Pizzul et al. 2006) and pesticides (De Schrijver and De Mot 1999). They grow rapidly on semi-selective, environmentally benign substrates, build a high biomass rapidly, and achieve high rates of pesticide degradation (Pogell 1995). Finally, strains can be genetically manipulated for future adaptation to new environments and new pollutants (Pogell 1995).

Pesticide degradation by actinobacteria

Many actinobacteria can degrade pesticides. Compared with Gram-negative bacteria, the metabolic pathway for pesticide degradation by actinobacteria has not been studied extensively. A common feature of the aerobic actinobacteria is the presence of many types of monooxygenases and dioxygenases (De Schrijver and DeMot 1999, Larkin et al. 2005). The genera *Arthrobacter*, *Clavibacter*, *Nocardia*, *Rhodococcus*, *Nocardioides* and *Streptomyces* are the most representative pesticide-degrading actinobacteria (De Schrijver and DeMot 1999). For example, *Streptomyces* spp. have the capacity to grow on and degrade several classes of pesticides, including atrazine, alachlor, carbofuran, cypermethrin, diuron and lindane, among others (Sette et al. 2004, 2005, Vancov et al. 2005, Castillo et al. 2006, Jayabarath et al. 2010, Fuentes et al. 2011, Lin et al. 2011). Detailed information about pesticide degradation by isolated actinobacterias is given in Table 2.

In a study performed by Benimeli et al. (2003), 93 indigenous colonies belonging to the actinobacteria group were isolated from wastewater sediment samples of a copper filter plant and tested against 11 organochlorine pesticides (OP), including aldrin, chlordane, DDD, DDE, DDT, dieldrin,

Table 2. Actinobacteria isolated for the pesticide degradation. The main obtained results and observations.

Chemical class/ pesticide	Microorganisms studied	Place of isolation	Condition of work	Response and observations	Reference
Pyrethroid					
-Cypermethrin	*Streptomyces* sp. HU-S-01	Wastewater sludge	-Liquid minimal medium. -50 mg L^{-1} of pesticide.	-92.1% degradation within 24 h and 100% after 30 h. -Optimum degradation condition at temperature of 26–28° and pH 7.5.	Lin et al. (2011)
Carbamate					
-Carbofuran	*Streptomyces alanonosinicus, Streptoverticillium album, Nocardia farcinia, S. atratus, Nocardia vaccini, Nocardia amarae and Micromonospora chalcea*	Saline soil	-Degradation as carbon source in minimal broth, and by co-metabolism in broth glycerol asparagine. -20 µg mL^{-1} of carbofuran.	-After 10 days, between 6.7 to 65.5% of pesticide was degraded as sole carbon source and between 69.8 to 95.3% was degraded as co-metabolism.	Jayabarath et al. (2010)
Organochlorine					
Aldrin	*Streptomyces* sp. Strain M7.	Wastewater sediment of a copper filter plant	-Liquid medium with 48 µg L^{-1} of aldrin; 96 h of incubation.	-After 72 h of incubation, 90% of the pesticide was degraded.	Benimeli et al. (2003)
Lindane	*Streptomyces* sp., *Micromonospora* sp.	Contaminated soil with organochlorine pesticide.	-Liquid minimal medium with 1.66 mg L^{-1} of lindane	-Dry weight of 0.02–0.09 g L^{-1} and residual lindane of 0.30–0.72 mg L^{-1}, after 7 days.	Fuentes et al. (2010)

Table 2. contd....

Table 2. contd....

Chemical class/ pesticide	Microorganisms studied	Place of isolation	Condition of work	Response and observations	Reference
Organochlorine					
	Streptomyces sp.	Contaminated soil with organochlorine pesticide.	-Liquid minimal medium with 1.66 mg L^{-1} lindane as sole carbon source. -Pure and mixed culture. -96 h of incubation	-Pure culture presented specific dechlorinase activity. Dechlorinase activity was improved by mixed cultures. Mixed cultures with two, three and four strains showed maximum removal of 46–68%.	Fuentes et al. (2011)
	Streptomyces sp. Strain M7	Pesticide-contaminated sediment.	-Liquid medium with 10 μg L^{-1} of lindane or 6 g L^{-1} of glucose. -96 h of incubation.	-Both, glucose and lindane were simultaneously consumed. Glucose improved lindane degradation and biomass.	Benimeli et al. (2007)
Chlordane	*Streptomyces* sp., *Micromonospora* sp.	Contaminated soil with organochlorine pesticide.	-Liquid minimal medium with 1.66 mg L^{-1} of chlordane.	-Dry weight of 0.01–0.23 g L^{-1}. No residual pesticide was detected after 7 days.	Fuentes et al. (2010)
Methoxychlor	*Streptomyces* sp., *Micromonospora* sp.	Contaminated soil with organochlorine pesticide.	-Liquid minimal medium with 1.66 mg L^{-1} of methoxychlor.	-Dry weight of 0.01–0.12 g L^{-1} and residual methoxichlor of 0.01–1.20 mg L^{-1}, after 7 days.	Fuentes et al. (2010)

Urea					
-Diuron	*Streptomyces* strains	Non-agricultual soil and agricultural soil.	-ISP2 medium supplemented with 4 mg L^{-1} of diuron.	-Between 50-70% of diuron was degraded. -*Streptomyces albidoflavus* showed 95% of degradation after 5 days and no residues after 10 days. -The strains isolated from agricultural soil exhibited higher degradation.	Castillo et al. (2006)
Chloroacetanilide					
-Alachlor	*Streptomyces* sp. LS166, LS177, and LS182.	Contaminated soil samples with alachlor.	-Pure culture; liquid medium supplanted with 144 mg L^{-1} of alachlor; 14 days of incubation.	-All strains were able to grow at studied concentration, and degrade between 60 to 75% in 14 days.	Sette et al. (2004)
	Streptomyces sp. LS151, LS143, LS153	Soil treated with alachlor.	- Mineral salt medium with 72 mg L^{-1} of alachlor	Six strains were able to grow and degrade ≥50% of pesticide.	Sette et al. (2005)
-Metolachlor	Actinobacteria no identified.	Metolachlor-contaminated soil.	-Liquid medium with 50 mg L^{-1} of pesticide.	-Over 50% metolachlor disappeared in medium with 0.4% of sucrose and 6% disappeared in medium with 0.05% sucrose.	Krause et al. (1985)

Table 2. contd....

Table 2. contd....

Chemical class/ pesticide	Microorganisms studied	Place of isolation	Condition of work	Response and observations	Reference
Organophosphorus					
-Chlorpyrifos	*Streptomyces radiopugnans*	Soil sprayed extensively with chlorpyrifos.	-Aislamiento through soil enrichment in mineral salt medium supplemented with chlorpyrifos as sole carbon source. -Chlorpyrifos removal as sole carbon source was evaluated in liquid medium.	-After 7 days of incubation an 25% of chlorpyrifos was degraded.	Sasikala et al. (2012)
	Streptomyces sp. Strain AC5 and AC7	Agricultural soil sprayed extensively with chlorpyrifos.	-Liquid medium with 25 and 50 mg L^{-1} of pesticide.	-After 24 h of incubation close to 90% of chlorpyrifos degradation was observed. Different concentration of degradation metabolite was produced by both studied strains.	Briceño et al. (2012)
-Fenamiphos	*Microbacterium* sp.	Soil which had received fenamiphos as the active component during 12 years.	-Soil extract agar mediun supplemented with 100 mg L^{-1} of pesticide.	-After 7 and 14 days post bacterium inoculation between 29–74% and 4–46% fenamiphos was found, respectively. Only one microbacteriun isolated was unable to degrade the pesticide.	Cabrera et al. (2010)

heptachlor, and heptachlor epoxides, lindane and methoxychlor. Among these strains, four, identified as *Streptomyces* sp., were selected based on multi-OP-tolerance. *Streptomyces* sp. M7, which was isolated from pesticide-contaminated sediment from Tucumán, Argentina, used lindane as the only carbon source (Benimeli et al. 2007). The presence of lindane induced the synthesis of dechlorinase by *Streptomyces* sp. M7, a characteristic that makes this strain useful in the bioremediation of soil or as a biosensor (Cuozzo et al. 2009). Briceño et al. (2012) isolated actinobacteria from a organophosphorus-contaminated agricultural soil used for blueberry crops and evaluated the ability of two selected strains, identified as *Streptomyces* sp. AC5 and *Streptomyces* sp. AC7, to grow and degrade the insecticide chlorpyrifos. The results indicated that both strains were able to rapidly degrade chlorpyrifos with approximately 90% degradation after 24 hrs of incubation. A different pattern of degradation was observed when its main metabolite, 3,5,6-trichloro-2-pyridinol, was evaluated. A maximum concentration of 0.46 mg L^{-1} of metabolite production by *Streptomyces* sp. strain AC5 was observed and its concentration decreased as a function of time. In contrast, production of the chlorpyrifos metabolites produced by *Streptomyces* sp. AC7 increased over time from 1.31 mg L^{-1} to 4.32 mg L^{-1}. These results were accompanied by microbial growth, pH modification and glucose consumption in the liquid medium. An actinobacteria consortium on lindane degradation was studied by Fuentes et al. (2011). In this study, consortia constituting two, three or four strains inoculated in liquid medium supplemented with lindane showed a greater removal of this compound compared with the individual strains, whereas no improvement in lindane removal was obtained with consortia constituting five or six strains.

Because actinobacteria produce extracellular ligninolytic enzymes (Godden et al. 1992, Kirby 2006), they may also degrade pesticides co-metabolically. Esposito et al. (1998) showed that actinobacteria strain CCT 4916 presented relatively good growth in soil contaminated with diuron (100 mg kg^{-1} soil) and degraded up to 37% of the herbicide in the *in vitro* assay. In this study, the degradation of diuron was found to be related to the production of manganese peroxidase-an enzyme known for the ability to degrade diverse pesticides (Pizzul et al. 2009).

Soil bioaugmentation can be performed through the addition of single strains or microbial consortia or by genetically modified indigenous bacteria. Most experiments have been performed using Gram-negative bacteria, such as the degradation of atrazine by *Pseudomonas* sp. in soil (Topp 2001), of chlorpyrifos by *Enterobacter* sp. (Singh et al. 2004), of 3,5,6-trichloro-2-pyridinol by *Alcaligenesfaecalis* (Yang et al. 2005) and of diazinon by *Serratia* sp. and *Pseudomonas* sp. (Cycon et al. 2009). Among fungi, *Trametesversicolor* was studied for atrazine bioremediation in calcareous soil (Bastos and Magan 2009), and *Antracophylum discolor* was investigated for degradation of pentachlorophenol (Rubilar et al. 2011). Actinobacteria have been less studied, and their use as a single strain in soil bioaugmentation is limited, despite their being known as good degraders of several pesticides. They are generally applied as a component of microbial consortia (Sasikala et al. 2012).

A loam soil (pH 7.30; 3.5% organic matter; 48% sand, 43% silt and 9% clay) was used to study atrazine degradation and mineralization in a soil microcosm by *Pseudomonas* sp. strain ADP, a *Pseudaminobacter* sp. and actinobacterias, *Nocardioides* sp. The results showed that only the *Pseudaminobacter* and *Nocardioides* accelerated atrazine dissipation. The *Pseudaminobacter* mineralized atrazine rapidly and without a lag, whereas atrazine was mineralized in the *Nocardioides* inoculated soil, but only after a lag of several days. Moreover, this study showed that the half-life of atrazine was approximately 10 days in the uninoculated and *Pseudomonas*-inoculated soil, approximately 5 days in the *Pseudaminobacter*-inoculated soil, and approximately 3 days in the *Nocardioides*-inoculated soil (Topp 2001). This work clearly showed the ability of an actinobacteria to degrade atrazine compared to another bacterium of soil. Benimeli et al. (2008) studied the lindane bioremediation ability of *Streptomyces* sp. M7 in soil samples and the pesticide effects on maize plants seeded in lindane-contaminated soil previously inoculated with *Streptomyces* sp. M7. Their results showed that *Streptomyces* sp. increased the biomass and concomitantly decreased residual lindane. The activity of this strain was not inhibited by natural soil microbial flora, and its growth was not inhibited by large amounts of pesticide. The optimum *Streptomyces* sp. M7 inoculum size, selected in sterile soil spiked with lindane (100 mg kg^{-1} of soil), was 2 g kg^{-1} soil, where a 56% removal was obtained. Lindane concentrations of 100–400 mg kg^{-1} soil did not affect the percentage of germination of maize plants, but the microorganism was inoculated under the same conditions in which a better vigor index, i.e., faster germination, was observed and in which 68% of lindane was removed. Finally, in a recent work, Fuentes et al. (2011) evaluated lindane biodegradation by an actinobacteria consortia were inoculated in soil (inoculum size 2.0 g kg^{-1} of soil). After 4 weeks, 32.6, 33.1 and 31.4% of the lindane applied in the soil (1.66 mg kg^{-1}) was degraded by the *Streptomyces* consortia composed of two, three and four strains, respectively.

Studies of pesticide degradation performed in soils (Table 3) have indicated that actinobacteria strains have the potential for application in bioremediation. However, as mentioned previously, despite the knowledge generated in the laboratory, there is a lack of successful applications in the field, and the use of immobilized inoculums is a step forward in filling this gap for actinobacteria.

Table 3. Degradation of pesticides by actinobacteria inoculated in soils.

Microorganisms	Pesticide	Condition	Observation	Reference
Actinobacteria strain CCT 4916	Diuron	Diuron was applied at concentration of 100 mg kg^{-1} / Soil used was a commercial potting soil (Terra do Paraiso, Brazil)/Soil plus diuron were added on top of the mycelia actinobacteria growth on the slant/In vitro assay was incubated for 7 d at 30°C.	37% of diuron degradation in 7 days was observed. The studied strain showed protease and urease activity. Moreover, manganese peroxidase activity was related with diuron degradation.	Esposito et al. (1998)
Pseudomonas sp. strain ADP, *Pseudaminobacter* sp., and *Nocardioides* sp.	Atrazine	A loam soil (pH 7.30; 3.5% organic matter; 48% sand, 43% silt and 9% clay) from London, Ontario, Canada was used./ The soil was under sod, and had no atrazine treatment history./Initial atrazine concentration was 1 mg/kg dry soil/ Bacteria were inoculated at concentrations of 1×10^5 or 1×10^7.	At inoculum densities of 10^5 cells/g soil, only the *Pseudaminobacter* and *Nocardioides* accelerated atrazine dissipation. The *Pseudaminobacter* mineralized atrazine rapidly and without a lag, whereas atrazine was mineralized in the *Nocardioides* inoculated soil but only after a lag of several days.	Topp (2001)
Streptomyces sp. M7	Lindane	In glass pots, 200 g non-sterile soil was taken and spiked with lindane at 100 mg kg^{-1}ww soil./Soil was inoculated using a microbial concentration of 2.0 g kg^{-1}ww soil. Incubation at room temperature for 2 weeks.	With the studied condition an optimal bioremediation of lindane was observed (56% removal).	Benimeli et al. (2008)
Streptomyces consortia	Lindane	Surface soil samples from experimental site, San Miguel de Tucuman, Argentina was used./The study was performed in glass pots with 200 g of sterile soil contaminated with 1.66 mg lindane kg^{-1} and inoculated with microbial concentration of 2.0 g kg^{-1} soil. Incubation at 30°C, 4 weeks.	Three mixed cultures were assayed for bioremediation of lindane polluted soil. All consortia exhibited good microbial growth. Higher duplication time was observed in treatment with lindane. Three consortia were able to remove about of 32% of lindane in 28 d.	Fuentes et al. (2011)

The immobilization of actinobacteria has been studied extensively for secondary metabolite production by *Streptomyces* sp. (Anisha and Prema 2008, Saudagar et al. 2008, Shrivastava et al. 2008, Kattimani et al. 2009), for chromium removal by *Streptomyces griseus* (Poopal and Laxman 2008) and for PAH removal by *Rhodoccocus* sp. (Quek et al. 2006) and only a small amount of work has been performed on immobilized pesticide-degrading actinobacteria (Table 4). Most studies have been performed with the aim of testing the ability of the immobilized bacteria to degrade a contaminant, and they have been performed in liquid media instead of soil. Vancov et al. (2005) studied atrazine degradation by encapsulated *Rhodococcus erythropolis* NI86/21. The three carriers tested-bentonite, powdered activated carbon and skimmed milk, favored a slow release of the strain and the degradation of the herbicide. However, bentonite-amended beads formulated with 1% skimmed milk did not provide an adequate number of cells to degrade atrazine in either liquid or soil compared to other studies. *Streptomyces rochei* 303 immobilized in polycaproamide fibers was studied for its ability to degrade individual chlorophenols and their mixtures. In a fermentation study with continuous substrate and air flow, a greater efficiency of chlorophenol degradation was obtained using immobilized cells compared with free cells (Golovlena et al. 1993). Cho et al. (2000) studied the degradation of *p*-nitrophenol, a major metabolite from the microbial degradation of parathion or methyl parathion, by freely suspended and calcium-alginate-immobilized *Nocardioides* sp. NSP41. The main results showed that the use of immobilized bacteria gave a high volumetric degradation rate of the compound and that it was possible to reuse the immobilized cells 12 times. Reusability for an extended period is one of the advantages of inoculation with immobilized cells.

Actinobacteria and biobeds

As previously discussed, actinobacteria can effectively degrade a wide range of organic pollutants, including pesticides. Additionally, due to the large amount of favorable results obtained in the field, the use of biobeds to prevent pesticide pollution is growing and many countries are adopting and adapting the system to local conditions and needs. Considering these facts, we discuss here the potential use of actinobacteria in biobeds.

Biobeds have been widely accepted by farmers and environmental authorities, in part because of their simplicity and low cost and the use of local materials and natural organisms (Castillo et al. 2008). To be feasible, inoculation, which implies the addition of external resources and an additional step in the construction of the biobed, should provide a considerable enhancement to biobed performance. Inoculation of

Table 4. Degradation of pesticides by immobilized actinobacteria.

Microorganisms	Pesticide	System of immobilization	Observations	References
Streptomyces rochei 303	2,4,6-trichlorophenol (TCP)	Inert carriers: ceramic large-pore carrier, vermiculite, glass fibre, nylon brush and polycaproamide fibre, and ion exchange carriers: copolymers of metacrylic acid and triehyleneglycol imethacrylate were tested.	Polycaproamide fiber was chosen as the optimal carrier for immobilization. The cells immobilized degraded high concentrations of individual chlorophenols and their mixtures including pentachlorophenol. Immobilized cells have not loss of activity during 2.5 months.	Golovlena et al. (1993)
Nocardioides sp. NSP41	*p*-nitrophenol (PNP) (a major metabolite resulting from the microbial degradation of parathion or methyl parathion	Beads of calcium alginate. Evaluation in liquid culture.	A high volumetric PNP and phenol degradation rate was achieved by immobilization because of the high cell concentration. When the immobilized cells were reused in the simultaneous degradation of PNP and phenol, they did not lose their PNP- and phenol-degrading activity for 12 times in semi-continuous cultures.	Cho et al. (2000)
Rhodococcus erythropolis NI86/21	Atrazine	Alginate encapsulation. Bentonite, powered activated carbon, skimmed milk and trehalose were added to the alginate mixture. Evaluation in liquid medium.	All beads types demonstrated capacity to degrade atrazine in basal minimal nutrient buffer whilst continually realizing viable bacterial cells. Skimmed milk sustained cell viability in bead formulations. Reductions of skimmed milk result in faster rates of atrazine degradation in both liquid medium and soil.	Vancov et al. (2005)

microorganisms into biobeds has not been a frequent practice, but the few studies related to fungal inoculation that have been reported are promising. The amount of total extractable isoproturon decreased by 78% after 28 days, and >99% had disappeared after 100 days in biobeds inoculated with the white-rot fungus *Phanerochaete chrysosporium.* A decrease of 76% was observed after only 100 days in the non-inoculated control (Wirén-Lehr et al. 2001), although part of the isoproturon could have remained as non-recovered residues. Several white-rot fungi individually inoculated into sterile biobed material degraded metalaxyl, atrazine, terbuthylazine, diuron and iprodione to different extents, depending on the fungus and the compound (Bending et al. 2002). In another study, inoculation with *Anthracophyllum discolors* Sp4 immobilized in lignocelluloses material increased the degradation of pentachlorophenol in two biological systems: biobeds and fixed-bed columns (Diez and Tortella 2008).

In the original Swedish biobed, the presence of peat in the biomix (25 vol%) produces a low pH that favors the growth of lignin-degrading fungi on the straw, which results in the production of ligninolytic enzymes and the subsequent degradation of pesticides. However, in some countries, the peat is replaced by compost for economic and/or environmental reasons (Fogg et al. 2004, Vischetti et al. 2004, Coppola et al. 2007), and the pH in the system can increase to values that are not favorable for the growth of certain fungi (Rousk et al. 2009). Under these conditions, the activity of bacteria plays a more important role. Among bacteria, actinobacteria possess several characteristics that make them the best replacement for fungi in the biobeds: a) the ability to degrade lignin and to produce phenoloxidases (Godden et al. 1992, Berrocal et al. 1997); b) mycelial growth that allows efficient colonization of the biomixture and c) the production of spores under adverse conditions that can be important for survival in case of, for example, high fluctuations in moisture levels. Moreover, compared to white-rot fungi, actinobacteria can degrade lignin at high nitrogen levels and are therefore less dependent on the C/N ratio of the materials used for the biomixture.

The original biobed was designed for the Swedish climate, with an annual precipitation of approximately 500 mm. In countries with a higher level of precipitation, the water content in the biomixture can increase. In some cases, biobeds are used for the treatment of larger volumes of water from tank and equipment washing. When it is used for this purpose, the biobed is isolated from the ground using an impermeable liner to avoid pesticide leaching and the depth is increased to prolong the retention time of the pesticide (Castillo et al. 2008). However, the effect of the higher moisture content on fungal growth and activity, if any, is not known. In general, bacteria are more tolerant of environments with higher water content than are white-rot fungi (Krishna 2005), and under those conditions, the addition of actinobacteria is expected to enhance pesticide degradation. The creation of a tailored biobed adapted to the types of pesticides more commonly used at the farm through inoculation with different specific strains may be possible. As discussed in the previous section, actinobacteria degrade several pesticides metabolically, in addition to those that can be degraded cometabolically by phenoloxidases. In some farms, the risk of pesticide pollution is associated with the accumulation of high amounts of metals. In Italy, for example, biobeds (renamed biomass beds) are used to treat pesticide-contaminated water in vineyards, and despite the efficient retention and degradation of the pesticides, concerns have arisen about the accumulation of copper in the biomixture. One interesting solution could be inoculation with an actinobacterium such as *Amycolatopsis* sp. AB0, which is a copper-resistant strain with high copper-specific biosorption ability (Albarracín et al. 2008). If immobilized in a suitable carrier, bacterial cells could be recovered from the treated waters or the biomaterial for reusing or for final disposal of the accumulated copper. The immobilized strain could also be used as a biofilter for the pretreatment of contaminated water before collection in the biobed.

The use of actinobacteria in biobeds is an unexplored area, and there is a need for new knowledge. Currently, research on bioaugmentation of a biobed with actinobacteria is being performed in Chile. The candidate microorganism is a chlorpyrifos-degrading *Streptomyces* sp., which has shown a good viability when encapsulated in alginate and adsorbed on activated carbon (Briceño et al. 2011).

Concluding Remarks

General awareness exists of the detrimental effects of inadequate pesticide use on the environment. Natural and environmental friendly technologies such as bioremediation, are an attractive option for the decontamination of pesticide-polluted soils. Actinobacteria have the ability to degrade a wide range of compounds and possess morphological features that make them good soil colonizers. Numerous examples of immobilized actinobacteria with strong potential for their use in the field also exist.

Soil and water contamination from diffuse sources can be prevented to a large extent by good farming practices, but additional measures are required for point-source contamination. Biobeds are a feasible tool that has been proven to be effective. Because of the increasing adoption of this technique in different areas and for different purposes, adjustments in the microbial composition may be needed. Actinobacteria are an appropriate and interesting replacement for white-rot fungi under conditions not suitable for fungal growth. However, many questions remain to be answered and a considerable amount of research at the laboratory and field scale needs to be conducted to understand the dynamics and interactions in these biological systems.

Acknowledgments

The authors gratefully acknowledge the financial support of FONDECYT Postdoctoral project N° 3100118 and the "Program of Scientific International Cooperation CONYCYT/MINCYT" 2009-111.

References

Albarracín, V.E., A.L. Avila, M.J. Amoroso and C.M. Abate. 2008. Copper removal ability by *Streptomyces* strains with dissimilar growth patterns and endowed with cupric reductase activity. FEMS Microbiol. Lett. 288: 141–148.

Anisha, G.S. and P. Prema. 2008. Cell immobilization technique for the enhanced production of α-galactosidae by Streptomyces griseoloalbus. Bioresour. Technol. 99: 3325–3330.

Arias-Estévez, M., E. López-Periago, E. Martínez-Carballo, J. Simal-Gándara, J.C. Mejuto and L. García-Río. 2008. The mobility and degradation of pesticides in soils and the pollution of groundwater resources. Agr. Ecosyst. Environ. 123: 247–260.

Ball, A.S., W.B. Betts and A.J. McCarthy. 1989. Degradation of lignin-related compounds by actinomycetes. Appl. Environ. Microbiol. 55: 1642–1644.

Barcelo, D. 1991. Occurrence, handling and chromatographic determination of pesticides in the aquatic environment. Analyst. 116: 681– 689.

Bastos, A.C. and N. Magan. 2009. *Trametes versicolor*: Potential for atrazine bioremediation in calcareous clay soil, under low water availability conditions. Int. Biodeterior. Biodegradation. 63: 389–394.

Bazot, S. and T. Lebeau. 2008. Simultaneous mineralization of glyphosate and diuron by a consortium of three bacteria as free and/or immobilized-cells formulations Appl. Microbiol. Biotechnol. 77: 1351–1358.

Bending, G., D.M. Friloux and A. Walker. 2002. Degradation of contrasting pesticides by white rot fungi and its relationship with ligninolytic potential. FEMS Microbiol. Lett. 212: 59–63.

Benimeli, C.S., M.J. Amoroso, A.P. Chaile and G.R. Castro. 2003. Isolation of four aquatic streptomycetes strains capable of growth on organochlorine pesticides. Bioresour. Technol. 89: 133–138.

Benimeli, C.S., G. Castro, A. Chaile and M.J. Amoroso. 2007. Lindane uptake and degradation by aquatic Streptomyces sp. strain M7. Int. Biodeter. Biodegradation. 59: 148–155.

Benimeli, C.S., M.S. Fuentes, C.M. Abate and M.J. Amoroso. 2008. Bioremediation of lindane-contaminated soil by Streptomyces sp. M7 and its effects on *Zea mays* growth. Int. Biodeter. Biodegradation. 61: 233–239.

Bérdy, J. 2005. Bioactive Microbial Metabolites. J. Antibiot. 58: 1–26.

Berrocal, M.M., J. Rodríguez, A.S. Ball, M.I. Perez-Leblic and M.E. Arias. 1997. Solubilisation and mineralisation of [C14] lignocellulose from wheat straw by Streptomyces cyaneus CECT 3335 during growth in solid-state fermentation. Appl. Microbiol. Biotechnol. 48: 379–384.

Briceño, G., G. Palma and M.C. Diez. 2011. Immobilization of organophosphate degrading actinomycetes isolated from agricultural soils. In the Proceeding of the 3rd International Workshop. Advances in Science and Technology of Bioresources. Chile.

Briceño, G., M.S. Fuentes, M.A. Jorquera, G. Palma, M.J. Amoroso and M.C. Diez. 2012. Chlorpyrifos biodegradation and 3,5,6- trichloro-2-pyridinol production by actinobacterias isolated from soil. Int. Biodeter. Biodegradation. 73: 1–7.

Cabrera, J.A., A. Kurtz, R.A. Sikora and A. Schouten. 2010. Isolation and characterization of fenamiphos degrading bacteria. Biodegradation. 6: 1017–27.

Castillo, M.P., A. Andersson, P. Ander, J. Stenstron and L. Torstensson. 2001. Establishment of the white rot fungus Phanerochaete chrysosporium on unsterile straw of solid substrate fermentation systems intended for degradation of pesticides. World J. Microbiol. Biotechnol. 17: 627–633.

Castillo, M., N. Felis, P. Aragón, G. Cuesta and C. Sabater. 2006. Biodegradation of the herbicide diuron by Streptomycetes isolated from soil. Int. Biodeter. Biodegradation. 58: 196–202.

Castillo, M.P. and L. Tortensson. 2007. Effect of biobed composition, moisture, and temperature on the degradation of pesticides. J. Agric. Food Chem. 55: 5725–5733.

Castillo, M.P., L. Torstensson and J. Stenström. 2008. Biobeds for environmental protection from pesticide uses—a review. J. Agric. Food Chem. 56: 6206–6219.

Cassidy, M.B., H. Lee and J.T. Trevors. 1996. Environmental applications of immobilized microbial cells: a review. J. Ind. Microbiol. 16: 79–101.

Coppola, L., M.P. Castillo, E. Monaci and C. Vischetti. 2007. Adaptation of the biobed composition for chlorpyrifos degradation to southern Europe conditions. J. Agr. Food Chem. 55: 396–401.

Crecchio, C., M. Curci, M. Pizzigallo, P. Ricciuti and P. Ruggiero. 2001. Molecular approaches to investigate herbicide-induced bacterial community changes in soil microcosms. Biol. Fertil Soils. 33: 460–466.

Cuozzo, S.A., G. Rollán, C.M. Abate and M.J. Amoroso. 2009. Specific dechlorinase activity in lindane degradation by Streptomyces sp. M7. World J. Microbiol. Biotechnol. 25: 1539–1546.

Cycon, M., M. Wójcik and Z. Piotrowska-Seget. 2009. Biodegradation of the organophosphorus insecticide diazinon by Serratia sp. and Pseudomonas sp. and their use in bioremediation of contaminated soil. Chemosphere. 76: 494–501.

Chen, S., C. Edwards and S. Subler. 2001. A microcosm approach for evaluating the effects of the fungicides benomyl and captan on soil ecological process and plant growth. Appl. Soil Ecol. 18: 69–82.

Cho, Y-G., S-K. Rhee and S.T. Lee. 2000. Influence of phenol on biodegradation of p-nitrophenol by freely suspended and immobilized Nocardiodes sp. NSP41. Biodegradation. 11: 21–28.

De Schrijver, A. and R. De Mot. 1999. Degradation of pesticides by actinomycetes. Crit. Rev. Microbiol. 25:85–119.

Diez, M.C. and G.R. Tortella. 2008. Pentachlorophenol degradation in two biological systems: biobed and fixed-bed column, inoculated with the fungus Anthracophyllum discolor. The Proceeding of the ISMOM, Chile.

Diez, M.C. 2010. Biological aspects involved in the degradation of organic pollutants. J. Soil. Sci. Plant. Nutr. 10: 244–267.

El-Nahhal, Y., S. Nir, T. Polubesova, L. Margulies and B. Rubin. 1997. Organo-clay formulations of alachlor: reduced leaching and improved efficacy. Proc. Brighton Crop Prot. Conf. Weeds 1: 21–26.

Ensign, J.C. Introduction to the actinomycetes. pp. 811–815. *In*: A. Ballows, H.G. Trüper, M. Dworkin, W. Harder and K.H. Schleifer [eds.]. 1992. The Prokaryotes. A handbook on the biology of bacteria: ecophysiology, isolation, identification, application. Springer-Verlag. New York.

Esposito, E., S.M. Paulillo and G.P. Manfio. 1998. Biodegradation of the herbicide diuron in soil by indigenous actinomycetes. Chemosphere. 37: 541–548.

Fantroussi, S.E. and S.N. Agathos. 2005. Is bioaugmentation a feasible strategy for pollutant removal and site remediation? Curr. Opin. Microbiol. 8: 268–275.

Fatoki, O.S. and O.R. Awofolu. 2005. Levels of organochlorine pesticide residues in marine, surface, ground and drinking waters from the Eastern Cape Province of South Africa. J. Environ. Sci. Heal. B. 39: 101–114.

Fogg, P., A.B. Boxall and A. Walker. 2003. Degradation of pesticides in biobeds: The effect of concentration and pesticide mixtures. J. Agr. Food Chem. 51: 5344–5349.

Fogg, P., A.B. Boxall, A. Walker and A. Jukes. 2004. Degradation and leaching potential of pesticides in biobed systems. Pest. Manag. Sci. 60: 645–654.

Fuentes, M.S., C.S. Benimeli, S.A. Cuozzo and M.J. Amoroso. 2010. Isolation of pesticide-degrading actinomycetes from a contaminated site: Bacterial growth, removal and dechlorination of organochlorine pesticides. Int. Biodeter. Biodegradation. 64: 434–441.

Fuentes, M.S., J.M. Sáez, C.S. Benimeli and M.J. Amoroso. 2011. Lindane biodegradation by defined consortia of indigenous Streptomyces strains. Water Air Soil Pollut. 222: 217–231.

Gilliom, R.J. 2007. Pesticides in U.S. streams and groundwater. Environ. Sci. Technol. 41: 3408– 414.

Godden B., A. Ball, P. Helvestein, A. McCarthy and M. Penninckx. 1992. Towards elucidation of the lignin degradation pathway in actinomycetes. J. Gen. Microbiol. 138:2441–2448.

Golovlena, L.A., O.E. Zaborina and A.Y. Arinbasarona. 1993. Degradation of 2,4,6-TCP and a mixture of isomeric chlorophenols by immobilized Streptomyces rochei 303. Appl. Microbiol. Biotechnol. 38: 815–819.

Goodfellow, M. and S. Williams. 1983. Ecology of actinomycetes. Annu. Rev. Microbiol. 37: 189–216.

Goodfellow, M., R. Davenport, F.M. Stainsby and T.P. Curtis. 1996. Actinomycete diversity associated with foaming in activated sludge plants. J. Ind. Microbiol. Biot. 17: 268–280.

Goss, M.J., D. Barry and D. Rudolph. 1998. Contamination in Ontario farmstead domestic wells and its association with agriculture: results from drinking water wells. J. Contam. Hydrol. 32: 267–293.

Hildebrandt, A., M. Guillamón, S. Lacorte, R. Tauler and D. Barceló. 2008. Impact of pesticides used in agriculture and vineyards to surface and groundwater quality (North Spain). Water Res. 42: 3315–3326.

Hosowaka, R., M. Nagai, M. Morikawa and H. Okuyama. 2009. Autochthonous bioaugmentation and its possible application to oil spills. World J. Microbiol. Biotechnol. 25: 1519– 1528.

Imada, C. 2005. Enzyme inhibitors and other bioactive compounds from marine actinomycetes. A. Van Leeuv. 87: 59–63.

Iwamoto, T. and M. Nasu. 2001. Current bioremediation practice and perspective. J. Biosci. Bioeng. 92: 1–8.

Jayabarath, J., S. Asma Musfira, R. Giridhar, S. Shyam Sundar and R. Arulmurugan. 2010. Biodegradation of carbofuran pesticide by saline soil actinomycetes. Int. J. Biot. Biochem. 6: 187–192.

Juwarkar, A.A., S.K. Singh and A. Mudhoo. 2010. A comprehensive overview of elements in bioremediation. Rev. Environ. Sci. Biotechnol. 9: 215–288.

Karasali, H., A. Hourdakis and H. Anagnostopoulos. 2002. Pesticide residues in thermal mineral water in Greece. J. Environ. Sci. Heal B. 37: 465–474.

Kattimani, L., S. Amena, V. Nandareddy and P. Mujugond. 2009. Immobilization of Streptomyces gulbargensis in polyurethane foam: A promising technique for L-asparaginase production. Irian J. Biot. 7: 199–204.

Kirby, R. 2006. Actinomycetes and lignin degradation. Adv. Appl. Microbiol. 58: 125–168.

Krause, A., W.G. Hancock, R.D. Minard, A.J. Freyer and R.C. Honeycutt. 1985. Microbial transformation of the herbicide metolachlor by a soil actinomycete. J. Agric. Food Chem. 33: 584–589.

Krishna, C. 2005. Solid-State Fermentation Systems—An Overview. Crit. Rev. Biotech. 25: 1–30.

Laabs, V., W. Amelung, A. Pinto, M.J. Wantzen, C. da Silva and W. Zech. 2002. Pesticides in surface water, sediment, and rainfall of the northeastern Pantanal basin, Brazil. J. Environ. Qual. 31: 1636–1648.

Larkin, M.J., L.A. Kulakov and C. CR. Allen. 2005. Biodegradation and Rhodococcus—masters of catabolic versatility. Curr. Opin. Biotech. 16: 282–290.

Lin, Q.S., S.H. Chen, M.Y. Hu, M.R. Ul Haq, L. Yang and H. Li. 2011. Biodegradation of cypermethrin by a newly isolated actinomycetes HU-S-01 from wastewater sludge. Int. J. Environ. Sci. Tech. 8: 45–56.

Maanen, J.M.S., M.A.J. de Vaan, A.W.F. Veldstra and W.P.A. Hendrix. 2001. Pesticides and nitrate in groundwater and rainwater in the province of Limburg in the Netherlands. Environ. Monit. Assess. 72: 95–114.

Maloschik, E., A. Ernst, G. Hegedüs, B. Darvas and A. Székács. 2007. Monitoring water-polluting pesticides in Hungary. Microchem. J. 85: 88–97.

Mason, M.G., A.S. Ball, B.J. Reeder, G. Silkstone, P. Nicholls and M.T. Wilson. 2001. Extracellular heme peroxidases in actinomycetes: a case of mistaken identity. Appl. Environ. Microbiol. 67: 4512–4519.

McCarthy, A. and S. Williams. 1992. Actinomycetes as agents of biodegradation in the environment—a review. Gene. 115: 189– 192.

Monciardini, P., M. Sosio, L. Cavaletti, C. Chiocchini and S. Donadio. 2002. New PCR primers for the selective amplification of 16S rDNA from different groups of actinomycetes. FEMS Microbiol. Ecol. 42: 419–429.

Mrozik, A. and Z. Piotrowska-Seget. 2010. Bioaugmentation as a strategy for cleaning up of soils contaminated with aromatic compounds. Microbiol. Res. 165: 363–375.

Niels, H., A. Helweg and K. Heinrichson. 2006. Leaching and degradation of 21 pesticides in full-scale model biobeds. Chemosphere. 65: 2223–2232.

Park, J.O., K.A. El-Tarabily, E.L. Ghisalberti and K. Sivasithamparam. 2002. Phatogenesis of Streptoverticillium albireticuli on Caenorhabditis elegans and its antagonism to soil-borne fungal pathogens. Lett. Appl. Microbiol. 35: 361–365.

Pasti, M.B., A.L. Pometto, M.P. Nuti and D.L. Crawford. 1990. Lignin-solubilizing ability of actinomycetes isolated from termite (Termitidae) gut. Appl. Environ. Microbiol. 56: 2213–2218.

Pimentel, D. 1995. Amounts of pesticides reaching target pests: Environmental impacts and ethics. J. Agr. Environ. Ethic. 8: 17–29.

Pizzul, L., M.P. Castillo and J. Stenström. 2006. Characterization of selected actinomycetes degrading polyaromatic hydrocarbons in liquid culture and spiked soil. World J. Microbiol. Biotechnol. 22: 745–752.

Pizzul, L., M.P. Castillo and J. Stenström. 2009. Degradation of glyphosate and other pesticides by ligninolytic enzymes. Biodegradation. 20: 751–759.

Planas, C., J. Caixach, F.J. Santos and J. Rivera. 1997. Occurrence of pesticides in Spanish surface waters. Analysis by high-resolution gas chromatography coupled to mass spectrometry. Chemosphere. 34: 2393–2406.

Plangklang, P. and A. Reungsang. 2009. Bioaugmentation of carbofuran residues in soil using Burkholderia cepacia PCL3 adsorbed on agricultural residues. Int. Biodeter. Biodegradation. 63: 515–522.

Pogell, B.M. Bioremediation of pesticides and herbicides by Streptomycetes. 38–46. *In*: M. Moo-Young, W.A. Anderson and A.M. Chakrabarty [eds.]. 1995. Environmental Biotechnology: Principles and Applications. Kluwer Academic Publisher, Netherlands.

Poopal, A.C. and R.S. Laxman. 2008. Hexavalent chromate reduction by immobilized Streptomyces griseus. Biotechnol. Lett. 30: 1005–1010.

Quek, E., Y-P. Ting and H.M. Tan. 2006. Rhodococcus sp. F92 immobilized on polyurethane foam shows ability to degrade various petroleum products. Bioresour. Technol. 97: 32–38.

Ravel, J., M.J. Amoroso, R.R. Colwell and R.T. Hill. 1998. Mercury-resistant actinomycetes from the Chesapeake Bay. FEMS Microbiol. Lett. 162: 177–184.

Raymond, J.W., T.N. Rogers, D.R. Shonnard and A.A. Kline. 2001. A review of structure-based biodegradation estimation methods. J. Hazard. Mater. 84: 189–215.

Rousk, J., P. Brookes and E. Bååth. 2009. Contrasting soil pH effects on fungal and bacterial growth suggest functional redundancy in carbon mineralization. Appl. Environ. Microbiol. 6: 1589–1596.

Rubilar, O., G. Tortella, M. Cea, F. Acevedo, M. Bustamante, L. Gianfreda and M.C. Diez. 2011. Bioremediation of a Chilean Andisol contaminated with pentacholophenol (PCP) by solid substrate cultures of white-rot fungi. Biodegradation. 22: 31–41.

Sasikala, C., S. Jiwal, P. Rout and M. Ramya. 2012. Biodegradation of chlorpyrifos by bacterial consortium isolated from agriculture soil. World J. Microbiol. Biotechnol. 28:1301–1308.

Saudagar, P.S., N.S. Shaligram and R.S. Singhal. 2008. Immobilization of Streptomyces clavuligerus on loofah sponge for the production of clavulanic acid. Bioresour. Technol. 99: 2250–2253.

Schipper, P.N.M., M.J.M. Vissers and A.M.A. van der Linden. 2008. Pesticides in groundwater and drinking water wells: overview of the situation in the Netherlands. Water Sci. Technol. 57: 1277–1286.

Sette, L., L. Mendonca Alves da Costa, A. Marsaiolo and G. Manfio. 2004. Biodegradation of alachlor by soil streptomycetes. Appl. Microbiol. Biotechnol. 64: 712–717.

Sette, L., V. de Oliveira and G. Manfio. 2005. Isolation and characterization of alachlor-degrading actinomycetes from soil. A. Van Leeuw. 87: 81–89.

Seong, C.N., Y.S. King, K.S. Baik, S.D. Lee, Y.C. Hah, S.B. Kim and M. Goodfellow. 1999. Mycolic acid-containing actinomycetes associated with activated sludge foam. J. Microbiol. 73: 66–72.

Seong, C., J. Choi and K-S. Baik. 2001. An improved selective isolation of rare actinomycetes from forest soil. J. Microbiol. 39: 17–23.

Shrivastava, S., S.F. Souza and P.D. Desai. 2008. Production of indole-3-acetic acid by immobilized actinomycete (Kitasatospora sp.) for soil application. Curr. Sci. India. 94: 1595–1604.

Singh, B., A. Walker, J. Morgan and D. Wright. 2004. Biodegradation of chlorpyrifos by Enterobacter strain B-14 and its use in bioremediation of contaminated soils. Appl. Environ. Microbiol. 70: 4855–4863.

Steger, K. 2006. Competition of microbial communities in composts. A tool to assess process development and quality of the final product. Thesis, Swedish University of Agricultural Science.

Teixeira, H., P. Proença, M. Alvarenga, M. Oliveira, E.P. Marques and D.N. Vieira. 2004. Pesticide intoxications in the Centre of Portugal: three years analysis. Forest Sci. Int. 16: 199–204.

Thompson, I.P., C.J. van der Gast, L. Ciric and A.C. Singer. 2005. Bioaugmentation for bioremediation: the challenge of strain selection. Environ. Microbiol. 7: 909–915.

Topp, E. 2001. A comparison of three atrazine-degrading bacteria for soil bioremediation. Biol. Fertil. Soils 33: 529–534.

Torstensson, L. and M.P. Castillo. 1997. Use of biobeds in Sweden to minimize environmental spillages from agricultural spraying equipment. Pesticide Outlook. 8: 24–27.

Vancov, T., K. Jury and L. Van Zwieten. 2005. Atrazine degradation by encapsulated Rhodococcus erythropolis NI86/21. J. Appl. Microbiol. 99: 767–775.

Vidali, M. 2001. Bioremediation. An overview. Pure Appl. Chem. 73: 1163–1172.

Vischetti, C., E. Capri, M. Trevisan, C. Casucci and P. Perucci. 2004. Biomassbed: a biological system to reduce pesticide point contamination at farm level. Chemosphere. 55: 823–828.

Vischetti, C., L. Coppola, E. Monaci, A. Cardinali and M.P. Castillo. 2007. Microbial impact of the pesticide chlorpyrifos in Swedish and Italian biobeds. Agron. Sustain. Dev. 27: 267–272.

Vischetti, C., E. Monaci, A. Cardinali, C. Casucci and P. Perucci. 2008. The effect of initial concentration, co-application and repeated applications on pesticide degradation in a biobed mixture. Chemosphere. 72: 1739–1743.

Vogel, T.M. and M.V. Walter. Bioaugmentation. pp. 952–959. *In*: C.J. Hurst, R.L. Crawford, G.R. Knudsen, M.J. McInerney, and L.D. Stetzenbach [eds]. 2001. Manual of Environmental Microbiology. American Society for Microbiology Press, Washington, DC, USA.

Wirén-Lehr, S., M.P. Castillo, L. Torstensson and I. Scheunert. 2001. Degradation of isoproturon in biobeds. Biol. Fert. Soils. 33: 535–540.

Wiren-Lehr, S., I. Scheunert and U. Dorfler. 2002. Mineralization of plant-incorporated residues of 14C-isoproturon in arable soils originating from different farming systems. Geoderma 105: 351–366.

Westbom, R., A. Hussen, N. Megersa, N. Negussie Retta, L. Mathiasson and E. Björklund. 2008. Assessment of organochlorine pesticide pollution in Upper Awash Ethiopian state farm soils using selective pressurized liquid extraction. Chemosphere. 72: 1181–1187.

Yang, L., Y. Zhao, B. Zhang, C. Yang, and X. Zhang. 2005. Isolation and characterization of a chlorpyrifos and 3,5,6-trichloro-2-pyridinol degrading bacterium. FEMS Microbiol. Lett. 251: 67–73.

CHAPTER 11

Lindane Removal Using *Streptomyces* Strains and Maize (*Zeas mays*) Plants

Analía Álvarez,[1,2,*] Luciano Matías Yañez[1,3,a] and María Julia Amoroso[1,3,4,b]

Introduction

Lindane (γ-hexachlorocyclohexane) is an organochlorine pesticide (OP) that has been used for crop protection worldwide and control of vector-borne diseases (Manickam et al. 2008). Lindane is a potential carcinogen and listed as a very well known pollutant by the US EPA (Walker et al. 1999). Although nowadays it's use is restricted or banned completely in most countries, residues of lindane are found all over the world in soil, water, air, plants, agricultural products, animals and humans (Piñero González et al. 2007, Kidd et al. 2008, Herrero-Mercado et al. 2010, Fuentes et al. 2011). Since toxicity associated with lindane is well-known, it is imperative to develop

[1]Planta Piloto de Procesos Industriales y Microbiológicos (PROIMI), CONICET, Av. Belgrano y Pasaje Caseros, 4000 Tucumán, Argentina.
[a]Email: lumaya12@hotmail.com
[b]Email: amoroso@proimi.org.ar
[2]Facultad de Ciencias Naturales e Instituto Miguel Lillo, Universidad Nacional de Tucumán, Miguel Lillo 205, Tucumán.
[3]Facultad de Bioquímica, Química y Farmacia, Universidad Nacional de Tucumán, Ayacucho 471, Tucumán.
[4]Universidad de Norte Santo Tomás de Aquino, 9 de Julio 165, Tucumán.
*Corresponding author: alvanalia@gmail.com

methods to remove it from the environment. Bioremediation technologies, which use plants and/or microorganisms to degrade toxic contaminants, have become the focus of interest.

Actinobacteria, the main group of bacteria presents in soils and sediments, have a great potential for bioremediating toxic compounds, since these Gram-positive microorganisms are already adapted to this habitat. In addition to their potential metabolic diversity, strains of *Streptomyces* may be well suited for soil inoculation as a consequence of their mycelial growth habit, relatively rapid rates of growth, colonization of semi-selective substrates, and their ability to be manipulated genetically (Shelton et al. 1996). However, little information is available on the ability of biotransformation of OPs by Gram-positive microorganisms, particularly actinobacteria (Lal et al. 2010).

Recent studies demonstrate enhanced dissipation and/or mineralization of OPs at the root-soil interface (Kidd et al. 2008). This rhizosphere effect is generally attributed to an increase in microbial density and/or metabolic activity due to the release of plant root exudates (REs). REs contain water soluble, insoluble, and volatile compounds including sugars, amino acids, organic acids, nucleotides, flavonones, phenolic compounds and certain enzymes (Chen et al. 2002, Kuiper et al. 2004, Chaudhry et al. 2005). A summary of potential root zone carbon sources is given in Table 1. Since REs are complex mixtures of substrates, they, not only provide a nutrient-rich habitat for pollutant degraders but can potentially enhance biodegradation in different ways. They may facilitate the co-metabolic transformation of pollutants with similar structures, induce catabolic enzymes involved in the degradation process and/or enhance the contaminant bioavailability. Some components of REs such as citric acid might increase the availability of xenobiotics in soil (Kidd et al. 2008, Gao et al. 2010). In addition, REs may induce contaminant degradation directly by root-driven extracellular enzymes (Barriada-Pereira et al. 2005, Gao et al. 2010). In this context, the

Table 1. Chemical compounds observed in plant root exudates and extracts.

Compound	**Examples**	**References**
Sugars	Glucose, xylose, mannitol, maltose, oligosaccharides	Pandya et al. (1999) Curl and Truelove (1986)
Aminoacids	Glutamate, isoleucine, methionine, tryptophan	Pandya et al. (1999)
Aromatics	Benzoate, phenols, *l*-carvone, limonene, *p*-cymene	Hegde and Fletchner (1996) Tang and Young (1982)
Organic acids	Acetate, citrate, malate, propionate	Curl and Truelove (1986)
Enzymes	Nitroreductase, dehalogenase, laccase	Schnoor et al. (1995)

phytostimulation of OP-degrading microorganisms by means of REs is therefore likely to be a successful strategy for the remediation of lindane-contaminated environments. However, limited research has evaluated this issue. Successful application of maize to the remediation of xenobiotics was reported previously (Luo et al. 2006, Gao et al. 2010) on the basis of which an active role of this plant in lindane degradation was established.

Previously, wild types of *Streptomyces* strains, which were able to remove lindane from different samples (Benimeli et al. 2003, Fuentes et al. 2010) were isolated and selected. Four of these strains showed promising results regarding their application for remediating polluted environments contaminated with OPs. The main objective of this chapter is to study the effect of maize REs on lindane removal by *Streptomyces* sp. strains already isolated in the lab.

Growth of *Streptomyces* sp. in Presence of Maize Root Exudates and Pesticide Removal

Pure cultures growing on minimal medium

During growth of plants, roots release a range of organic compounds which enhance biodegradation of xenobiotics potentially in different ways. Stimulation of bacterial growth, provided they have the corresponding metabolic abilities, is one of these ways. In the present work, four native *Streptomyces* sp. strains were cultured on minimal medium (MM) (Hopwood et al. 1985) amended with maize REs or glucose (1 g L^{-1}) as sole carbon source and spiked with lindane (1.66 mg L^{-1}). *Streptomyces* sp. A5, M7, A11 and A2 were isolated from sediments and soil samples from Argentina and contaminated with several OPs (Benimeli et al. 2003, Fuentes et al. 2010). Taxonomic identification of these strains have been confirmed by amplification and partial sequencing of their 16S rDNA genes [GenBank IDs: AY45953 (M7) (Benimeli et al. 2007), GQ867055 (A11), GQ867050 (A5) and GU085103 (A2) (Fuentes et al. 2010)]. To obtain spore suspensions, strains were plated on starch-casein medium (SC agar) at 30ºC, prior to being grown on MM at 30°C, for seven days. Microbial biomass was estimated after culture centrifugation by washing the pellets and drying to constant weight at 105ºC. Similar experiments were carried out without lindane, REs and/or glucose as controls.

REs were obtained from maize plants that were grown on nutrient solution as described by Luo et al. (2006). The solution in the culture flasks was replaced twice daily with distilled and sterilized water in the morning and fresh nutrient solution in the evening, for two weeks. The nutrient solution collected in the evening from each flask was used as the source of

REs. Exudates were lyophilized, then diluted in water and filter sterilized (0.22 μ) prior to use.

All assayed strains were able to grow on MM supplemented with REs as sole carbon source (Figs. 1A, B, C and D). Maximum biomass was reached by *Streptomyces* sp. A5 when it was cultured in the presence of REs-lindane (Fig.1A). There was no evidence of microbial growth in MM without added carbon sources (control cultures, data not shown).

This result indicates on one hand, that our *Streptomyces* sp. strains are competitive at the rhizosphere level and on the other hand, that REs represent a convenient carbon and energy source and possible nitrogen, since there was no evidence of microbial growth in MM without adding either REs or other carbon source (data not shown).

Lindane removal

Residual lindane was detected in centrifuged culture supernatants, using a gas chromatograph with electron micro-capture detector (GC/μECD). As expected, all *Streptomyces* sp. strains were able to consume lindane from the culture medium and/or degrade it because the residual pesticide values detected were less than the initial concentration. Pesticide removal, calculated as percentage of initial lindane minus percentage of residual lindane, varied from 8.55 to 55.00 percent (Figs. 1A, B, C and D). Furthermore, in assays with *Streptomyces* sp. A5 and M7, the obtained residual lindane concentration was more than half of the amount in MM supplemented with glucose-lindane, compared to MM supplemented with REs-lindane. These results suggest that REs could be a more appropriate carbon source as electron donors to support aerobic dehalogenation of the pesticide (Cuozzo et al. 2009). In this connection, Benimeli et al. (2007) found that removal of different OPs by *Streptomyces* sp. M7 was more efficient when other carbon sources were present in the medium.

It is noteworthy that the highest percentage of lindane removal obtained in the current study (55.0 percent), correspond to *Streptomyces* sp. A5 cultivated with REs-lindane (Fig. 1A), whereas lindane removal by *Streptomyces* sp. M7 and A2 was approximately half in the same culture conditions (Figs. 1B and D). The strong biological effect of *Streptomyces* sp. A5 is very relevant taking into account the feasibility of its application: *Streptomyces* strains are already adapted to the habitat, and maize plants are well adapted to acidic conditions as generated during lindane degradation (Benimeli et al. 2008). Additionally, maize might create particularly good environmental conditions for soil microorganisms (Lin et al. 2008).

No decrease in pesticide concentration was observed by *Streptomyces* sp. A11 after incubation with REs-lindane, although biomass registered was elevated (0.70 ± 0.08 g L^{-1}) (Fig. 1C). This is not surprising considering that

A

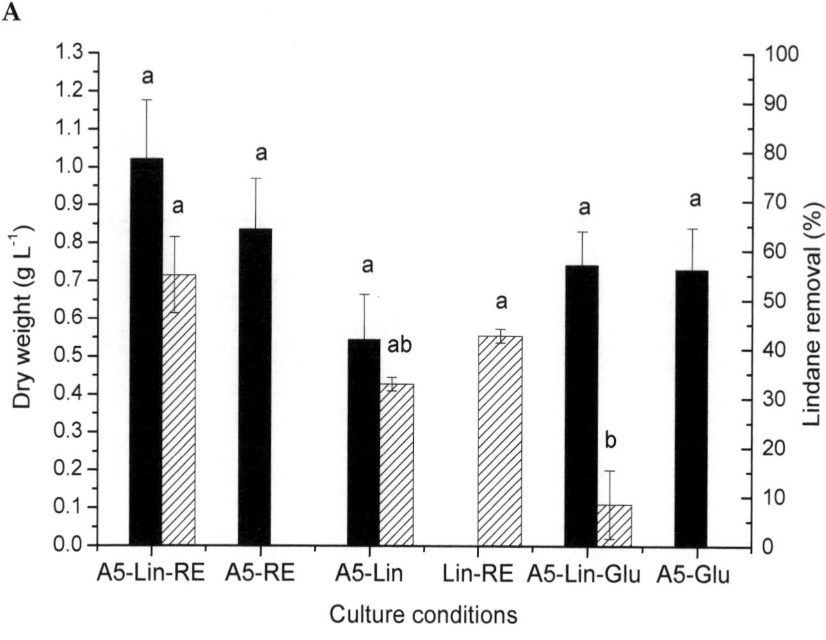

B

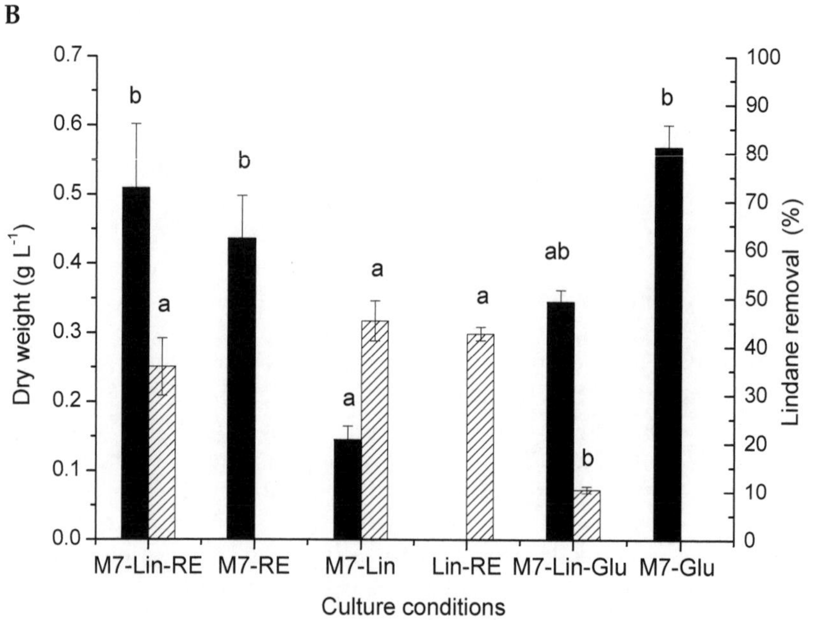

Figure 1. contd....

C

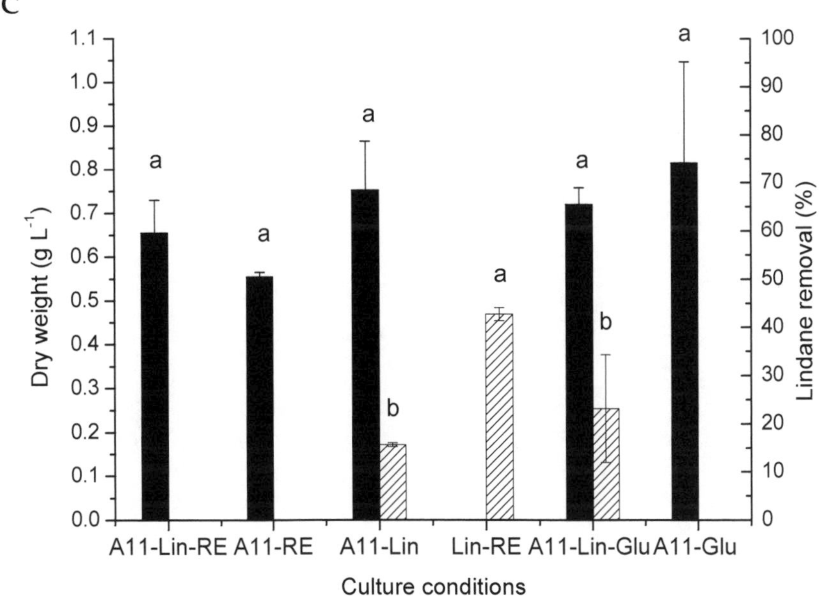

D

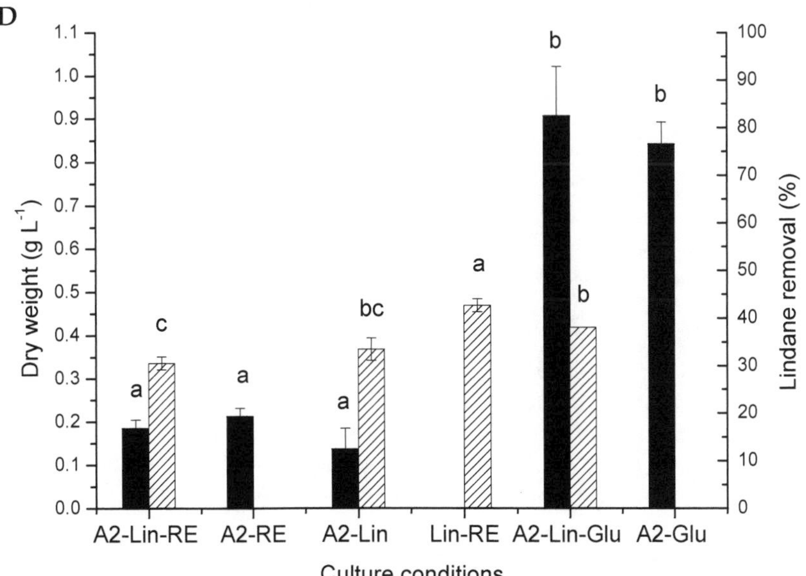

Figure 1. Comparison of growth on different carbon sources (■) and lindane removal (▧) by *Streptomyces* sp. strains cultivated with maize root exudates (RE), glucose (Glu) or lindane (Lin). (A) *Streptomyces* sp. A5. (B) *Streptomyces* sp. M7. (C) *Streptomyces* sp. A11. (D) *Streptomyces* sp. A2. Bars sharing the same letter were not significantly different ($P > 0.05$, Tukey post-test).

REs are a complex mixture of substrates (Table 1) and some of them could be repressing lindane-degrading activity of the microorganism. Similar results were obtained by Rentz et al. (2004) and Louvel et al. (2011), who studied the repression of phenanthrene-degrading activity of *Pseudomonas putida* in the presence of REs of different plants species.

Root exudates composition

The protein concentration of REs was determined according to the Bradford method (Bradford 1976) and carbohydrates were determined by the dinitrosalicylic acid (DNS) method described by Miller (1959). Regarding this issue, Personeni et al. (2007) found that glucose is a very common sugar in REs of maize growth in hydroponics. In fact, we found 193.50 ± 16.00 (μg mL^{-1}) of total proteins and 0.80 ± 0.02 (g L^{-1}) of carbohydrates in concentrate REs of five hundred maize plants. It is known that hydroponics statics cultures underestimate amounts of secretions by plants because these are absorbed by roots (Personeni et al. 2007). However, in our experiments, the nutritive growing solution was renewed daily, minimizing re-uptake. Hence, amounts of sugars and proteins detected could be considerate to reflect the amounts presents in the rhizosphere of maize.

Specific dechlorinase activity

Plants secrete enzymes that may also contribute to degradation of xenobiotics (phytodegradation) (Gao et al. 2010, Van Aken et al. 2010). For instance, Magee et al. (2008) reported recently about dechlorination of polychlorinated biphenyls (PCBs) by crude extract of nitrate reductase from *Medicago sativa* and a pure commercial nitrate reductase from maize.

Specific dechlorinase activity (SDA) of the REs was indirectly determined using a colorimetric assay, a modification of Phillips et al. (2001), in which phenol red sodium salt was added to the supernatant at a ratio of 1 per 10 as pH indicator. The change in color from red to orange to yellow in the presence of chlorides in the supernatant was indicative of lindane dechlorination and, therefore, a positive result. One enzymatic unit was defined as the amount of chloride ions released (micromoles) in 1 hr (EU=μmol Cl^- per hr) and the SDA was defined as EU per milligram of protein. In this study, we detect 10.80 ± 0.40 μmol Cl per hr of SDA in maize REs, which could explain the elevated lindane removal (42.60 ± 1.40 percent) in REs-lindane assay, without microorganisms (Figs. 1A, B, C and D). Similar results were obtained by Barriada-Pereira et al. (2005) who analyzed data obtained from the bulk and rhizosphere soils from *Avena sativa* and *Cytisus striatus* and concluded that both plant species tend to reduce levels of all hexachlorocyclohexane (HCH) isomers in the rhizosphere due to secreted enzymes able to dechlorinate.

Surfactant activity

Surfactants are employed in primary mechanisms for bioremediation of petroleum and other hydrocarbon pollutants from the environment (Calvo et al. 2009). These compounds are able to emulsify hydrophobic pollutants and enhance their water solubility. In this connection, REs may enhance pesticide biodegradation by increasing it's bioavailability, as reported by Calvelo-Pereira et al. (2006). The authors showed an increase in the aqueous solubility of different HCH isomers (included lindane) and a reduction in their concentration in the rhizosphere of *Avena sativa* and *Cytisus striatus* in relation to the bulk the soil. In this context, we studied the bioemulsifier activity of maize REs. The emulsification index (EI) was determined by mixing equal volumes of a hydrocarbon (kerosene) and REs and the EI was calculated as the percentage of the height of the emulsified layer (mm) divided by the total height of the liquid column (mm) (Cooper and Goldenberg 1987). Our experiment showed that REs have a bioemulsifier activity, since the EI calculated was 46.20 ± 9.80 percent (Fig. 2). This activity could contribute to increase lindane removal, making this more available

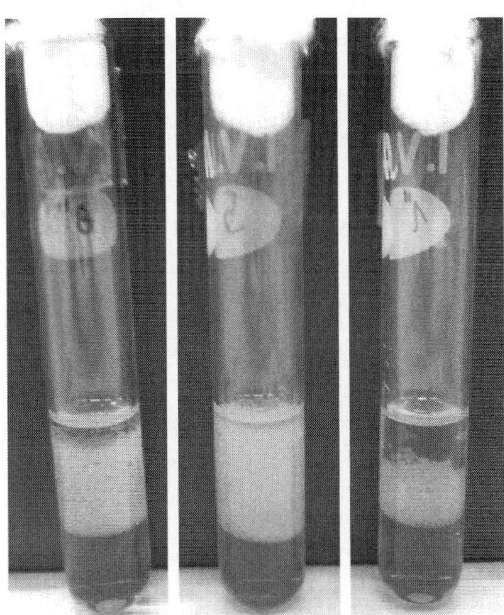

Figure 2. Emulsion of kerosene and maize root exudates. Each tube corresponds to one replicate.

for microbial uptake, as shown in A5-lindane-REs assay (Fig. 1A). Similar results were obtained by Gao et al. (2010), who observed that artificial REs promote the release of polycyclic aromatic hydrocarbons (PAHs) from soil.

On the other hand, in aqueous solution, surfactants facilitate the formation of emulsions between liquids of different polarities, and, at a given concentration, surfactant molecules aggregate to form structures such as bilayers, vesicles, or micelles (Miller 1995). Regarding this approach, we observed a micelles formation in REs-lindane assay (without microorganisms), which could also contribute to pesticide reduction (42.60 percent of removal) in addition to enzyme activity. We hypothesize that lindane binds to organic acids and other carbonaceous components of REs to form a complex matrix in which pesticide is being held and consequently it reduces in culture supernatant.

The potential applications of REs as emulsifiers in bioremediation could be highly promising, as capable of combining marked enhancing effects on the process with a complete biodegradability.

Mixed cultures growing on soil extract medium

Mixed cultures are considered potential agents in biodegradation of recalcitrant compounds because, in some cases, they have proven to be more efficient than pure cultures (Hamer 1997). This may be due to an increased number of catabolic pathways available to degrade the contaminants (Nyer et al. 2002, Rahman et al. 2002a,b, Siripattanakul et al. 2009). From an applied perspective, to import to a contaminated site, a microbial consortium rather than a pure culture for bioremediation is more advantageous. Considering that no growth inhibition was observed among any of the four studied *Streptomyces* strains (data not shown) they were cultured as mixed culture to assess their potentiating effect on lindane removal. Moreover, as it is relevant for the use in bioremediation approaches, microorganisms were cultured in soil extract medium (SEM) rather than in standard MM medium to obtain data relevant for applications in the field. SEM was prepared according to Benimeli et al. (2007). Briefly, soil (OPs free) was mixed with the same amount of water and autoclaved for 1 hr. After 24 hrs, the solution was centrifuged and the supernatant was employed as SEM after being filtered through filter paper. Sterilized SEM was amended with maize REs and spiked with lindane (1.66 mg L^{-1}).

As we expected, strains were able to grow on SEM supplemented with different carbon sources, although the biomass obtained was rather poor (maximun reached was 0.20 g $L^{-1} \pm 0.01$) (Fig. 3) compared to growth obtained on MM. In this connection, lindane removal did not exceed a 20

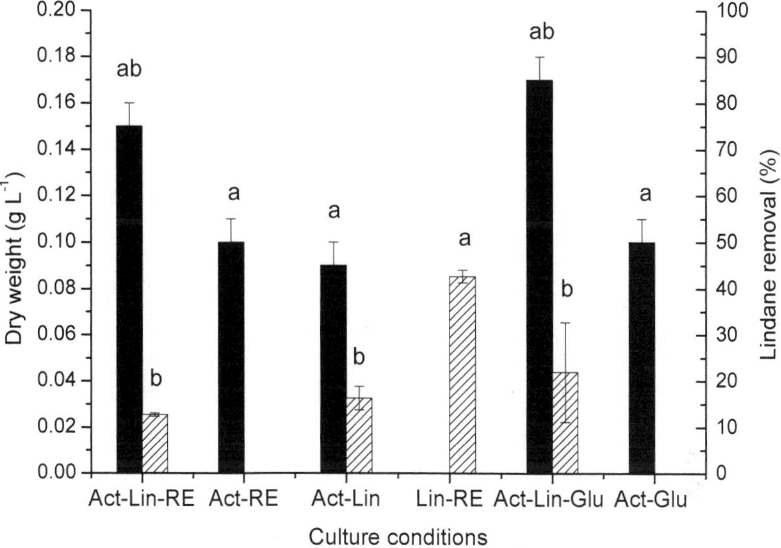

Figure 3. Comparison of growth (■) and lindane removal (▧) by a mixed culture of *Streptomyces* sp. strains cultivated on soil extract medium (SEM) amended with maize root exudates (RE), glucose (Glu) or lindane (Lin). Bars sharing the same letter were not significantly different ($P > 0.05$, Tukey post-test).

percent in the presence of microorganisms. However, considering that SEM is a nutritionally poor medium, it is noteworthy that strains were able to grow in this within a reasonable time, suggesting that microbial consortium would have conditions to be grown in the field.

Mixed cultures growing on maize plant

In situ phytoremediation involves placement of live plants in contaminated water, soil or sediment for the purpose of remediation. In this regard, the *Streptomyces* consortium was cultivated on SEM (2 g L^{-1}) spiked with lindane (1.66 g L^{-1}) where maize plants were grown individually. As shown in Fig. 4, maximum lindane removal (85.2 ± 0.2 percent) was reached when mixed cultures were grown in the presence of maize plants. However, pesticide removal was also very high without microorganisms (81.0 ± 1.4 percent) suggesting that the maize plant itself has an active role in the process of remediation. In fact, there were no statistically significant differences in lindane removal with and or without the bacteria ($P > 0.05$). Since the high hydrophobicity of lindane make their uptake and translocation within the plant unlikely ($\log K_{OW}$ 3.7–4.1; Willett et al. 1998), the high pesticide removal

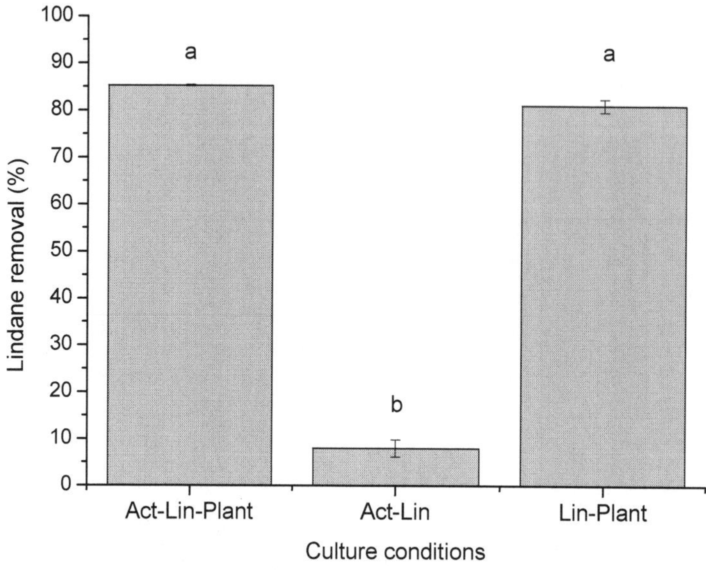

Figure 4. Lindane removal by a mixed culture of *Streptomyces* sp. strains cultivated on soil extract medium (SEM) implanted with maize. Bars sharing the same letter were not significantly different (P >0.05, Tukey post-test).

would be due by the enzymes release in REs which probably remain more actives in live plants that in extracted REs. Susarla et al. (2002) discussed this approach emphasizing that the length of time that enzymes remain active for breakdown a contaminant is one important factor to consider when this technology is applied. In addition, due to physicochemical characteristics of HCH isomers, lindane tends to sorbs to organic material (Rodríguez Garrido 2003). It is possible that lindane sticks to the plant root and therefore it is reduced. This result is very relevant taking into account that *in situ* phytoremediation either one or the other mechanism is generally the least expensive phytoremediation strategy (Susarla et al. 2002).

The results presented here show that maize plants and/or their REs influenced the removal of lindane markedly by *Streptomyces* sp. strains. Phytostimulation of lindane-degrading actinobacteria by this approach is therefore likely to be a successful strategy for the remediation of lindane-contaminated environments. Further studies evaluating the soil-plant-microbe system and its influence on lindane biodegradation are necessary so as to explore and exploit an undoubtedly huge potential.

Acknowledgements

This work was supported by Consejo de Investigaciones de la Universidad Nacional de Tucumán (CIUNT), Agencia Nacional de Promoción Científica y Tecnológica (ANPCyT), Consejo Nacional de Investigaciones Científicas y Técnicas (CONICET) and Fundación Bunge y Born.

References Cited

Barriada-Pereira, M., M.J. Gonzalez-Castro, S. Muniategui-Lorenzo, P. Lopez-Mahia, D. Prada-Rodriguez and E. Fernandez-Fernandez. 2005. Organochlorine pesticides accumulation and degradation products in vegetation samples of a contaminated area in Galicia (NW Spain). Chemosphere. 58: 1571–1578.

Benimeli, C.S., G. Castro, A. Chaile and M.J. Amoroso. 2007. Lindane uptake and degradation by aquatic *Streptomyces* sp. strain M7. Int. Biodeterior. Biodegradation. 59: 148–155.

Benimeli, C.S., M.J. Amoroso, A. Chaile and G. Castro. 2003. Isolation of four aquatic streptomycetes strains capable of growth on organochlorine pesticides. Bioresour. Technol. 89: 133–138.

Benimeli, C.S., M.S. Fuentes, C.M. Abate and M.J. Amoroso. 2008. Bioremediation of-lindane contaminated soil by *Streptomyces* sp. M7 and its effects on *Zea mays* growth. Int. Biodeterior. Biodegradation. 61: 233–239.

Bradford, M.M. 1976. A rapid and sensitive method for quantitation of microgram quantities of protein utilizing the principle of protein-dye binding. Anal. Biochem. 72: 248–254.

Calvelo Pereira, R., M. Camps-Arbestain, B. Rodríguez Garrido, C. Macías and F. Monterroso. 2006. Behaviour of α-, β-, γ-, and δ-hexachlorocyclohexane in the soil-plant system of a contaminated site. Environ. Pollut. 144: 210–217.

Calvo, C., M. Manzanera, G.A. Silva-Castro, I. Uad and J. González-López. 2009. Application of bioemulsifiers in soil oil bioremediation processes. Future prospects. Sci. Total Environ. 407: 3634–3640.

Chaudhry, Q., M. Blom-Zandstra, S. Gupta and E.J. Joner. 2005. Utilising the synergy between plants and rhizosphere microorganisms to enhance breakdown of organic pollutants in the environment. Environ. Sci. Pollut. R. 12: 34–48.

Chen, Y., Y. Guo, S. Han, C. Zou, Y. Zhou and G. Cheng. 2002. Effect of root derived organic acids on the activation of nutrients in the rhizosphere soil. J. Forest Res. 13: 115–118.

Cooper, D.G. and B.G. Goldenberg. 1987. Surface-active agents from two Bacillus species. Appl. Environ. Microbiol. 53: 224–229.

Cuozzo, S.A., G.G. Rollán, C.M. Abate and M.J. Amoroso. 2009. Specific dechlorinase activity in lindane degradation by *Streptomyces* sp. M7. World J. Microb. Biot. 25: 1539–1546.

Curl, E.A. and B. Truelove. 1986. The Rhizosphere. Springer-Verlag, Heidelberg, Berlin.

Fuentes, M.S., C.S. Benimeli, S.A. Cuozzo and M.J. Amoroso. 2010. Isolation of pesticide-degrading actinomycetes from a contaminated site: Bacterial growth, removal and dechlorination of organochlorine pesticides. Biodeterior. Biodegradation. 64: 434–441.

Fuentes, M.S., J.M. Sáez, C.S. Benimeli and M.J. Amoroso. 2011. Lindane Biodegradation by Defined Consortia of Indigenous *Streptomyces* Strains. Water Air Soil Poll. 222: 217–231.

Gao, Y., L. Ren, W. Ling, S. Gong, B. Sun and Y. Zhang. 2010. Desorption of phenanthrene and pyrene in soils by root exudates. Bioresour. Technol. 101: 1159–1165.

Hamer, G. 1997. Microbial consortia for multiple pollutant biodegradation. Pure Appl. Chem. 69: 2343–2356.

Hegde, R.S. and J.S. Fletcher. 1996. Influence of plant growth stage and season on the release of root phenolics by mulberry as related to development of phytoremediation technology. Chemosphere. 32: 2471–2479.

Herrero-Mercado, M., S.M., Waliszewski, R. Valencia-Quintana, M. Caba, F. Hernández-Chalate and E. García-Aguilar. 2010. Organochlorine pesticide levels in adipose tissue of pregnant women in Veracruz, Mexico. B. Environ. Contam. Tox. 84: 652–656.

Hopwood, D.A., M.J. Bibb, K.F. Chater, T. Kieser, C.J. Bruton, H.M. Kieser, D.J. Lydiate, C.P. Smith, J.M. Ward and H. Schrempf. 1985. Genetic Manipulation of *Streptomyces*. A Laboratory Manual. John Innes Foundation, Norwich.

Kidd, P., A. Prieto-Fernández, C. Monterroso and M.J. Acea. 2008. Rhizosphere microbial community and hexachlorocyclohexane degradative potential in contrasting plant species. Plant Soil. 302: 233–247.

Kuiper, I., E.L. Lagendijk, G.V. Bloemberg and B.J. Lugtenberg. 2004. Rhizoremediation: a beneficial plant-microbe interaction. Mol. Plant Microbe In. 17: 6–15.

Lal, R., G. Pandey, P. Sharma, K. Kumari, S. Malhotra and R. Pandey. 2010. Biochemistry of microbial degradation of hexachlorocyclohexane and prospects for bioremediation. Microbiol. Mol. Biol. R. 74: 58–80.

Lin, Q., K.-L. Shen, H.-M. Zhao and W.-H. Li. 2008. Growth response of *Zea mays* L. in pyrene-copper co-contaminated soil and the fate of pollutants. J. Hazard. Mater. 150: 515–521.

Louvel, B., A. Cébron and C. Leyval, C. 2011. Root exudates affect phenanthrene biodegradation, bacterial community and functional gene expression in sand microcosms. Biodeterior. Biodegradation. 65: 947–953.

Luo, L., S. Zhang, X. Shan and Y. Zhu. 2006. Oxalate and root exudates enhance the desorption of p,p-DDT from soils. Chemosphere. 63: 1273–1279.

Magee, K., A. Michael, H. Ulla and S.K. Dutta. 2008. Dechlorination of PCB in the presence of plant nitrate reductase. Environ. Toxicol. Phar. 25: 144–147.

Manickam, N., M. Reddy, H. Saini and R. Shanker. 2008. Isolation of hexa-chlorocyclohexane degrading *Sphingomonas* sp. by dehalogenase assay and characterization of genes involved in g-HCH degradation. J. Appl. Microbiol. 104: 952–960.

Miller, G.L. 1959. Use of dinitrosalicylic acid reagent for determination of reducing sugar. Anal. Chem. 31: 426–428.

Miller, R. 1995. Biosurfactant-facilitated Remediation of Metal-contaminated Soils. Environ. Health Perspect. 103: 59–62.

Nyer, E.K., F. Payne and S. Suthersan. 2002. Environment vs. bacteria or let's play 'name that bacteria'. Ground Water Monit. Remediat. 23: 36–45.

Pandya, S., P. Iyer, V. Gaitonde, T. Parekh and A. Desai. 1999. Chemotaxis of Rhizobiumsp. S2 towards *Cajanus cajan* root exudates and its major components. Curr. Microbiol. 38: 205–209.

Personeni, E., C. Nguyen, P. Marchal and L. Pages. 2007. Experimental evaluation of an efflux–influx model of C exudation by individual apical root segments. J. Exp. Bot. 58: 2091–2099.

Phillips, T., A. Seech, H. Lee and J. Trevors. 2001. Colorimetric assay for lindane dechlorination by bacteria. J. Microbiol. Meth. 47: 181–188.

Piñero González, M., P. Izquierdo Córser, M. Allara Cagnasso and A. García Urdaneta. 2007. Residuos de plaguicidas organoclorados en 4 tipos de aceites vegetales. Archivos Latinoamericanos de Nutrición. 57: 397–401.

Rahman, K.S.M., J. Thahira-Rahman, P. Lakshmanaperumalsamy and I.M. Banat. 2002b. Towards efficient crude oil degradation by a mixed bacterial consortium. Bioresour. Technol. 85: 257–261.

Rahman, K.S.M., I.M. Banat, J. Thahira, T. Thayumanavan and P. Lakshmanaperumalsamy. 2002a. Bioremediation of gasoline contaminated soil by a bacterial consortium amended with poultry litter, coir pith, and rhamnolipid biosurfactant. Bioresour. Technol. 81: 25–32.

Rentz, J.A., P.J.J. Alvarez and J.L. Schnoor. 2004. Repression of *Pseudomonas putida* phenanthrene-degrading activity by plant root extracts and exudates. Environ. Microbiol. 6: 574–583.

Rodríguez Garrido, B. 2003. Isómeros HCH: Retención en Suelo y Deshalogenación Reductiva en Medio Abiótico. Thesis, University of Santiago de Compostela, Espain.

Schnoor, L., L.A. Licht, S.C. McCutcheon, N.L. Wolfe and L.H. Carreira. 1995. Phvtoremediation of Oraanic and Nutrient Contaminants. Environ. Sci. Technol. 29: 318–323.

Shelton, D.R., S. Khader, J.S. Karns and B.M. Pogell. 1996. Metabolism of twelve herbicides by *Streptomyces*. Biodegradation. 7: 129–136.

Siripatanakul, S.W. Wirojanagud, J. McEvoy, T. Limpiyakorn and E. Khan. 2009. Atrazine degradation by stable mixed cultures enriched from agricultural soil and their characterization. J. Appl. Microbiol. 106: 986–992.

Susarla, S., F. Medina and S.C. McCutcheon. 2002. Phytoremediation: An ecological solution to organic chemical contamination. Ecol. Eng. 18: 647–658.

Tang, C.S. and C. Young. 1982. Collection and identification of allelopathic compounds from the undisturbed root system of Bigalta limpograss (*Hemarthia altissima*). Plant Physiology. 69: 155–160.

Van Aken, B., P. Correa and J. Schnoor. 2010. Phytoremediation of Polychlorinated Biphenyls: New Trends and Promises. Environ. Sci. Technol. 44: 2767–2776.

Walker, K., D.A. Vallero and R.G. Lewis. 1999. Factors influencing the distribution of lindane and other hexachlorocyclohexanes in the environment. Environ. Sci. Technol. 33: 4373–4378.

Willett, K.L., E.M. Ulrich and R.A. Hites. 1998. Differential toxicity and environmental fates of hexachlorocyclohexane isomers. Environ. Sci. Technol. 32: 2197–2207.

CHAPTER 12

Actinobacteria Consortia as Lindane Bioremediation Tool for Liquid and Soil Systems

María S. Fuentes,[1,*] Juliana M. Sáez,[1] Claudia S. Benimeli[1,2] and María J. Amoroso[1,2,3]

Introduction

The very same properties that once made certain chemical compounds valuable in industry, are today resulting in serious health hazards in humans and wildlife (Colborn et al. 1993, Mocarelli et al. 2008). Over the past three decades, authorities have become increasingly aware of the potential adverse health effects of chemical pollutants and studies addressing this environmental matter have indicated that persistent organic pollutants (POPs) are of particular concern. Persistent organic pollutants, such as polychlorinated biphenyls (PCBs) and organochlorine pesticides (OCPs), are synthetic compounds with great chemical stability and are commonly found in aquatic and terrestrial organisms (Jones and Voogt 1999). POPs constitute a large group of lipophilic chemicals that tend to bioaccumulate in animals and humans, thus contributing to a long-term toxic exposure and these are listed in the Stockholm Convention on POPs (UNEP 2011).

[1]Planta Piloto de Procesos Industriales y Microbiológicos (PROIMI), CONICET, Av. Belgrano y Pasaje Caseros, 4000 Tucumán, Argentina.
[2]Universidad de Norte Santo Tomás de Aquino, 9 de Julio 165, Tucumán, Argentina.
[3]Facultad de Bioquímica, Química y Farmacia, Universidad Nacional de Tucumán, Ayacucho 471, Tucumán, Argentina.
*Corresponding author: soledadfs@gmail.com

Since their introduction in the environment, POPs have been detected in air, sediment, fish, animals, and humans (blood, adipose tissue and breast milk) in industrialized countries, as well as in very isolated and pristine areas around the world (Bates et al. 2004, Patterson et al. 2009). Some of these pollutants are highly toxic and have a large variety of chronic effects, including endocrine dysfunction, mutagenesis and carcinogenesis (Tanabe 2002).

Organochlorine pesticides were put on the market in the 1940s and 50s and initially played an important role in the control of certain pests and disease vectors (WHO 1990). Unfortunately, OCPs accumulate in animal tissues and most of them are extremely stable and persist in the environment. Consequently, they can enter the food chain, not only directly via target organisms, but also indirectly via water intake and plants eaten by herbivores (Vega et al. 2007). On the other hand, OCP residues have been detected in Arctic and Antarctic regions, although they had never been used or produced there. They spread to these areas due to migration via various routes (WHO 2003).

Since the early 1970s, most countries have banned or severely restricted the production, handling, and disposal of OCPs, due to their high persistence in the environment and their proven or suspected clinical effects at doses traditionally considered safe, including reproductive disorders, teratogenicity, endocrine disruption, and carcinogenicity (Olea et al. 2001a, b, Porta et al. 2008, UNEP 2002). Nevertheless, it has been reported that these xenobiotics are still used in several countries, either legally (e.g., in vector control campaigns) or illegally (Roberts et al. 1997, UNEP 2002, Porta et al. 2008). Exposure to environmental pollutants is of particular concern in the populations of developing countries, because of inadequate legislation on this matter, the increasing presence of manufacturers, and the lack of trained personnel and equipment. In South American countries, the lack of regular scientific biomonitoring studies means that scant data is available on human exposure to persistent organic pollutants, and several populations might be especially at risk. Adipose tissue offers a good measure of cumulative internal exposure to OCPs, accounting for all routes and sources of exposure (Kohlmeier and Kohlmeier 1995, Pearce et al. 1995). Lipid-adjusted values provide a more accurate picture of the total burden of these residues in adipose tissue throughout the body (Patterson et al. 1988). Serum levels of OCPs reflect the recirculation of xenobiotics released from adipose tissue due to lipolysis and current exposures (Crinnion 2009).

Among OCPs most commonly used is lindane, the gamma isomer of hexachlorocyclohexane (γ-HCH). This compound is a halogenated organic insecticide that has been used worldwide, in spite of being banned in first world countries. Lindane has been used for crop protection and prevention of vector-borne diseases for many decades. Negative impacts of lindane on the environment and human health have been reported worldwide (Quintero et al. 2005, Camacho-Pérez et al. 2012). Lindane residues persist in the environment, undergo volatilization under tropical conditions, migrate long distances with air currents, deposit in colder regions and cause widespread contamination.

The γ-HCH residues enter the human body through the food chain and get biomagnified at each trophy level. γ-HCH is a lipophilic compound and therefore it tends to accumulate and concentrate in human body fat (Johri et al. 2000). The low aqueous solubility and chlorinated nature of lindane contribute to its persistence and resistance to degradation by microorganisms (Phillips et al. 2005). γ-HCH residues have been reported in soil, water, air, plants, agricultural products, animals, food, microbial environments and humans (Botella et al. 2004, Brilhante et al. 2006, Piñero González et al. 2007, Herrero-Mercado et al. 2010).

Since γ-HCH toxicity is well-known, it is imperative to develop methods to remove lindane from the environment. One of the strategies adopted is bioremediation using microorganisms with degrading potential. Many Gram-negative bacteria have been reported to have metabolic abilities to attack lindane. *Sphingobium japanicum* UT26, *Sphingobium francense* SpC, *Sphingobium indicum* B90A, *Sphingobium ummariense* among other strains belonging to the Sphingomonadaceae family (Dogra et al. 2004, Boltner et al. 2005, Lal et al. 2006, Mohn et al. 2006, Nagata et al. 2007, Lal et al. 2008, Singh and Lal 2009) and fungi such as *Pleurotus eryngii, Trametes hirsutus, Cyathus bulleri* and *Phanerochaete sordida* (Singh and Kuhad 2000, Quintero et al. 2008) have been reported to degrade the pesticide.

However, little information is available on the ability of biotransformation of organochlorine pesticides by Gram-positive microorganisms and particularly by actinobacteria species, the main group of bacteria present in soils and sediments (De Schrijver and De Mot 1999, Lal et al. 2010). These Gram-positive microorganisms have a great potential for biodegradation of organic and inorganic toxic compounds and several studies have demonstrated oxidation and partial dechlorination and dealkylation of aldrin, DDT and herbicides like metolachlor and atrazine by actinobacteria, particularly those belonging to the *Streptomyces* genus (Liu et al. 1991, Radosevich et al. 1995). Benimeli et al. (2003, 2006, 2007) isolated and selected wild type *Streptomyces* strains which were able to tolerate and remove lindane from culture media and soil. Cuozzo et al. (2009) detected dechlorinase activity and lindane catabolism products as a result of microbial lindane degradation by *Streptomyces* sp. M7, isolated in Tucumán, Argentina.

In addition to their potential metabolic diversity, *Streptomyces* strains may be well suited for soil inoculation as a consequence of their mycelial growth, relatively rapid growth rate, colonization of semi-selective substrates and their ability to be genetically manipulated (Shelton et al. 1996). One additional advantage is that the vegetative hyphal mass of these microorganisms can differentiate into spores that assist in spreading and persistence; the spores constitute a semi-dormant stage in the life cycle that can survive in the soil for long periods and are resistant to low nutrient concentrations and water availability (Karagouni et al. 1993).

Microbial mixed cultures have been shown to be more suitable for bioremediation of recalcitrant compounds than pure cultures. It has been suggested that growth rates and substrate utilization are frequently higher in enriched mixed cultures than those in pure cultures isolated from the mixture. The rationale is that their biodiversity can enhance environmental survival and increase the number of catabolic pathways available for contaminant biodegradation (Smith et al. 2005). For example, in atrazine degradation assays bacterial consortia appeared to be more common and more effective than individual species (Mandelbaum et al. 1993, Assaf and Turco 1994) and some consortia were reported for their metabolic cooperative actions by examining individual contribution in atrazine degradation (Smith et al. 2005, Yang et al. 2010). However, there are no reports on γ-HCH degradation by consortia of actinobacteria. Efforts have been made in our laboratory to isolate efficient indigenous actinobacteria with the ability to degrade lindane when they are grown together forming different consortia.

Selection of Actinobacteria Strains Able to Grow in the Presence of Lindane

Thirteen actinobacteria strains were cultured in Minimal Medium (MM) (Hopwood 1967) supplemented with 1.66 mg L^{-1} lindane as the only carbon and energy source, for 7 days. Eleven of them had been previously isolated from a contaminated environment in Santiago del Estero, Argentina, where about 30 tons of organochlorine pesticides were found in 1994. The isolates were identified as *Streptomyces* genera, except one of them designed as strain A10, which belonged to the *Micromonospora* genus (Fuentes et al. 2010). *Streptomyces coelicolor* A3, obtained from the German Collection of Microorganisms and Cell Cultures (DSMZ) and *Streptomyces* sp. M7, isolated at our laboratory from a pesticides contaminated site (Benimeli et al. 2003) were used as well.

All assayed actinobacteria showed growth in the presence of lindane and pesticide removal. Biomass ranged from 51.1 to 93.3 mg L^{-1} and residual lindane from 0.37 to 0.67 mg L^{-1}.

It is important to notice that these actinobacteria strains were not able to grow in MM, which contains L-asparagine as nitrogen source, without the addition of any carbon source such as glucose (Benimeli et al. 2003). On the other hand, there was no evidence of microbial growth in control cultures in MM without added lindane. Besides, all strains were able to consume lindane from the culture medium and/or degrade it because the residual pesticide values detected were less than the initial concentration. This favorable performance of the strains assayed may be due to selective pressure of the environment.

Because no linear relationship was found between the residual lindane concentration and microbial growth, we decided to examine the ratio between the two parameters in order to select the best strains for further experiments, following the methodology previously described by Benimeli et al. (2007). Microorganisms with a minimal ratio were chosen because of the low concentrations of residual lindane and high biomass production. Thus, six strains (*Streptomyces* sp. M7, *Streptomyces coelicolor* A3, *Streptomyces* sp. A2, A5, A8 and A11) were found to be the most efficient, considering their growth capacity in the presence of lindane and pesticide removal ability (Table 1).

The actinobacteria isolated from the contaminated samples showed efficient growth with lindane as the only carbon source, indicating that these microorganisms could survive in contaminated environments either due to the tolerance to the pesticides or their ability to degrade them. Microbial communities living in contaminated ecosystems tend to be dominated by organisms capable of using and/or tolerating toxic

Table 1. Selection of six actinobacteria strains individually cultured in the presence of 1.66 mg L^{-1} lindane (Adapted from Fuentes et al. 2010)

Strains	RL/B x 10^3
Streptomyces sp. A1	9.34
Streptomyces sp. A2	7.29
Streptomyces sp. A3	7.80
Streptomyces sp. A5	7.36
Streptomyces sp. A6	7.84
Streptomyces sp. A7	8.81
Streptomyces sp. A8	5.29
Micromonospora sp. A10	7.80
Streptomyces sp. A11	5.63
Streptomyces sp. A12	9.63
Streptomyces sp. A14	9.88
Streptomyces sp. M7	6.65
S. coelicolor A3	6.00

RL: residual lindane (mg L^{-1}); B: biomass (dry weight, mg L^{-1})

S. coelicolor: Streptomyces coelicolor

compounds (MacNaughton et al. 1999). Any environmental or nutritional condition that reduces the likelihood of growth or cell replication will result in an immediate reaction at the metabolic level and secondly at the genetic level (Hubert et al. 1999, Whiteley and Bailey 2000).

Specific Dechlorinase Activity (SDA) and Lindane Removal in Pure Cultures of *Streptomyces*

The key reaction during microbial degradation of halogenated compounds is the removal of the halogen atom, i.e., dehalogenation of the organic molecule. During this step, the halogen atom(s), which is (are) usually responsible for the toxic and xenobiotic character of the compound is (are) most commonly replaced by hydrogen or a hydroxyl group. Halogen removal reduces both recalcitrance to biodegradation and the risk of forming toxic intermediates during subsequent metabolic steps (Camacho-Pérez et al. 2012). Each lindane molecule has six chlorine atoms and hence, dechlorination is a very significant step in its degradation process.

The degradation pathway of γ-HCH has been studied extensively in *Sphingobium japonicum* UT26, identifying different types of dehalogenases, including LinA and LinB, which are involved in the first steps of degradation of this pesticide (Nagata et al. 1999, Nagata et al. 2007).

This pathway was also studied by Manickam et al. (2006) in the Gram (+) bacterium *Microbacterium* sp. ITRC1, which was able to degrade HCH isomers. The authors amplified gene sequences belonging to LinB and LinC enzymes. It is possible that this strain has a similar xenobiotic degradation pathway to the one studied by Nagata et al. (2007) in Gram (–) bacteria.

Cuozzo et al. (2009) detected the presence of the first two metabolites (γ-pentachlorocyclohexene and 1,3,4,6-tetrachloro-1,4-cyclohexadiene) produced during dechlorination of lindane in cell free extract of *Streptomyces* sp. M7 after 96 hrs of incubation, using the pesticide as only carbon source.

In order to find out whether actinobacteria have the ability to dechlorinate lindane, specific dechlorinase activity was determined in cell free extracts of pure cultures of the six selected strains (*Streptomyces* sp. A2, A5, A8, A11, M7 and *Streptomyces coelicolor* A3). Previously, these strains were cultured twice in MM supplemented with lindane (1.66 mg L^{-1}) as carbon source, for 96 hrs. Microorganisms cultivated in the presence of lindane were washed and then lysed in a French Press obtaining cell free extracts (CFE). Specific dechlorinase activity of the CFE was determined indirectly using a colorimetric assay, a modification of Phillips et al. (2001). The results obtained confirmed the presence of dechlorinase activity in all assayed strains (Table 2).

Table 2. Specific dechlorinase activity in cell free extracts of *Streptomyces* pure cultures (Adapted from Fuentes et al. 2011)

Strains	Specific dechlorinase activity (EU mg^{-1} protein)
Streptomyces sp. A2	5.88 ± 0.03
Streptomyces sp. A5	18.29 ± 0.03
Streptomyces sp. A8	10.13 ± 0.01
Streptomyces sp. A11	8.15 ± 0.03
Streptomyces sp. M7	16.06 ± 0.05
S. coelicolor A3	5.14 ± 0.03

Enzimatic unit (EU) was defined as the amount of chloride ions released (micromoles) in 1 hr (EU=μmol Cl^{-} h^{-1}).

To assay pesticide removal ability of the six selected strains, residual lindane concentration was measured in cultures supernatants by gas chromatography with electron microcapture detector (GC-μECD). Maximum removal was observed when MM was inoculated with *Streptomyces* sp. M7 (37%), whereas *Streptomyces* sp. A2 showed lowest removal (23%). No decrease in lindane concentration was observed in abiotic controls after 96 hrs of incubation (Fig. 1).

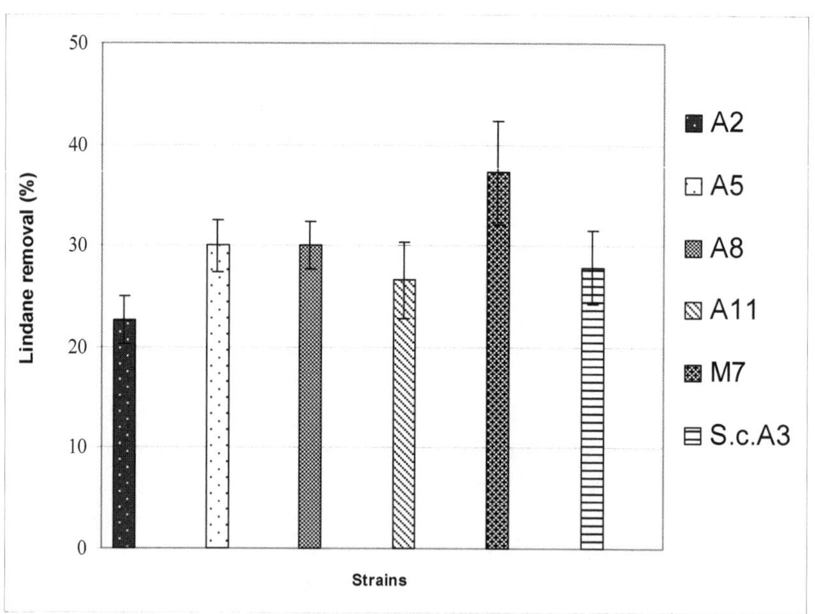

Figure 1. Lindane removal by pure cultures of *Streptomyces* sp. A2, A5, A8, A11, M7 and Streptomyces coelicolor A3 (Adapted from Fuentes et al. 2011).

Antagonism between *Streptomyces* Strains

In order to find mixed cultures with lindane removal ability, antagonism among any one of the six strains was evaluated, according a methodology described by Bell et al. (1980).

When individual strains were confronted with each other on solid MM supplemented with 1.66 mg L^{-1} lindane, no growth inhibition was observed (Fig. 2). So, there were no significant differences in responses to antagonism between the assayed strains. Although this *"in vitro"* screening for biological antagonists among *Streptomyces* strains is a simplistic approach, it should yield useful information about the behavior of these microorganisms. The next step was to grow several strains together in order to obtain a microbial consortium with enhanced dechlorinase activity to accelerate lindane degradation in environmental samples.

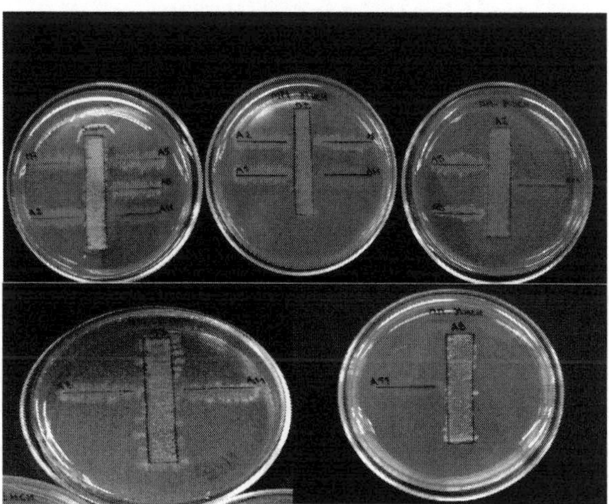

Figure 2. Compatibility test among *Streptomyces* sp. A2, A5, A8, A11, M7 and *Streptomyces coelicolor* A3 (Adapted from Fuentes et al. 2011).

Specific Dechlorinase Activity (SDA) and Lindane Removal in Mixed Cultures of *Streptomyces*

Mixed cultures are considered potential agents in biodegradation of recalcitrant compounds, because in some cases they have proven to be more efficient than pure cultures (Hamer 1997). This may be due to an increased number of catabolic pathways available to degrade the contaminants (Siripattanakul et al. 2009).

Typical aerobic pathways of γ-HCH are well known for a few selected microbial strains, however less is known about actinobacteria consortia where the possibility of synergism, antagonism, and mutualism can lead to more particular routes and more effective degradation of this pesticide.

In view of the results obtained in the antagonism assay, 57 different combinations of the studied strains were cultured to assess their enhancer effect with the aim to select a mixed culture with higher SDA and pesticide removal than any of the pure cultures.

The six *Streptomyces* strains were precultured in the presence of lindane and inoculated in MM with lindane (1.66 mg L^{-1}), with all possible mixed combinations of two, three, four, five and six strains. Mixed cultures were incubated for 96 hrs and SDA in cell free extracts was determined as described above. The results were compared with the expected activity, which was calculated as the average of the specific activity of each pure culture. The results are presented in Table 3.

Fifty two mixed cultures showed SDA values higher than the expected results, suggesting that pesticide degradation could be improved using microbial consortia. These mixed cultures could also be able to degrade rather high lindane concentrations, because more efficient degradation of the pesticide requires the use of mutually compatible strains. Similar results have been observed by Guevara and Zambrano (2006), who studied carboxymethyl cellulase activity in supernatants of mixed cultures, obtaining an increase in activity in some cases when compared with pure cultures.

The combination of four strains consisting of *Streptomyces* sp. A2, A5, M7 and *S. coelicolor* A3 presented the highest specific enzyme activity. This is reflected in the relative activity (observed/expected) of 12.62. It is interesting to emphasize that all consortia with high SDA values are combinations that included *Streptomyces* sp. A5, the strain that presented highest SDA when grown individually.

To assay pesticide removal ability of the 57 actinobacteria consortia, residual lindane concentration was measured in cultures supernatants by GC-μECD. Lindane removal by mixed cultures of two strains (Fig. 3a) was higher than that obtained with the corresponding pure cultures. Combinations of *Streptomyces* sp. A2–A8 and *Streptomyces* sp. A2-M7 showed lindane removal of 54 and 50%, respectively, whereas *Streptomyces* sp. A2 in pure culture presented the lowest removal capacity (23%). These results show the enhancement in pesticide removal between microorganisms when acting together.

A

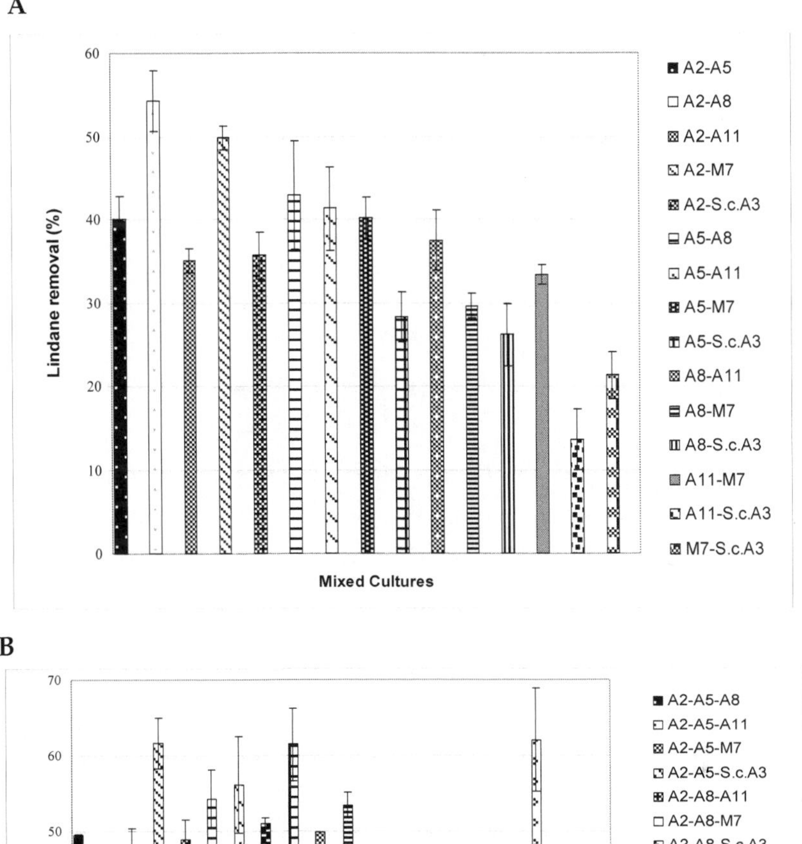

B

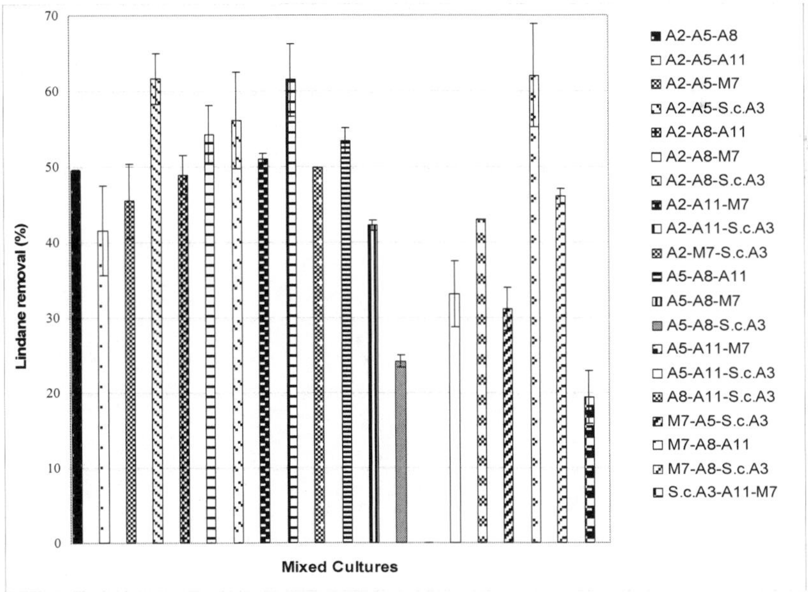

70
60
50
40
30
20
10
0
Lindane removal (%)
Mixed Cultures
A2-A5-A8
A2-A5-A11
A2-A5-M7
A2-A5-S.c.A3
A2-A8-A11
A2-A8-M7
A2-A8-S.c.A3
A2-A11-M7
A2-A11-S.c.A3
A2-M7-S.c.A3
A5-A8-A11
A5-A8-M7
A5-A8-S.c.A3
A5-A11-M7
A5-A11-S.c.A3
A8-A11-S.c.A3
M7-A5-S.c.A3
M7-A8-A11
M7-A8-S.c.A3
S.c.A3-A11-M7

Figure 3. contd....

Figure 3. contd....

C

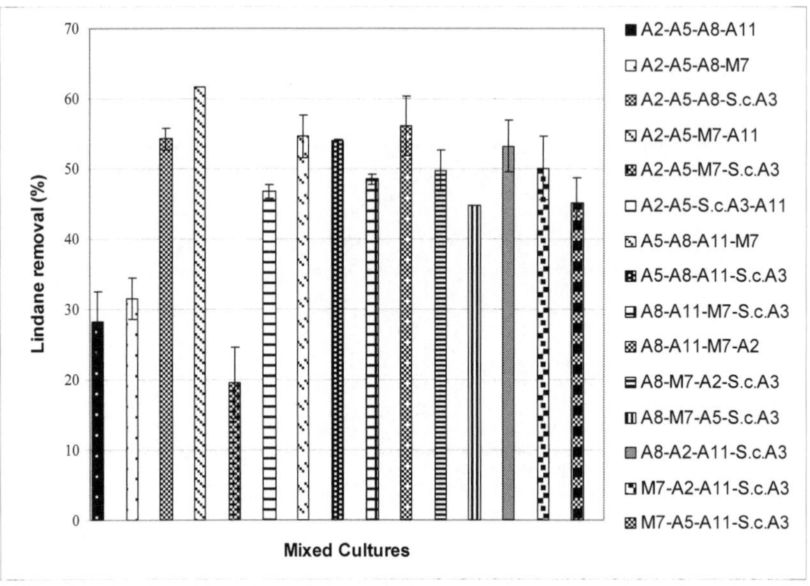

D

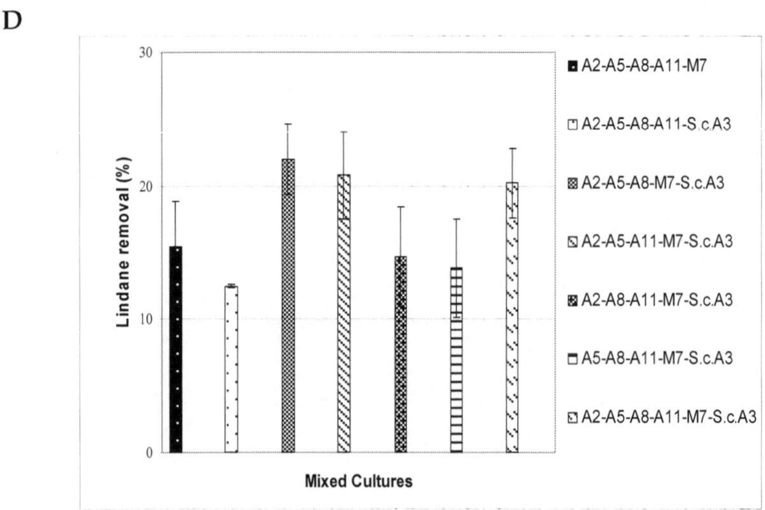

Figure 3. Biodegradation of lindane by mixed cultures of *Streptomyces*: (A) two strains; (B) three strains; (C) four strains; (D) five and six strains (Adapted from Fuentes et al. 2011).

Table 3. Specific dechlorinase activity in cell free extracts of *Streptomyces* mixed cultures (Adapted from Fuentes et al. 2011).

	Specific dechlorinase activity (EU mg^{-1} protein)		Relative activity
Mixed culture	**Observed**	**Expected**	**Obs/Exp**
	Two strains		
A2-A5	69.44 ± 0.21	12.09 ± 0.06	5.74 ± 0.05
A2-A8	32.47 ± 0.01	8.01 ± 0.04	4.05 ± 0.02
A2-A11	44.77 ± 0.26	7.02 ± 0.06	6.38 ± 0.09
A2-M7	13.95 ± 0.12	10.97 ± 0.08	1.27 ± 0.02
A2-*Sc* A3	16.42 ± 0.25	5.51 ± 0.06	2.98 ± 0.08
A5-A8	32.17 ± 0.25	14.21 ± 0.04	2.26 ± 0.02
A5-A11	34.48 ± 0.26	13.22 ± 0.06	2.61 ± 0.03
A5-M7	82.39 ± 0.24	17.18 ± 0.08	4.80 ± 0.04
A5-*Sc* A3	32.53 ± 0.25	11.72 ± 0.06	2.78 ± 0.04
A8-A11	72.43 ± 0.26	9.14 ± 0.04	7.92 ± 0.06
A8-M7	21.38 ± 0.19	13.10 ± 0.06	1.63 ± 0.02
A8-*Sc*A3	28.67 ± 0.26	7.64 ± 0.04	3.75 ± 0.05
A11-M7	12.58 ± 0.21	12.11 ± 0.08	1.04 ± 0.02
A11-*Sc* A3	40.98 ± 0.24	6.65 ± 0.05	6.16 ± 0.08
M7-*Sc* A3	34.10 ± 0.39	10.60 ± 0.08	3.22 ± 0.06
	Three strains		
A2-A5-A8	70.26 ± 0.05	11.43 ± 0.07	6.15 ± 0.04
A2-A5-A11	31.17 ± 0.09	10.77 ± 0.09	2.89 ± 0.03
A2-A5-M7	26.09 ± 0.21	13.41 ± 0.11	1.95 ± 0.03
A2-A5-*Sc* A3	24.85 ± 0.02	9.77 ± 0.09	2.54 ± 0.03
A2-A8-A11	33.58 ± 0.08	8.05 ± 0.07	4.17 ± 0.05
A2-A8-M7	10.65 ± 0.04	10.69 ± 0.09	1.00 ± 0.01
A2-A8-*Sc* A3	29.00 ± 0.07	7.05 ± 0.07	4.11 ± 0.05
A2-A11-M7	27.07 ± 0.02	10.03 ± 0.11	2.70 ± 0.03
A2-A11-*Sc* A3	6.88 ± 0.09	6.39 ± 0.09	1.08 ± 0.03
A2-M7-*Sc* A3	13.70 ± 0.06	9.03 ± 0.11	1.52 ± 0.02
A5-A8-A11	9.52 ± 0.08	12.19 ± 0.07	0.78 ± 0.01
A5-A8-M7	14.56 ± 0.08	14.83 ± 0.09	0.98 ± 0.01
A5-A8-*Sc* A3	19.35 ± 0.08	11.19 ± 0.07	1.73 ± 0.02
A5-A11-M7	24.41 ± 0.05	14.17 ± 0.11	1.72 ± 0.02
A5-A11-*Sc* A3	41.33 ± 0.03	10.53 ± 0.09	3.92 ± 0.04
A8-A11-*Sc* A3	5.49 ± 0.02	7.81 ± 0.07	0.70 ± 0.01
M7-A5-*Sc* A3	31.42 ± 0.02	13.16 ± 0.11	2.39 ± 0.02
M7-A8-A11	29.34 ± 0.05	11.45 ± 0.09	2.56 ± 0.02
M7-A8-*Sc* A3	16.44 ± 0.03	10.44 ± 0.07	1.57 ± 0.01
Sc A3-A11-M7	38.31 ± 0.01	9.78 ± 0.11	3.92 ± 0.04

Table 3. contd....

Table 3. contd....

	Specific dechlorinase activity (EU mg^{-1} protein)		Relative activity
Mixed culture	**Observed**	**Expected**	**Obs/Exp**
	Four strains		
A2-A5-A8-A11	100.37 ± 0.03	10.61 ± 0.10	9.46 ± 0.09
A2-A5-A8-M7	75.45 ± 0.03	12.59 ± 0.12	5.99 ± 0.06
A2-A5-A8-*Sc* A3	18.14 ± 0.06	9.86 ± 0.10	1.84 ± 0.02
A2-A5-M7-A11	61.87 ± 0.07	12.10 ± 0.14	5.11 ± 0.06
A2-A5-M7-*Sc* A3	143.10 ± 0.10	11.34 ± 0.14	12.62 ± 0.06
A2-A5-*Sc* A3-A11	44.77 ± 0.03	9.37 ± 0.12	4.78 ± 0.06
A5-A8-A11-M7	22.15± 0.06	13.16 ± 0.12	1.68 ± 0.02
A5-A8-A11-*Sc* A3	22.51 ± 0.02	10.43 ± 0.10	2.16 ± 0.02
A8-A11-M7-*Sc* A3	36.33 ± 0.03	9.87 ± 0.12	3.68 ± 0.05
A8-A11-M7-A2	27.50 ± 0.11	10.06 ± 0.12	2.73 ± 0.04
A8-M7-A2-*Sc* A3	14.40 ± 0.08	9.30 ± 0.12	1.55 ± 0.03
A8-M7-A5-*Sc* A3	55.87 ± 0.01	12.41 ± 0.12	4.5 ± 0.04
A8-A2-A11-*Sc* A3	17.53 ± 0.11	7.33 ± 0.10	2.39 ± 0.05
M7-A2-A11-*Sc* A3	31.43 ± 0.11	8.81 ± 0.14	3.57 ± 0.07
M7-A5-A11-*Sc* A3	24.06 ± 0.11	11.91 ± 0.14	2.02 ± 0.03
	Five strains		
A2-A5-A8-A11-M7	22.91 ± 0.06	11.70 ± 0.15	1.96 ± 0.03
A2-A5-A8-A11-*Sc* A3	35.14 ± 0.02	9.52 ± 0.13	3.69 ± 0.05
A2-A5-A8-M7-*Sc* A3	94.79 ± 0.10	11.10 ± 0.15	8.54 ± 0.12
A2-A5-A11-M7-*Sc* A3	25.96 ± 0.11	10.70 ± 0.17	2.43 ± 0.05
A2-A8-A11-M7-*Sc* A3	45.48 ± 0.06	9.07 ± 0.15	5.01 ± 0.09
A5-A8-A11-M7-*Sc* A3	47.19 ± 0.01	11.55 ± 0.15	4.09 ± 0.05
	Six strains		
A2-A5-A8-A11-M7-*Sc* A3	62.17 ± 0.11	10.61 ± 0.18	5.86 ± 0.11

Sc A3: *Streptomyces coelicolor* A3. Obs/Exp: ratio Observed/Expected
Enzimatic unit (EU) was defined as the amount of chloride ions released (micromoles) in 1 hr (EU=μmol Cl^{-} h^{-1})

Maximum removal was more than 60% by combinations of three strains: *Streptomyces* sp. A2-A5-*S. coelicolor* A3, *Streptomyces* sp. A2-A11-*S. coelicolor* A3 and *Streptomyces* sp. M7-A8-A11 (Fig. 3b). Nevertheless, in the *Streptomyces* sp. A5-A11-M7 consortium no lindane removal was observed.

Microbial consortia of four strains (Fig. 3c) showed that in most cases lindane removal was higher than 40%, with highest degradation by *Streptomyces* sp. A2-A5-M7-A11 (61.7%).

Consortia of five and six strains showed a decrease in pesticide removal ability. Degradation ranged from 12.5 to 22%, which discards these consortia for lindane bioremediation (Fig. 3d). There was no evident diminution in the residual lindane concentrations in uninoculated culture media after 96 hrs of incubation.

Several researchers have studied biodegradation of other pesticides by mixed cultures obtaining similar results. Carrillo-Pérez et al. (2004) studied DDT degradation with a mixed culture of native strains isolated from a DDT-contaminated area, detecting that only 43% of this pesticide was used for bacterial growth after 80 hrs of incubation. Hirano et al. (2007) examined anaerobic biodegradation of chlordane and hexachlorobenzenes by mixed cultures of indigenous bacteria isolated from river sediment and they observed 22 to 33% removal of *cis*- and *trans*-chlordane; hexachlorobenzene degradation was much faster and an effectiveness of 60% removal was measured.

De Souza et al. (1998) reported that a mixed culture produced higher atrazine biodegradation than the pure atrazine-degrading culture, *Clavibacter michiganese* ATZ1. Kumar and Philip (2007) studied endosulfan-contaminated soil in a pilot-scale reactor, which was bioaugmented with a mixed bacterial culture. The authors observed that 87% of the pesticide was removed after 56 days of incubation. Later, Siripattanakul et al. (2009) studied atrazine degradation by a mixed bacterial culture, observing an efficiency of 33–51%.

The mixed bacterial cultures consisting of two, three and four *Streptomyces* strains have proven to be more efficient than the individual strains for lindane removal, which make these consortia suitable to be used in bioremediation of lindane contaminated environments.

Selection of the Most Efficient Microbial Consortium for Lindane Biodegradation

Analyzing the above results, no linear relationship was found between specific enzyme activity of mixed cultures and their respective percentages of lindane removal. Therefore, two criteria for the selection of the microbial consortium with highest capacity for lindane biodegradation were established. The first criterion was the average residual lindane concentration of the 57 mixed cultures. Mixed cultures with residual lindane concentrations higher than average (0.79 mg L^{-1}) were rejected.

The second criterion was the ratio between the residual lindane concentration and specific dechlorinase activity of the remaining mixed cultures. Mixed cultures which presented the lowest value for this ratio were chosen. According to the first criterion, 33 microbial consortia were selected for their potential to degrade the pesticide; mixed cultures with five and six strains were discarded because no improvement about lindane degradation and dechlorinase activity were obtained in these two types of mixed cultures. Probably in both cases a growth competition among the strains reduces the capacity to degrade this pesticide. More studies are needed in order to confirm these suppositions.

The mixed culture of *Streptomyces* sp. A2-A5-M7-A11 presented the lowest ratio between residual lindane concentration and specific enzyme activity (Table 4). This culture was found to be the most effective one for

Table 4. Selection of the most efficient mixed cultures in the lindane biodegradation (Adapted from Fuentes et al. 2011).

Mixed culture	**RL/SDA x 10^2**
A2-A5	1.12 ± 0.16
A2-A8	1.83 ± 0.37
A2-M7	4.68 ± 0.18
A5-A8	2.31 ± 0.27
A5-A11	2.21 ± 0.19
A5-M7	0.94 ± 0.05
A2-A5-A8	0.93 ± 0.01
A2-A5-A11	2.44 ± 0.26
A2-A5-M7	2.72 ± 0.19
A2-A5-*Sc* A3	2.00 ± 0.72
A2-A8-A11	1.98 ± 0.12
A2-A8-M7	5.58 ± 1.14
A2-A8-*Sc* A3	1.96 ± 0.28
A2-A11-M7	2.35 ± 0.04
A2-A11-*Sc* A3	7.28 ± 0.37
A2-M7-*Sc* A3	4.75 ± 0.09
A5-A8-A11	6.35 ± 0.27
A5-A8-M7	5.15 ± 0.10
A8-A11-*Sc* A3	13.47 ± 0.23
M7-A8-A11	1.68 ± 0.31
M7-A8-*Sc* A3	4.26 ± 0.07
A2-A5-A8-*Sc* A3	3.28 ± 0.12
A2-A5-M7-A11	0.8 ± 0.02
A2-A5-*Sc* A3-A11	1.55 ± 0.02
A5-A8-A11-M7	2.67 ± 0.19
A5-A8-A11-*Sc* A3	2.65 ± 0.05
A8-A11-M7-*Sc* A3	1.84 ± 0.03
A8-A11-M7-A2	2.08 ± 0.22
A8-M7-A2-*Sc* A3	4.54 ± 0.12
A8-M7-A5-*Sc* A3	1.28 ± 0.02
A8-A2-A11-*Sc* A3	3.47 ± 0.19
M7-A2-A11-*Sc* A3	2.06 ± 0.11
M7-A5-A11-*Sc* A3	2.97 ± 0.21

Sc A3: *Streptomyces coelicolor* A3. RL: Residual lindane concentration SDA: Specific dechlorinase activity

pesticide biodegradation because it showed one of the lowest lindane concentration values of the 57 mixed cultures and exhibited significant specific dechlorinase activity.

On the other hand, the *Streptomyces* sp. A5-M7 and *Streptomyces* sp. A2-A5-A8 combinations also presented very low ratios and hence they showed a high efficiency of lindane biodegradation as well. These microbial consortia and the four-strain consortium previously selected could be used for future studies.

Halogenated organic compounds constitute one of the main groups of contaminants due to their persistence in the environment. The microbial consortia assayed and selected in the current study, based on their pesticide degradation ability, could be used effectively to eliminate these toxic compounds from contaminated soils, sediments and wastewaters.

Bioremediation of Lindane Contaminated Soil

Because of the results obtained, the three mixed cultures with lowest ratios, *Streptomyces* sp. A5-M7, *Streptomyces* sp. A2-A5-M7-A11 and *Streptomyces* sp. A2-A5-A8, were later assayed for bioremediation of soil microcosms polluted with lindane (1.66 mg Kg^{-1}) at 30°C during 28 days.

Surface soil samples were taken from an experimental site northwest of San Miguel de Tucumán, Argentina, without organochlorine pesticide contamination and then were sterilized (three successive sterilizations at 100°C, 1 hr each, with 24 hrs in between). For inoculation of the samples, actinobacterias strains were precultured individually in Trypticase Soy Broth for three days. Soil pots with and without lindane were inoculated with the precultured strains belonging to the consortia at a final microbial concentration of 2.0 g kg^{-1} of soil.

All three consortia exhibited excellent bacterial growth when inoculated in lindane-polluted soil and, interestingly, no growth inhibition was observed. The fact that the original environment where the strains were isolated was highly polluted with organochlorine pesticides could be explanation for this. The duplication time of the selected cultures was higher than the control for mixed cultures with two and three strains and similar to that mixed culture with four strains (Table 5). Chromatography analysis demonstrated that the three consortia were able to remove about 32% of the pesticide after 28 days of incubation (Table 5).

In this sense, Cycoń et al. (2009) showed that a consortium consisting of *Serratia* and *Pseudomonas* strains was more efficiently degrading diazinon in sterilized soil as compared to soils inoculated with a single bacterial strain and non-sterilized soil without bacteria inoculum.

Table 5. Microbial growth and lindane removal by *Streptomyces* consortia in soil samples contaminated with lindane (1.66 mg kg^{-1} soil) (Adapted from Fuentes et al. 2011).

Microbial consortia	Lindane removal (%)				Duplication time (d)	
	Day 7	Day 14	Day 21	Day 28	With lindane	Without lindane
A5-M7	25.5 ± 0.1	27.2 ± 0.1	28.1 ± 0.1	32.6 ± 0.1	2.48 ± 0.02	1.35 ± 0.05
A2-A5-A8	24.7 ± 0.1	32.7 ± 0.1	29.9 ± 0.1	33.1 ± 0.1	2.46 ± 0.01	1.35 ± 0.01
A2-A5-A11-M7	28.2 ± 0.1	28.5 ± 0.1	31.5 ± 0.1	31.4 ± 0.1	2.29 ± 0.01	2.16 ± 0.03

Conclusion

Lindane removal improved when defined actinobacteria consortia of two, three or four strains were inoculated in liquid medium supplemented with this toxic compound compared with the corresponding individual strains. No improvement in lindane removal was obtained with mixed cultures of five or six actinobacteria. Our results reveal that microbial consortia of *Streptomyces* strains can be used effectively to degrade this xenobiotic from contaminated soils, sediments and wastewaters improving the biodegradation process in a considerable way.

It can be concluded that typical anaerobic and aerobic pathways of γ-HCH are well known for a few selected microbial strains, although less is known for aerobic consortia where the possibility of synergism, antagonism and mutualism can lead to more particular routes and more effective degradation of γ-HCH. Enzyme and genetic characterization of the molecular mechanisms involved are in their early infancy; more work is needed to elucidate them in the future.

Acknowledgements

This work was supported by Consejo de Investigaciones de la Universidad Nacional de Tucumán (CIUNT), Agencia Nacional de Promoción Científica y Tecnológica (ANPCyT) and Consejo Nacional de Investigaciones Científicas y Técnicas (CONICET).

References Cited

Assaf, N.A. and R.F. Turco. 1994. Accelerated biodegradation of atrazine by a microbial consortium is possible in culture and soil. Biodegradation. 5: 29–35.

Bates, M.N., S.J. Buckland, N. Garrett, H. Ellis, L.L. Needham, D.G. Patterson, W.E. Turner and D.G. Russell. 2004. Persistent organochlorines in the serum of the non-occupationally exposed New Zealand population. Chemosphere. 54: 1431–1443.

Bell, D.K., H.D. Wells and C.R. Markham. 1980. *In vitro* Antagonism of *Trichoderma* species Against Six Fungal Plant Pathogens. Phytopathology. 72: 379–382.

Benimeli, C.S., M.J. Amoroso, A.P. Chaile and G.R. Castro. 2003. Isolation of four aquatic streptomycetes strains capable of growth on organochlorine pesticides. Bioresour. Technol. 89: 348–357.

Benimeli, C.S., G.R. Castro, A.P. Chaile and M.J. Amoroso. 2006. Lindane removal induction by *Streptomyces* sp. M7. J. Basic Microbiol. 46: 348–357.

Benimeli, C.S., G.R. Castro, A.P. Chaile and M.J. Amoroso. 2007. Lindane uptake and degradation by aquatic *Streptomyces* sp. strain M7. Int. Biodeterior. Biodegradation. 59: 148–155.

Boltner, D., S. Moreno-Morillas and J.L. Ramos. 2005. 16S rDNA phylogeny and distribution of lin genes in novel hexachlorocyclohexane degrading *Sphingomonas* strains. Environ. Microbiol. 7: 1329–1338.

Botella, B., J. Crespo, A. Rivas, I. Cerrillo, M.F. Olea-Serrano and N. Olea. 2004. Exposure of women to organochlorine pesticides in Southern Spain. Environ. Res. 96: 34–40.

Brilhante, O.M. and R. Franco. 2006. Exposure pathways to HCH and DDT in Cidade dos Meninos and its surrounding districts of Amapa, Figueiras and Pilar, metropolitan regions of Rio de Janeiro, Brazil. Int. J. Environ. Health. Res. 16: 205–217.

Camacho-Pérez, B., E. Ríos-Leal, N. Rinderknecht-Seijas and H.M. Poggi-Varaldo. 2012. Enzymes involved in the biodegradation of hexachlorocyclohexane: A mini review. J. Environ. Manage. 95: S306–S318.

Carrillo-Pérez, E., A. Ruiz-Manríquez and H. Yeomans-Reina. 2004. Aislamiento, identificación y evaluación de un cultivo mixto de microorganismos con capacidad para degradar DDT. Revista Internacional de Contaminación Ambiental. 20: 69–75.

Colborn, T., F.S.V. Saal and A.M. Soto. 1993. Developmental effects of endocrinedisrupting chemicals in wildlife and humans. Environ. Health Perspect. 101: 378–384.

Crinnion, W.J. 2009. Chlorinated pesticides: threats to health and importance of detection. Altern. Med. Rev. 14: 347–349

Cuozzo, S.A., G.C. Rollán, C.M. Abate and M.J. Amoroso. 2009. Specific dechlorinase activity in lindane degradation by *Streptomyces* sp. M7. World J. Microbiol. Biotechnol. 25: 1539–1546.

Cycoń, M., M. Wójcik and Z. Piotrowska-Seget. 2009. Biodegradation of the organophosphorus insecticide diazinon by *Serratia* sp. and *Pseudomonas* sp. and their use in bioremediation of contaminated soil. Chemosphere. 76: 494–501

De Schrijver, A. and R. De Mot. 1999. Degradation of pesticides by actinomycetes. Crit. Rev. Microbiol. 25: 85–119.

De Souza, M.L., D. Newcombe, S. Alvey, D.E. Crowley, A. Hay, M.J. Sadowsky and L.P. Wackett. 1998. Molecular basis of a bacterial consortium: interspecies catabolism of atrazine. Appl. Environ. Microbiol. 64: 178–184.

Dogra, C., V. Raina, R. Pal, M. Suar, S. Lal, K.H. Gartemann, C. Holliger, J.R. van der Meer and R. Lal. 2004. Organization of lin genes and IS6100 among different strains of hexachlorocyclohexanedegrading *Sphingomonas paucimobilis*: evidence for horizontal gene transfer. J. Bacteriol. 186: 2225–2235.

Fuentes, M.S., C.S. Benimeli, S.A. Cuozzo and M.J. Amoroso. 2010. Isolation of pesticide-degrading actinomycetes from a contaminated site: Bacterial growth, removal and dechlorination of organochlorine pesticides. Int. Biodeterior. Biodegradation. 64: 434–441.

Fuentes, M.S., J.M. Sáez, C.S. Benimeli and M.J. Amoroso. 2011. Lindane biodegradation by defined consortia of indigenous *Streptomyces* strains. Water Air Soil Pollut. 222: 217–231.

Guevara, C. and M.M. Zambrano. 2006. Sugarcane cellulose utilization by a defined microbial consortium. FEMS Microbiol. Lett. 255: 52–58.

Hamer, G. 1997. Microbial consortia for multiple pollutant biodegradation. Pure and Applied Chemistry. 69: 2343–2356.

Herrero-Mercado, M., S.M. Waliszewski, R. Valencia-Quintana, M. Caba, F. Hernández-Chalate, E. García-Aguilar and R. Villalba. 2010. Organochlorine pesticide levels in adipose tissue of pregnant women in Veracruz, Mexico. Bull Environ. Contam. Toxicol. 84: 652–656.

Hirano, T., T. Ishida, K. Oh, and R. Sudo. 2007. Biodegradation of chlordane and hexachlorobenzenes in river sediment. Chemosphere. 67: 428–434.

Hopwood, D.A. 1967. Genetic analysis and genome structure in *Streptomyces coelicolor.* Bacteriological Reviews. 31: 373–403.

Hubert, C.C., Y. Shen and G. Voordouw. 1999. Composition of toluene-degrading microbial communities from soil at different concentrations of toluene. Appl. Environ. Microbiol. 65: 3064–3070.

Johri, A.K., M. Dua, D.M. Saxena and N. Sethunathan. 2000. Enhanced degradation of hexachlorocyclohexane isomers by *Sphingomonas paucimobilis.* Curr. Microbiol. 41: 309–311.

Jones, K.C. and P. Voogt. 1999. Persistent organic pollutants (POPs): state of the science. Environ. Pollut. 100: 209–221.

Karagouni, A.D., A.P. Vionis, P.W. Baker, and E.M.H. Wellington. 1993. The effect of soil moisture content on spore germination, mycelium development and survival of a seeded streptomycete in soil. Microbial Releases. 28: 47–51.

Kohlmeier, L. and M. Kohlmeier. 1995. Adipose tissue as a medium for epidemiologic exposure assessment. Environ. Health Perspect. 103: 99–106.

Kumar, M. and L. Philip. 2007. Biodegradation of endosulfan-contaminated soil in a pilot-scale reactor-bioaugmented with mixed bacterial culture. J. Environ. Sci. Health B. 42: 707–715.

Lal, R., M. Dadhwal, K. Kumari, P. Sharma, A. Singh, H. Kumari, S. Jit, S.K. Gupta, A. Nigam, D. Lal, M. Verma, J. Kaur, K. Bala and S. Jindal. 2008. *Pseudomonas* sp. to *Sphingobium indicum*: a journey of microbial degradation and bioremediation of hexachlorocyclohexane. Indian J. Microbiol. 48: 3–18.

Lal, R., C. Dogra, S. Malhotra, P. Sharma and R. Pal. 2006. Diversity, distribution and divergence of lin genes in hexachlorocyclohexanedegrading sphingomonads. Trends Biotechnol. 24: 121–130.

Lal, R., G. Pandey, P. Sharma, K. Kumari, S. Malhotra, R. Pandey, V. Raina, H.P. Kohler, C. Holliger, C. Jackson and J.G. Oakeshott. 2010. Biochemistry of microbial degradation of hexachlorocyclohexane and prospects for bioremediation. Microbiol. Mol. Biol. Rev. 74: 58–80.

Liu, S.U., A.J. Freyer and J.M. Bollag. 1991. Microbial dechlorination of the herbicide metolachlor. J. Agric. Food Chem. 39: 631–636.

MacNaughton, S.J., J.R. Stephen, A.D. Venosa, G.A. Davis, Y.J. Chang and D.C. White. 1999. Microbial population changes during bioremediation of an experimental oil spill. Appl. Environ. Microbiol. 65: 3566–3574.

Mandelbaum, R.T., L.P. Wackett and D.L. Allan. 1993. Mineralization of the s-triazine ring of atrazine by stable bacterial mixed cultures. Appl. Environ. Microbiol. 59: 1695–1701.

Manickam, N., M. Mau and M. Schloemann. 2006. Characterization of the novel HCH-degrading strain *Microbacterium* sp. ITRC1. Appl. Microbiol. Biotechnol. 69: 580–588.

Mocarelli, P., P.M. Gerthoux, D.G. Patterson, S. Milani, G. Limonta, M. Bertona, S. Signorini, P. Tramacere, L. Colombo, C. Crespi, P. Brambilla, C. Sarto, V. Carreri, E.J. Sampson, W.E Turner and L.L. Needham. 2008. Dioxin exposure, from infancy through puberty, produces endocrine disruption and affects human semen quality. Environ. Health Perspect. 116: 70–77.

Mohn, W.W., B. Mertens, J.D. Neufeld, W. Verstraete and V. de Lorenzo. 2006. Distribution and phylogeny of hexachlorocyclohexane degrading bacteria in soils from Spain. Environ. Microbiol. 8: 60–68.

Nagata, Y., R. Endo, M. Itro, Y. Ohtsubo and M. Tsuda. 2007. Aerobic degradation of lindane (γ-hexachlorocyclohexane) in bacteria and its biochemical and molecular basis. Appl. Microbiol. Biotechnol. 76: 741–752.

Nagata, Y., A. Futamura, K. Miyauchi and M. Takagi. 1999. Two different types of dehalogenases, Lin A and Lin B, involved in γ-hexachlorocyclohexane degradation in *Sphingomonas paucimobilis* UT26 are localized in the periplasmic space without molecular processing. J. Bacteriol. 181: 5409–5413.

Olea, N., M.F. Fernandez and P. Martin-Olmedo. 2001a. Endocrine Disrupters. The case of oestrogenic xenobiotics. Revista de Salud Ambiental. 1: 6–11.

Olea, N., M.F. Fernandez and P. Martin-Olmedo. 2001b. Endocrine Disrupters. The case of oestrogenic xenobiotics II: synthetic oestrogens. Revista de Salud Ambiental. 1: 64–72.

Patterson, Jr., D.G., L.L. Needham, J.L. Pirkle, D.W. Roberts, J. Bagby, W.A. Garrett, J.S. Andrews Jr., H. Falk, J.T. Bernert and E.J. Sampson. 1988. Correlation between serum and adipose tissue levels of 2.3.7.8-tetrachlorodibenzo- p-dioxin in 50 persons from Missouri. Arch. Environ. Contam. Toxicol. 17: 139–143

Patterson, D.G., L.Y. Wong, W.E. Turner, S.P. Caudill, E.S. Dipietro, P.C. McClure, T.P. Cash, J.D. Osterloh, J.L. Pirkle, E.J. Sampson and L.L. Needham. 2009. Levels in the US population of those persistent organic pollutants (2003–2004) included in the stockholm convention or in other long-range transboundary air pollution agreements. Environ. Sci. Technol. 43: 1211–1218.

Pearce, N., S. de Sanjose, P. Boffetta, M. Kogevinas, R. Saracci and D. Savitz. 1995. Limitations of biomarkers of exposure in cancer epidemiology. Epidemiology. 6: 190–194.

Phillips, T.M., A.G. Seech, H. Lee and J.T. Trevors. 2001. Colorimetric assay for lindane dechlorination by bacteria. J. Microbiol. Methods. 47: 181–188.

Phillips, T.M., A.G. Seech, H. Lee and J.T. Trevors. 2005. Biodegradation of hexachlorocyclohexane (HCH) by microorganisms. Biodegradation. 16: 363–392.

Piñero González, M., P. Izquierdo Córser, M. Allara Cagnasso and A. García Urdaneta. 2007. Residuos de plaguicidas organoclorados en 4 tipos de aceites vegetales. Arch. Latinoam. Nutr. 57: 397–401.

Porta, M., E. Puigdomenech, F. Ballester, J. Selva, N. Ribas-Fito, S. Llop and T. Lopez. 2008. Monitoring concentrations of persistent organic pollutants in the general population: the international experience. Environ. Int. 34: 546–561.

Quintero, J.C., M.T. Moreira, G. Feijoo and J.M. Lema. 2005. Anaerobic degradation of hexachlorocyclohexane isomers in liquid and soil slurry systems. Chemosphere. 61: 528–536.

Quintero, J.C., M.T. Moreira, G. Feijoo and J.M. Lema. 2008. Screening of white rot fungal species for their capacity to degrade lindane and other isomers of hexachlorocyclohexane (HCH). Ciencia e Investigación Agraria. 35: 159–167.

Radosevich, M., S.J. Traina, Y.L. Hao and O.H. Tuovinen. 1995. Degradation and mineralization of atrazine by a soil bacterial isolate. Appl. Environ. Microbiol. 61: 297–302.

Roberts, D.R., L.L. Laughlin, P. Hsheih and L.J. Legters. 1997. DDT, global strategies, and a malaria control crisis in South America. Emerg. Infect. Dis. 3: 295–302.

Shelton, D.R., S. Khader, J.S. Karns and B.M. Pogell. 1996. Metabolism of twelve herbicides by *Streptomyces*. Biodegradation. 7: 129–136.

Singh, A. and R. Lal. 2009. *Sphingobium ummariense* sp. nov., a novel hexachlorocyclohexane degrading bacterium, isolated from HCH contaminated soil. Int. J. Syst. Evol. Microbiol. 59: 162–166.

Singh, B.K. and R.C. Kuhad. 2000. Degradation of the insecticide lindane (γ-HCH) by white-rot fungi *Cyathus bulleri* and *Phanerochaete sordida*. Pest. Manag. Sci. 56: 142–146.

Siripattanakul, S., W. Wirojanagud, J. Mc.Evoy, T. Limpiyakorn and E. Khan. 2009. Atrazine degradation by stable mixed cultures enriched from agricultural soil and their characterization. J. Appl. Microbiol. 106: 986–992.

Smith, D., S. Alvey and D.E. Crowley. 2005. Cooperative catabolic pathways within an atrazine-degrading enrichment culture isolated from soil. FEMS Microbiol. Ecol. 53: 265–273.

Tanabe, S. 2002. Contamination and toxic effects of persistent endocrine disrupters in marine mammals and birds. Mar Pollut. Bull. 45: 69–77.

UNEP. 2002. United Nations Environment Program Regionally Based Assessment of Persistent Toxic Substances. Easternand Western South America Regional Report. /http://www.bvsde.paho.org/bvsacops/i/fulltext/ewsamer.pdfS.

UNEP. 2011. Draft revised guidance on the global monitoring plan for persistent organic pollutants, UNEP/POPS/COP.5/INF/27, United Nations Environment Programme, UNEP Chemicals Geneva, Switzerland.

Vega, F.A., E.F. Covelo and M.L. Andrade. 2007. Accidental organochlorine pesticide contamination of soil in Porriño, Spain. J. Environ. Qual. 36: 272–279.

Whiteley, A.S. and M.J. Bailey. 2000. Bacterial community structure and physiological state within an industrial phenol bioremediation system. Appl. Environ. Microbiol. 66: 2400–2407.

WHO. 1990. Public health impact of pesticides used in agriculture. WHO and UNEP, Geneva.

WHO. 2003. Health risks of persistent organic pollutants from long-range trans-boundary air pollution, Chapter 3. Hexachlorocyclohexanes. Joint WHO/Convention Task Force on the Health Aspects of Air Pollution. Geneva, Switzerland, pp. 61–85.

Yang, C., Y. Li, K. Zhang, X. Wang, C. Ma, H. Tang and P. Xu. 2010. Atrazine degradation by a simple consortium of *Klebsiella* sp. A1 and *Comamonas* sp. A2 in nitrogen enriched medium. Biodegradation. 21: 97–105.

CHAPTER 13

Chlordane Biodegradation Under Aerobic Conditions by Actinobacteria Strains

Natalia Bourguignon,[1] Sergio A. Cuozzo,[1,2,5,*] María S. Fuentes,[1,5] Claudia S. Benimeli[1,4] and María J. Amoroso[1,3,4]

Introduction

Persistent organic pollutants (POPs) are organic compounds that remain intact in the environment for long periods, favoring their wide geographical distribution. They accumulate in the fatty tissue of living organisms and their toxic effect on humans and wildlife has been proven (Hirano et al. 2007). Chlordane (C10H6Cl8, 1,2,4,5,6,7,8,8a-octachloro-2,3,3a,4,7,7a-hexahydro-4,7-methanoindene) is an organochlorine pesticide that has been used worldwide, especially during the early 1980s, on farmlands, lawns in houses and gardens and also as a termiticide for foundations in houses (Dearth and Hites 1990, Colt et al. 2009).

[1]Planta Piloto de Procesos Industriales Microbiológicos (PROIMI), CONICET, Avenida Belgrano y Pasaje Caseros, T40001MVB, Tucumán, Argentina.
[2]Facultad de Ciencias Naturales e Instituto Miguel Lillo, Universidad Nacional de Tucumán, Miguel Lillo 205, Tucumán.
[3]Facultad de Bioquímica, Química y Farmacia, Universidad Nacional de Tucumán, Ayacucho 471, Tucumán.
[4]Universidad del Norte Santo Tomás de Aquino (UNSTA), 9 de Julio 165, Tucumán.
[5]Universidad de San Pablo-Tucumán (USP-T). Av. Solano Vera y Camino a Villa Nougues, San Pablo, Tucumán.
*Corresponding author: scuozzo@proimi.org.ar

Commercially available technical-grade chlordane (CLD) is a mixture of over 140 different but related compounds; the three most common components among them are α-(cis-) chlordane, γ-(trans-) chlordane and trans-nonachlor (Dearth and Hites 1990).

Many of the chlordane components and their metabolites are ubiquitous and persistent and have a tendency for biomagnification. They have been shown to be toxic in higher animals; they are suspected to be carcinogenic and may have estrogenic activities (Chia et al. 2010, Liu et al. 2010). Technical chlordane, along with many other organochlorine pesticides, has been phased out from the market under the UNEP (United Nations Environmental Program) and has been included in the group of persistent organic pollutants (UNEP 2000). Although chlordane ranks among the "dirty dozen" priority pollutants established during the Stockholm Convention, a global treaty to protect human health and the environment that came into force in May 2004 (Wong et al. 2005), this pesticide is still detected in food and environmental samples around the world (Barber et al. 2006, Hageman et al. 2006, Yago et al. 2006, Bempah and Donkor 2010, Jia et al. 2010, Lal et al. 2010). There is great concern about the adverse effects on the ecosystem. It is not only important to monitor sites where chlordane was used and where pesticide stockpiles were managed but it is also crucial to assess the risk of chlordane remaining in the environment.

To date, little fundamental research on chlordane biodegradation has been carried out and there are only very few studies on the degradation pathways or degradation products of this compound (Yamada et al. 2008). Actinobacteria have a great potential for biodegradation of organic and inorganic toxic compounds (Ravel et al. 1998). There exist studies demonstrating how these microorganisms have been able to oxidize and partially dechlorinate and dealkylate organochlorine pesticides such as aldrin, DDT, metolachlor and atrazine (Liu et al. 1990, 1991, Radosevich et al. 1995). In our laboratory, Benimeli et al. (2003, 2006, 2007) and Fuentes et al. (2010, 2011) isolated and selected wild actinobacteria strains, which were tolerant to lindane and able to remove it from culture media and soil. Cuozzo et al. (2009) detected dechlorinase activity and lindane catabolism products as a result of microbial lindane degradation by *Streptomyces* sp. M7, isolated in Tucumán, Argentina. However, to our knowledge, there exists only one report about an actinobacterium strain (*Nocardiopsis* sp.) isolated from soil that has been able to metabolize pure cis- and trans-chlordane in a pure culture (Beeman and Matsumura 1981). The aim of this work was to evaluate whether indigenous actinobacteria strains isolated from contaminated environments in Argentina would be able to remove and degrade technical-grade chlordane from culture medium and soil samples under aerobic conditions.

Bacterial Growth and Technical-grade Chlordane Removal by *Streptomyces* Strains in Minimal Medium

Figure 1 shows the growth profiles of the six *Streptomyces* strains cultured in Minimal Medium (MM) (Hopwood et al. 1967) in the presence of 1.66 mg L^{-1} technical-grade chlordane as the sole carbon source. The microorganisms showed dissimilar behaviors; different biomass values were obtained under the same culture conditions. After one day of incubation, it was observed that the highest biomass values were achieved by *Streptomyces* sp. A5, A13 and *S. coelicolor* A3. However, *Streptomyces* sp. A13 and *S. coelicolor* A3 showed a decrease in the biomass production after 7 days incubation, which could be due to the presence of toxic metabolites produced during microbial growth.

Analysis of the growth kinetics of the six microorganisms exhibited duplication times (DT) from 6.77 to 18.78 hr, with *Streptomyces* sp. A5 showing the lowest value (Table 1).

Technical-grade Chlordane added to culture medium is a mixture of over 140 different but related compounds, and the three most common components are α-chlordane (15%), γ-chlordane (15%) and trans-nonachlor (9.7%) (Dearth and Hites 1990). Residual γ-CLD in the culture media was

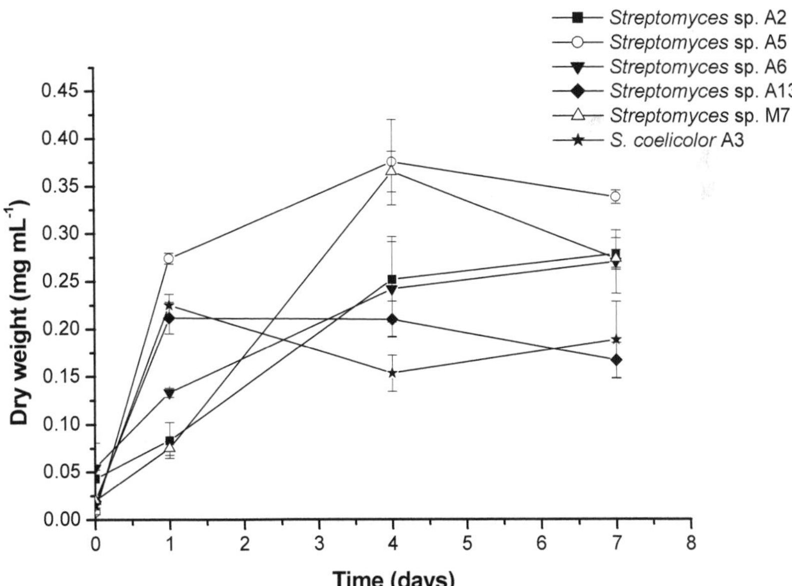

Figure 1. Time-course of bacterial growth at 30°C, initial pH 7 and at 200 rpm of the six *Streptomyces* strains assayed in the presence of 1.66 mg L^{-1} technical-grade chlordane. Error bars represent standard deviation (Cuozzo et al. 2011).

Table 1. Duplication times (DT) of *Streptomyces* in Minimal Medium supplemented with 1.66 mg L^{-1} technical-grade chlordane (Cuozzo et al. 2011).

Strains	*Streptomyces* sp. A2	*Streptomyces* sp. A5	*Streptomyces* sp. A6	*Streptomyces* sp. A13	*Streptomyces* sp. M7	*S. coelicolor* A3
DT (h)	13.06	6.77	18.78	6.98	12.58	8.71

determined by gas chromatography (GC/µECD) (Fig. 2). All *Streptomyces* strains showed high average rates of γ-CLD removal, from 97 to 99.8% after 24 hr of incubation, and *Streptomyces* sp. A5 was the most efficient strain (99.8%). From then until seven days of incubation, no significant change in γ-CLD concentration was observed indicating that the metabolism of the strains was highly adaptable and previous adaptation to the pesticide was not necessary.

The six selected *Streptomyces* strains showed release of chloride ions into supernatants. Maximum values were reached within 24 hr of incubation, indicating microbial dechlorination activity in MM and the use of CLD as sole carbon source (Fig. 3). The strains with highest Cl^- release were *Streptomyces* sp. M7 (Cuozzo et al. 2009) and *Streptomyces* sp. A6, reaching ΔA_{540} values of 0.12 approximately.

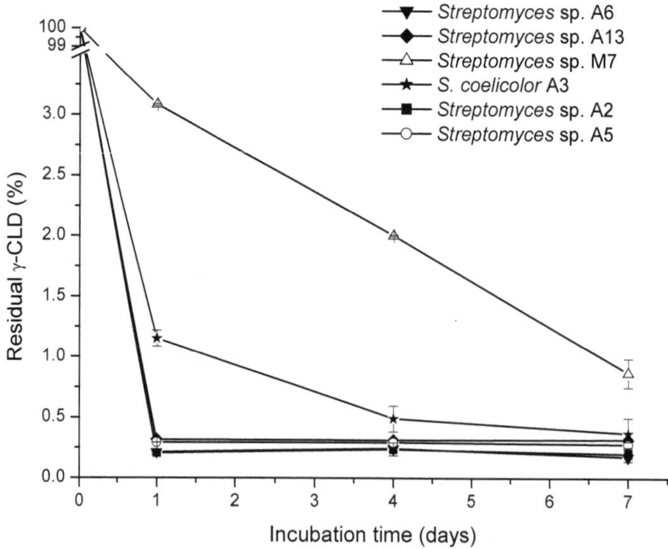

Figure 2. γ-Chlordane removal in Minimal Medium by the six *Streptomyces* strains assayed. Error bars represent standard deviation (Cuozzo et al. 2011).

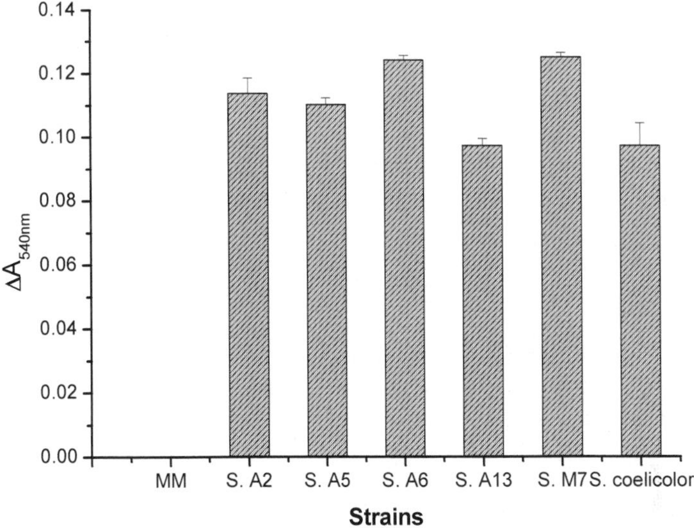

Figure 3. Chloride ion released from chlordane after 24 hr of incubation by the six *Streptomyces* strains cultured in minimal medium. Error bars represent standard deviation (Cuozzo et al. 2011).

Selection of the Most Efficient Chlordane-degrading *Streptomyces* Strain

Based on previous results, three criteria were established for selection of the most efficient strain for chlordane biodegradation: (i) biomass production and duplication time during microbial growth; (ii) residual chlordane concentration and release of chloride ions (measured as ΔA_{540}) in culture supernatants; (iii) comparison of growth kinetics in the presence of glucose (1 g L^{-1}) or chlordane (1.66 mg L^{-1}).

Streptomyces sp. A5 was selected based on the first two criteria, as it showed the highest growth potential, reaching the highest biomass of the strains assayed between one and seven days after incubation: 0.28 and 0.35 mg mL^{-1}, respectively (Fig. 1). The strain also exhibited the lowest duplication time (6.44 hr) in MM supplemented with 1.66 mg L^{-1} technical-grade chlordane (Table 1) and it removed 99.8% of γ-CLD from the medium (Fig. 2) with a high release of chloride ions (ΔA_{540} 0.11) (Fig. 3).

The biodegradation of organic chemicals generally requires acclimatization of the microbial population before detectable bioconversion rates occur. During this period they induce new enzymes, undergo genetic changes and exhaust preferential substrates (Yikmis et al. 2008). This acclimatization period usually appears as a lag period, in which little or no biodegradation is observed. However, in our case no lag period was observed for γ-CLD removal by the *Streptomyces* strains assayed (Fig. 1). This is because these microorganisms were isolated from sites contaminated with organochlorine pesticides, the acclimation period had most likely finished and the strains had probably acquired the ability to biodegrade the chemicals (Elcey and Kunhi 2010). Comparatively, Hirano et al. (2007) observed that anaerobic degradation of γ-CLD and α-CLD in river sediments occurred after an initial lag period of 4 weeks with residual pesticide concentrations of 67.0 and 88.0%, respectively.

Aerobic degradation of CLD was reported by Murray et al. (1997), but the authors observed only 9% pesticide removal after 21 days of incubation using indigenous bacteria from a banana farm.

It should be emphasized that the *Streptomyces* strains assayed in this study showed removal percentages that are similar or higher compared to those reported in other studies, and pesticide depletion from the medium was achieved in a shorter incubation period (less than 24 hrs) than those cited in the literature (Hirano et al. 2007, Baczynski et al. 2010).

It is well known that the elimination of halogens from halogenated xenobiotics is a key step in their degradation because the carbon-halogen bond is relatively stable (Fetzner and Lingens 1994). For example, Nagata et al. (2007) determined two different types of dehalogenases, which are involved in the early steps of lindane (γ-HCH) degradation by *Sphingobium japonicum* UT26: dehydrochlorinase and halidohydrolase. Because dehalogenation plays a central role in biodegradation of many chlorinated compounds, the current study examined the release of chloride ions to assess chlordane degradation by six actinobacteria strains. Manickam et al. (2008) used specific and rapid dechlorinase activity assays to screen bacteria from contaminated soils for HCH-degrading activity. Benimeli et al. (2006) reported on the release of chloride ions from lindane by a streptomycete strain. Cuozzo et al. (2009) demonstrated that synthesis of dechlorinase in *Streptomyces* sp. M7 was induced when the microorganism was grown in the presence of lindane as sole carbon source.

Influence of pH, Temperature and Concentration of Chlordane on its Degradation by *Streptomyces* sp. A5

Streptomyces sp. A5 was able to grow in the presence of CLD. The microbial growth, pesticide removal and degradation were higher when compared with other strains. No significant differences were observed in the growth kinetics of *Streptomyces* sp. A5 in the presence of the two alternative carbon sources, glucose and CLD (Fig. 4a), indicating that the pesticide was not

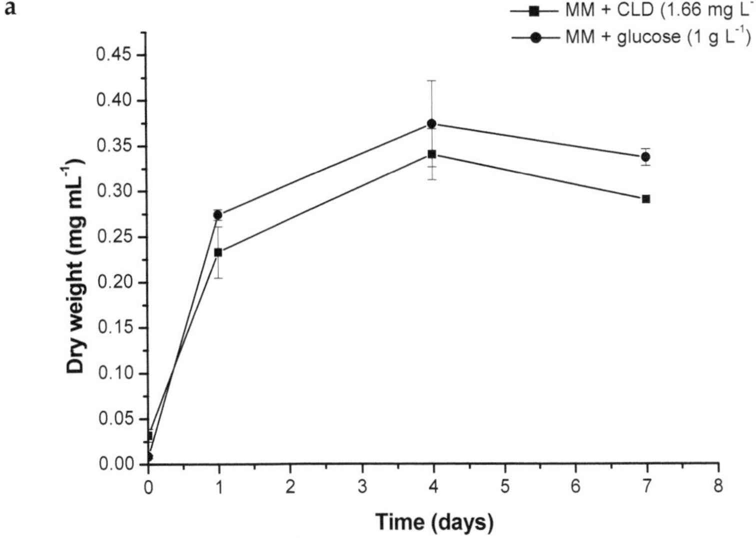

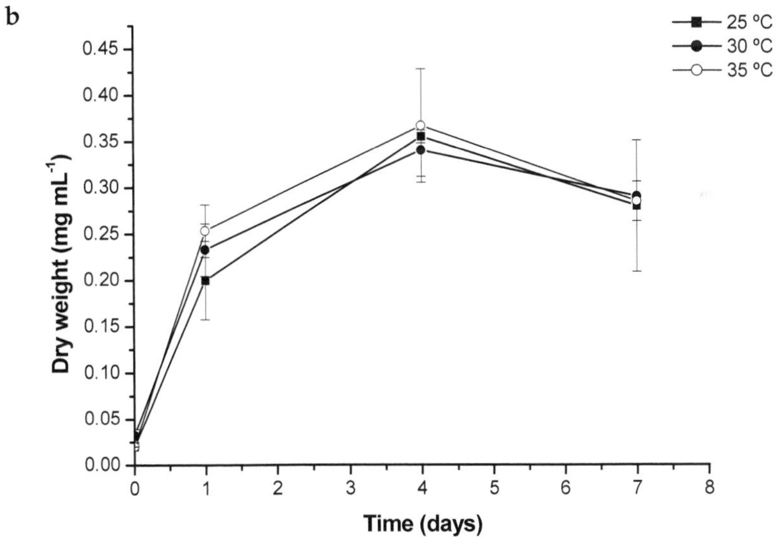

Figure 4. contd....

toxic to the cells at the assayed concentration (1.66 mg mL^{-1}) and probably the strain did not accumulate toxic metabolites that could produce growth inhibitory effects. Therefore, this strain could be highly promising for future experiments.

Figure 4. contd....

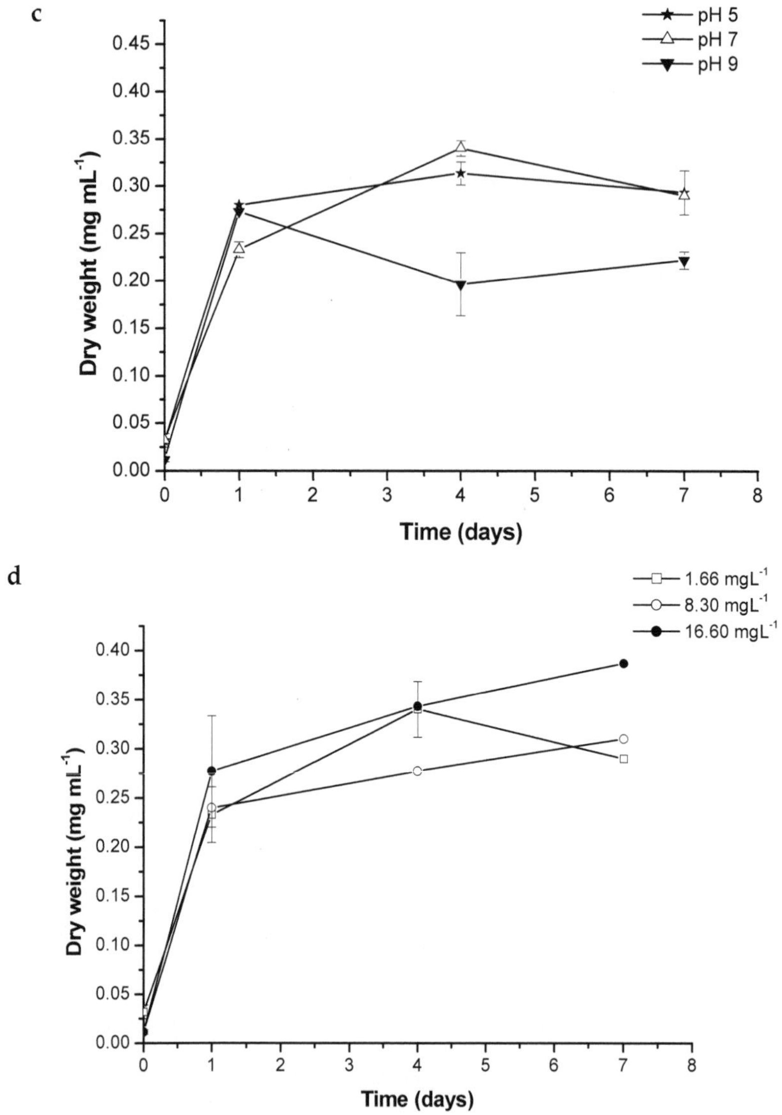

Figure 4. (a) Growth of *Streptomyces* sp. A5 in the presence of CLD or glucose as sole carbon source. (b) Effect of temperature on microbial growth of *Streptomyces* sp. A5 in MM in the presence of 1.66 mg L^{-1} CLD. (c) Effect of pH on microbial growth of *Streptomyces* sp. A5 in MM in the presence of 1.66 mg L^{-1} CLD. (d) Microbial growth of *Streptomyces* sp. A5 in MM supplemented with three different CLD concentrations: 1.66, 8.30 and 16.60 mg L^{-1}. Error bars represent standard deviation (Cuozzo et al. 2011).

The appropriate conditions for chlordane degradation by *Streptomyces* sp. A5 were examined by assaying pH, temperature and CLD concentration. At a fixed agitation speed (200 rpm), initial pH 7 and a CLD concentration of 1.66 mg L^{-1}, the biomass production of *Streptomyces* sp. A5 varied according to the temperature (Fig. 4b).

Duplication times of the strain at 25, 30 and 35°C were 7.17, 6.77 and 6.55 hr, respectively, and 35°C was considered the optimum temperature for microbial growth. However, no considerable difference was observed in chlordane removal at different temperatures; highest CLD removal was observed at 30 and 35°C with 98.8 and 99.8%, respectively. Highest ΔA_{540} value (0.11) was detected in the culture supernatant after seven d of incubation at 30°C, indicating a release of chloride ions.

Chlordane degradation by *Streptomyces* sp. A5 was also monitored in MM at different initial pH values (5, 7 and 9) at 200 rpm and at 30°C (Fig. 4c). The highest microbial growth (0.34 mg L^{-1}) was found at pH 7 and the lowest duplication time (6.36 hr) was observed at pH 9. However, no significant difference ($P < 0.05$) was observed for γ-CLD removal at different pH values: at pH 5, 7 and 9 removal was 99.2, 99.8 and 99.5%, respectively. Because of these results, pH 7 was used for further studies on chlordane degradation by *Streptomyces* sp. A5 in MM at 200 rpm and 30°C. It should be emphasized that a pH of 7 is generally considered optimal for metabolic activities of the genus *Streptomyces* (Cuozzo et al. 2009). In addition, Robinson et al. (2009) observed that an acidic environment was not favorable for aerobic bacterial dehalogenation. Chlordane degradation by *Streptomyces* sp. A5 in MM at pH 7 and at 30°C was assayed after addition of different CLD concentrations (Fig. 4d). Highest microbial biomass (0.39 mg L^{-1}) and lowest duplication time (5.25 hr) were obtained with increasing CLD concentration, but no considerable difference in pesticide removal was found at the different CLD concentrations. When the strain was grown in MM supplemented with an initial concentration of 1.66 mg L^{-1} technical-grade CLD (corresponding to 0.26 mg L^{-1} γ-CLD), maximum removal was achieved after one day of incubation, but at higher initial CLD concentrations (8.30 and 16.60 mg L^{-1}) maximum removal was observed after four days of incubation (data not shown).

At different degradation conditions, it was observed that the culture, temperature optimum was 30°C (Fig. 4b), which is comparable with previous results obtained by Benimeli et al. (2007), who studied lindane degradation by *Streptomyces* sp. M7 in medium with soil extract. Kennedy et al. (1990) observed 23% removal of γ-CLD under aerobic conditions in liquid medium at 39°C by a fungus, *Phanerochaete chrysosporium*, after 60 days of incubation. Comparatively, *Streptomyces* sp. A5 showed a greater ability to remove γ-CLD in a much shorter incubation time. As a result, the pesticide concentrations assayed were not toxic enough to inhibit microbial growth and pesticide degradation (Fig. 4d).

Bioremediation of Chlordane-contaminated Sterile Soil by *Streptomyces* sp. A5

Growth of *Streptomyces* sp. A5 in sterile soil samples contaminated with technical-grade chlordane was studied for 4 weeks. Simultaneously, CLD removal by *Streptomyces* sp. A5 was determined. Identical experiments without CLD were carried out as culture control. Growth profiles of contaminated and non-contaminated soil samples were similar (data not shown), which strengthens the hypothesis that chlordane present in soil is not toxic to *Streptomyces* sp. A5.

A decline in residual chlordane was observed after 2 weeks of incubation whereas the compound did not disappear from the noninoculated sterile control (Fig. 5).

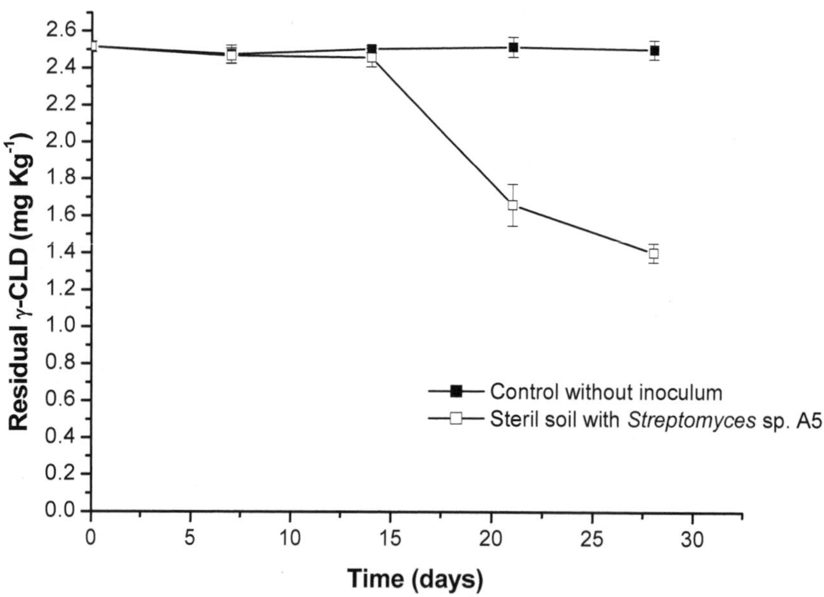

Figure 5. Removal of γ-chlordane by *Streptomyces* sp. A5 in sterile soil samples after 28 days of incubation. Error bars represent standard deviation (Cuozzo et al. 2011).

In contrast, Arisoy (1998) found that concentrations of heptachlor higher than 50 µM produced a toxic effect on the growth of the fungus *Phanerochaete chrysosporium*. Wang et al. (2006) observed that degradation of hexazinone by a mixed bacterial culture (*Pseudomonas* sp. and *Enterobacter cloacap*) clearly decreased concomitantly with increasing initial concentration of the herbicide. A hexazinone concentration of 150–200 mg L^{-1} was found to be toxic enough to inhibit its degradation totally.

It was found that the maximum pesticide removal of γ-CLD (44.0%) was observed after 4 weeks of incubation (Fig. 5). Similar results were found by Benimeli et al. (2008), who demonstrated a depletion of 56% of residual lindane in sterile soil spiked with lindane (100 µg kg^{-1}soil) and inoculated with *Streptomyces* sp. M7. Additional studies are necessary to optimize bioremediation under natural environmental conditions by stimulating bacterial growth and chlordane biodegradation ability. However, survival and activity of the inoculated strain are not always guaranteed and present another issue to be addressed in bioremediation studies.

Conclusions

Five actinobacteria strains isolated from organochlorine pesticide-contaminated soils showed chlordane-degrading ability in MM. Growth of *Streptomyces* sp. A5 in MM chlordane-supplemented and the ability of this microorganism to remove CLD and release chloride ions from the medium was higher than that of the other strains assayed. *Streptomyces* sp. A5 had the shortest duplication time and the highest microbial growth rate. Optimum conditions for the strain to degrade chlordane were pH 7, 30°C and agitation at 200 rpm. The microorganism probably uses the pesticide as carbon and energy source, because an increase in CLD concentration of 10 times still showed similar microbial growth still.

Besides, the concentrations assayed were not toxic to the bacterium, because it was able to remove CLD almost completely. A reduction of 56% in γ-chlordane was observed in soil after 28 days, which is much faster than that found in previous studies. Our results also provide the first evidence of aerobic biodegradation of CLD by regional actinobacteria strains. Although *Streptomyces* sp. A5 seems to be a potential agent for bioremediation of environments contaminated with chlordane, many studies have shown that results from laboratory studies can differ greatly from results in field studies due to a number of variables. Therefore, mechanisms of uptake, accumulation and biodegradation of chlordane by *Streptomyces* are presently being studied at our laboratory.

Acknowledgements

This research was financially supported by grants from Consejo de Investigaciones de la Universidad Nacional de Tucuman (CIUNT), Agencia Nacional de Promocion Cientifica y tecnologica (ANPCYT) and Consejo Nacional de Investigaciones Cientificas y Tecnicas (CONICET).

References Cited

Arisoy, M. 1998. Biodegradation of chlorinated organic compounds by white-rot fungi. Bull. Environ. Contam. Toxicol. 60: 872–876.

Baczynski, T.P., D. Pleissner and T. Grotenhuis. 2010. Anaerobic biodegradation of organochlorine pesticides in contaminated soil—Significance of temperature and availability. Chemosphere. 78: 22–28.

Barber, L.B., S.H. Keefe, R.C. Antweiler and H.E. Taylor. 2006. Accumulation of contaminants in fish from wastewater treatment wetlands. Environ. Sci. Technol. 40: 603–611.

Beeman, R.W. and F. Matsumura. 1981. Metabolism of *cis*- and *trans*-chlordane by a soil microorganism. J. Agric. Food Chem. 29: 84–89.

Bempah, C.K. and A.K. Donkor. 2010. Pesticide residues in fruits at the market level in Accra Metropolis, Ghana, a preliminary study. Environ. Monit. Assess. 175: 551–561.

Benimeli, C.S., M.J. Amoroso, A.P. Chaile and R.G. Castro. 2003. Isolation of four aquatic streptomycetes strains capable of growth on organochlorine pesticides. Bioresour. Technol. 89: 348–357.

Benimeli, C.S., R.G. Castro, A.P. Chaile and M.J. Amoroso. 2006. Lindane removal induction by *Streptomyces* sp. M7. J. Basic Microbiol. 46: 348–357.

Benimeli, C.S., G.R. Castro, A.P. Chaile and M.J. Amoroso. 2007. Lindane uptake and degradation by aquatic *Streptomyces* sp. strain M7. Int. Biodeter. Biodegr. 59: 148–155.

Benimeli, C.S., M.S. Fuentes, C.M. Abate and M.J. Amoroso. 2008. Bioremediation of lindane contaminated soil by *Streptomyces* sp. M7 and its effects on *Zea mays* growth. Int. Biodeter. Biodegradation. 61: 233–239.

Colt, J.S., N. Rothman, R.K. Severson, P. Hartge, J.R. Cerhan, N. Chatterjee, W. Cozen, L.M. Morton, A.J. De Roos, S. Davis, S. Chanock and S.S. Wang. 2009. Organochlorine exposure, immune gene variation, and risk of non-Hodgkin lymphoma. Blood. 113: 1899–1905.

Chia, V.M., Y. Li, S.M. Quraishi, B.I. Graubard, J.D. Figueroa, J.P. Weber, S.J. Chanock, M.V. Rubertone, R.L. Erickson and K.A. McGlynn. 2010. Effect modification of endocrine disruptors and testicular germ cell tumour risk by hormone-metabolizing genes. Int. J. Androl. 33: 588–596.

Cuozzo, S.A., M.S. Fuentes, N. Bourguignon, C.S. Benimeli and M.J. Amoroso. 2011. Chlordane biodegradation under aerobic conditions by indigenous *Streptomyces* strains. Int. Biodeter. Biodegradation. 66: 19–24.

Cuozzo, S.A., G.C. Rollán, C.M. Abate and M.J. Amoroso. 2009. Specific dechlorinase activity in lindane degradation by *Streptomyces* sp. M7. World J. Microb. Biot. 25: 1539–1546.

Dearth, M.A. and R.A. Hites. 1990. Highly chlordane dimethanofluorenes in technical chlordane and in human adipose tissue. J. Am. Soc. Mass Spectrom. 1: 99–103.

Elcey, D.C. and M.A.A. Kunhi. 2010. Substantially enhanced degradation of hexachlorocyclohexane isomers by a microbial consortium on acclimation. J. Agric. Food Chem. 58: 1046–1054.

Fetzner, S. and F. Lingens. 1994. Bacterial dehalogenases: biochemistry, genetics and biotechnological applications. FEMS Microbiol. Rev. 58: 641–685.

Fuentes, M.S., C.S. Benimeli, S.A. Cuozzo and M.J. Amoroso. 2010. Isolation of pesticide-degrading actinomycetes from a contaminated site: Bacterial growth, removal and dechlorination of organochlorine pesticides. Int. Biodeterior. Biodegradation. 64: 434–441.

Fuentes, M.S., J.M. Sáez, C.S. Benimeli and M.J. Amoroso. 2011. Lindane biodegradation by defined consortia of indigenous *Streptomyces* strains. Water Air Soil Pollut. 222: 217–231.

Hageman, K.J., S.L. Simonich, G.R. Wilson, D.H. Campbell and D.H. Landers. 2006. Atmospheric deposition of current-use and historic-use pesticides in snow at national parks in the western United States. Environ. Sci. Technol. 40: 3174–3180.

Hirano, T., T. Ishida, K. Oh and R. Sudo. 2007. Biodegradation of chlordane and hexachlorobenzenes in river sediment Chemosphere. 67: 428–434.

Hopwood, D.A. 1967. Genetic analysis and genome structure in *Streptomyces coelicolor*, Bacteriol. Rev. 31: 373–403.

Jia, H., Y. Chang, Y. Sun, D. Wang, X. Liu, M. Yang, D. Xu, B. Meng and Y.F. Li. 2010. Distribution and potential human risk of organochlorine pesticides in market mollusks from Dalian, China. Bull. Environ. Contam. Toxicol. 84: 278–284.

Kennedy, D.W., S.D. Aust and J.A. Bumpus. 1990. Comparative biodegradation of alkyl halide insecticides by the white rot fungus, *Phanerochaete chrysosporium* (BKM-F-1767). Appl. Environ. Microbiol. 8: 2347–2353.

Lal, R., G. Pandey, P. Sharma, K. Kumari, S. Malhotra, R. Pandey, V. Raina, H.P. Kohler, C. Holliger, C. Jackson and J.G. Oakeshott. 2010. Biochemistry of microbial degradation of hexachlorocyclohexane and prospects for bioremediation. Microbiol. Mol. Biol. Rev. 74: 58–80.

Liu, S.U., A.J. Freyer and J.M. Bollag. 1991. Microbial dechlorination of the herbicide metolachlor. J. Agric. Food Chem. 39: 631–636.

Liu, Z., H. Zhang, M. Tao, S. Yang, L. Wang, Y. Liu, D. Ma and Z. He. 2010. Organochlorine pesticides in consumer fish and mollusks of Liaoning province, China: distribution and human exposure implications. Arch. Environ. Contam. Toxicol. 59: 444–453.

Liu, S.Y., M.H. Liu and J.M. Bollag. 1990. Transformation of metolachlor in soil inoculated with *Streptomyces* sp. Biodegradation. 1: 9–17.

Manickam, N., M.K. Reddy, H.S. Saini and R. Shanker. 2008. Isolation of hexachlorocyclohexane-degrading *Sphingomonas* sp. by dehalogenase assay and characterization of genes involved in gamma-HCH degradation. J. Appl. Microbiol. 104: 952–960.

Murray, R., P. Phillips and J. Bender. 1997. Degradation of pesticides applied to banana farm soil: Comparison of indigenous bacteria and a microbial mat. Environ. Toxicol. Chem. 16: 84–90.

Nagata, Y., R. Endo, M. Itro, Y. Ohtsubo and M. Tsuda. 2007. Aerobic degradation of lindane (γ-hexachlorocyclohexane) in bacteria and its biochemical and molecular basis. Appl. Microbiol. Biotechnol. 76: 741–752.

Radosevich, M., S.J. Traina, Y.L. Hao and O.H. Tuovinen. 1995. Degradation and mineralization of atrazine by a soil bacterial isolate. Appl. Environ. Microbiol. 61: 297–302.

Ravel, J., M.J. Amoroso, R.R. Colwell and R.T. Hill. 1998. Mercury resistant actinomycetes from Chesapeake Bay. FEMS Microbiol. Lett. 162: 177–184.

Robinson, C., D.A. Barry, P.L. McCarty, J.I. Gerhard and I. Kouznetsova. 2009. pH control for enhanced reductive bioremediation of chlorinated solvent source zones. Sci. Total Environ. 407: 4560–4573.

UNEP, 2000. Preparation of an international legally binding instrument for implementing international action on certain persistent organic pollutants, UNEP/POPS/INC.4/INF/6; United Nations Environment Programme: Nairobi, Kenya.

Wang, X., S. Zhou, H. Wang and S. Yang. 2006. Biodegradation of hexazinone by two isolated bacterial strains (WFX-1 and WFX-2). Biodegradation. 17: 331–339.

Wong, M.H., A.O. Leung, J.K. Chan and M.P. Choi. 2005. A review on the usage of POP pesticides in China, with emphasis on DDT loadings in human milk. Chemosphere. 60: 740–752.

Yago, H., H. Murayama, T. Suzuki, N. Hatamoto, Y. Tominaga and N. Shibuya. 2006. Atmospheric concentration of persistent organic pollutants in Sado. J. Environ. Chem. 16: 71–80.

Yamada, S., Y. Naito, M. Funakawa, S. Nakai and M. Hosomi. 2008. Photodegradation fates of cis-chlordane, trans-chlordane, and heptachlor in ethanol. Chemosphere. 70: 1669–1675.

Yikmis, M., M. Arenskötter, K. Rose, N. Lange, H. Wernsmann, L. Wiefel and A. Steinbüchel. 2008. Secretion and transcriptional regulation of the latex-clearing protein, Lcp, by the rubber-degrading bacterium *Streptomyces* sp. strain K30. Appl. Environ. Microbiol. 74: 5373–5382.

CHAPTER 14

Metabolic Diversity and Flexibility for Hydrocarbon Biodegradation by *Rhodococcus*

Héctor M. Alvarez* and Roxana A. Silva

Introduction

The members of the genus *Rhodococcus* are aerobic, non-motile and non-sporulating bacteria belonging to the actinomycetes group. These microorganisms are taxonomically close to *Nocardia*, *Mycobacterium* and *Dietzia* genera. They are usually found in diverse environments, such as tropical and artic soil, deserts as well in marine and deep-sea sediments (Whyte et al. 1999, Heald et al. 2001, Peressutti et al. 2003, Alvarez et al. 2004, Luz et al. 2004, Peng et al. 2008). The huge metabolic repertoire of these microorganisms as well as their capability to adapt their metabolism to a wide range of nutritional conditions, are in part responsible for the occurrence of these actinobacteria in different natural ecosystems. The central metabolism of rhodococci provides all necessary intermediates for the biosynthesis of a wide range of molecules which are used for assembling complex macromolecules, secondary metabolites and cellular structures. This extraordinary biosynthetic capability of rhodococci is combined with a broad catabolic ability. These microorganisms seem to have the potential

Regional Centre for Research and Development (CRIDECIT), Faculty of Natural Science, University of Patagonia San Juan Bosco, Ruta Provincial Nro 1—Ciudad Universitaria, (9000) Comodoro Rivadavia, Chubut, Argentina.
*Corresponding author: halvarez@unpata.edu.ar

to use alternatively different glycolytic pathways, such as EMP or ED pathways, the PP pathway, and a partial or fully functional tricarboxylic cycle (TCA) in both, oxidative and reductive directions according to the experimental conditions (Alvarez 2010). Moreover, rhodococci posses the ability to produce variable amounts of storage compounds, such as polyhydroxyalkanoates (PHA), triacylglycerols (TAG), glycogen and polyphosphate, which probably permit cells to respond rapidly to changes in nutritional state and to balance metabolism under different environmental conditions (Hernández et al. 2008). One relevant feature of rhodococci is their ability to degrade and transform a wide range of pollutant compounds, such as hydrocarbons with diverse chemical structures, pesticides, xenobiotics and explosives (Warhurst and Fewson 1994, Larkin et al. 2005, Martínková et al. 2009). All these properties make such microorganisms promising candidates for bioremediation of polluted environments. In this context, some success has been achieved in bioremediation of contaminated soil, air and waters using rhodococci as bioaugmentation agents (Kuyukina and Ivshina 2010).

This chapter describes the ability of rhodococci to degrade a wide range of hydrocarbons and other pollutants under fluctuating conditions based on the high flexibility of their metabolism.

Rhodococcus are Frequently Found in Polluted Environments

Members of rhodococci have been isolated from diverse environments in different geographical regions of the world, frequently from polluted environments. The occurrence of members of *Rhodococcus* genus in polluted soil and water has been reported in a wide range of ecological regions, such as temperate and extreme cold or hot environments. Only a few examples on the wide distribution of rhodococci in polluted soil and water of the different geographical regions of the world will be included in this chapter.

Temperate and cold environments of Europe and Asia

Koronelli et al. (1987) reported the prevalent occurrence of rhodococci and mycobacteria in water near harbors of the Baltic sea and Kurshsky bay polluted after a fuel oil spill; whereas the isolation of a large number of alkane-degrading *Rhodococcus* strains from soil, water, snow or air in diverse regions of the former Soviet Union has been reported by Ivshina et al. (1994, 1995). Those strains have been identified as *R. erythropolis*, *R. fascians*, *R. rhodochrous* and *R. ruber*. In other work, *Rhodococcus* bacteria have been identified among the hydrocarbon assimilating-microorganisms from a temperate agricultural soil in France (Chaîneau et al. 1999). These microorganisms were isolated from a polluted soil collected in a field treated with oily cuttings resulting from the drilling of an onshore well. Hernandez-Raquet et al. (2006) investigated the biodiversity of microbial mats inhabiting the oil-contaminated lagoon Etang de Berre (France) by molecular approaches. They reported the occurrence of *Rhodococcus* members among the species involved in oil degradation, which was determined by combining culture-based approaches and DGGE. Andreoni et al. (2004) analyzed three soils samples (e.g., Belgian, German, and Italian) with different properties and hydrocarbon-pollution history with regard to their potential to degrade phenanthrene. The authors isolated phenanthrene-degrading strains from the Belgian soil, where they demonstrated a rapid decrease of polyaromatic hydrocarbons (PAH) and phenanthrene contents. From the fastest phenanthrene-degrading culture, a representative strain was identified as *Rhodococcus aetherovorans* (100%), among others belonging to *Achromobacter xylosoxidans*, *Methylobacterium* sp., *Rhizobium galegae*, *Stenotrophomonas acidaminiphila*, *Alcaligenes* sp. and *Aquamicrobium defluvium*. In addition, DGGE-profiles of culture also showed bands attributable to *Rhodococcus* members. In other study, Wagner-Döbler et al. (1998) employed a microcosm enrichment approach to isolate bacteria which are representative of long-term biphenyl-adapted microbial communities. In all of the microcosms, isolates identified as *Rhodococcus opacus* dominated the cultivable microbial community, therefore, they postulated that *R. opacus* is a promising candidate for development of effective long-term inocula for polychlorinated biphenyl bioremediation.

In other study, a multiapproach methodology was applied to investigate the bacterial communities in two different shoreline matrices, rocks and sand from the Costa da Morte, in northwestern Spain, during 12 months after being affected by the Prestige oil spill. Culture-dependent and culture-independent approaches suggested that the genus *Rhodococcus* could play a key role in the *in situ* degradation of the alkane fraction of the Prestige fuel together with other members of the suborder Corynebacterineae. Moreover, other members of this suborder, such as *Mycobacterium* spp., together with Sphingomonadaceae bacteria were related as well to the degradation of the aromatic fraction of the Prestige fuel (Alonso-Gutiérrez et al. 2009).

Margesin et al. (2003) investigated the prevalence of seven genotypes involved in the degradation of *n*-alkanes (*Pseudomonas putida* GPo1 *alk*B; *Acinetobacter* spp. *alk*M; *Rhodococcus* spp. *alk*B1; and *Rhodococcus* spp. *alk*B2), aromatic hydrocarbons (*P. putida xyl*E) and polycyclic aromatic hydrocarbons (*P. putida ndo*B and *Mycobacterium* sp. strain PYR-1 *nid*A) in 12 oil-contaminated and 8 pristine Alpine soils from Tyrol (Austria) by PCR hybridization analyses of total soil community DNA, using oligonucleotide primers and DNA probes specific for each genotype. The genotypes containing genes from gram-positive bacteria (*Rhodococcus alk*B1 and *alk*B2 and *Mycobacterium* nidA) were detected at a high frequency in both contaminated (41.7 to 75%) and pristine (37.5 to 50%) soils, indicating that they were already present in substantial number before the occurrence of contamination.

Finally, two strains of *R. erythropolis* were isolated from oil-polluted sites located in central Taiwan (Lin et al. 2005). These strains which were found to float and grow near the diesel layer on the surface were characterized by a high cell-surface hydrophobicity and cell-residue emulsification activity.

Arid environments of Asia, Australia and South America

The isolation of *Rhodococcus* members from soil of extreme conditions like deserts has been reported. Radwan et al. (1995) reported the occurrence of *Rhodococcus* bacteria, among other hydrocarbon-degrading microorganisms, in oil-polluted desert samples from Kuwait. In another interesting approach, Sorkhoh et al. (1995) utilized a natural microbial cocktail for remediating oil-polluted desert in the Arabian Gulf region. Oil-degrading microorganisms immobilized within dense cyanobacterial mats on oily coasts of the Gulf were successfully established in oil-contaminated sand. Members of the *Rhodococcus* genus predominated in the first few weeks, but after 22 weeks population of *Pseudomonas* spp. increased, sharing *Rhodococcus* in the predominance. *Rhodococcus* bacteria together with the genera *Agrobacterium, Arthrobacter, Pseudomonas* and *Gordonia,* predominated in the rhizosphere of two turf cover sorts: Bermuda grass and American grass (Mahmoud et al. 2011). Quantitative determinations revealed that predominant bacteria consumed crude oil and representative aliphatic (*n*-octadecane) and aromatic (phenanthrene) hydrocarbons efficiently. The authors postulated that phytoremediation by covering oil-polluted soil of Arabian deserts with turf cover minimized atmospheric pollution, increased the numbers of the oil-utilizing/nitrogen-fixing bacteria by about 20 to 46% thus, encouraging oil attenuation.

There are additional reports on the dominant occurrence of rhodococci in arid environments. Warton et al. (2001) reported a case of enhanced biodegradation of metham sodium soil fumigant on a farm located in Western Australia. They identified 11 strains belonging to *Rhodococcus* spp. among a total of 18 Gram-positive isolates, which were responsible of the fumigant degradation in soil. These strains were able to resist dry heat treatments and to recover their degrading ability following dehydration. In another study, Peressutti et al. (2003) reported the occurrence of *Rhodococcus* and *Gordonia* bacteria as relevant components of microbial communities of soil polluted with hydrocarbons in the semiarid Patagonia (Argentina). Some of the isolates belonging to *Rhodococcus* genus displayed a wide potential for hydrocarbon biodegradation and persistence in the oil Patagonian polluted soil. The soil of this region is characterized by low moisture contents, low nutrient levels, alkaline pH and cold and fluctuating temperatures. In this context, some native strains belonging to *R. erythropolis*, *R. fascians*, *Dietzia maris*, *Nocardia globerula*, *N. asteroides* and *N. restricta* were able to degrade hydrocarbons, principally *n*-alkanes, during cultivation under nitrogen starvation (Alvarez 2003). Recently, Silva et al. (2010) reported the isolation of an indigenous *Rhodococcus jostii* strain (strain 602) from a polluted soil sample collected in semiarid Patagonia (Argentina) with the ability to degrade PAH and transform naphthyl compounds in lipids under nitrogen-limiting conditions.

Extreme cold environments of Arctic and Antarctica

The occurrence of rhodococci in cold environments, such as the soils of the Arctic region as well as in Antarctica, has also been reported. Whyte et al. (1996) isolated a psychrotrophic strain Q15 identified as a *Rhodococcus* sp., which was able to mineralize the C28 *n*-paraffin octacosane, alkanes and diesel fuel at both 23 and 5°C. The psychrotrophic strain Q15 possessed two large plasmids of approximately 90 and 115 kb which were not required for alkane mineralization, although the 90-kb plasmid enhanced mineralization of some alkanes and growth on diesel oil at both 5 and 25°C (Whyte et al. 1998). *Rhodococcus* sp. strain Q15 showed interesting physiological adaptations to assimilate alkanes at low temperatures. During growth at 5°C on *n*-hexadecane or diesel fuel, strain Q15 produced a cell surface-associated biosurfactant(s) and exhibited increased cell surface hydrophobicity. Cells seemed to produce an extracellular polymeric substance (EPS) at 5°C, containing a complex mixture of glycoconjugates, during its growth on diesel fuel (Whyte et al. 1999). In addition, cells decreased the degree of saturation of membrane lipid fatty acids, but it did so to a lesser extent when it was grown on hydrocarbons at 5°C; suggesting that strain Q15 modulates membrane fluidity in response to the counteracting influences of low temperature and hydrocarbon toxicity.

In another study, Ruberto et al. (2005) isolated three *Rhodococcus* strains from polluted Antarctic soils, which were able to grow at a wide range of temperatures. Soil microcosm studies demonstrated that bioaugmentation with the Antarctic *Rhodococcus* sp. strain ADH improved biodegradation of hydrocarbons, either by themselves or mixed with an indigenous hydrocarbon-degrading *Acinetobacter* strain. The authors concluded that members of *Rhodococcus* genus are present in Antarctic soils and might play an important role in decontamination of polluted environments in those extreme environments.

The presence of rhodococci in diverse polluted cold environments has also been demonstrated by molecular approaches. Whyte et al. (2002a) investigated the prevalence of four alkane monooxygenase genotypes (*Pseudomonas putida* GPo1, Pp *alk*B; *Rhodococcus* sp. strain Q15, Rh *alk*B1 and Rh *alk*B2; and *Acinetobacter* sp. strain ADP1, Ac *alk*M) in hydrocarbon-contaminated and pristine soils from the Arctic and Antarctica, by both culture-independent (PCR hybridization analyses) and culture-dependent (colony hybridization analyses) molecular methods. The Rh *alk*B1 genotype was clearly more prevalent in culturable cold-adapted bacteria than in culturable mesophiles. Finally, the authors concluded that whereas *Pseudomonas* spp. may become enriched in polar soils following contamination events, *Rhodococcus* spp. may be the predominant alkane-degradative bacteria in both pristine and contaminated polar soils. Moreover, the alkane monooxygenase genes, Rh *alk*B1 and Rh *alk*B2, were the most frequently detected *alk* genes in soil contaminated with hydrocarbons in Brazil as well as in Antarctica (Luz et al. 2004).

Marcos et al. (2009) analyzed the PAH-degrading bacterial populations in cold marine ecosystems of Subantarctic marine sediments (Ushuaia Bay, Argentina). They used a degenerate primer set targeting genes encoding the alpha subunit of PAH-dioxygenases from Gram-positive bacteria to amplify gene fragments from metagenomic DNA isolated from such marine sediments. The authors were able to identify 14 distinct groups of genes showing significant relatedness with dioxygenases from *Rhodococcus* genus, in addition to *Mycobacterium*, *Nocardioides*, *Terrabacter* and *Bacillus* genera.

Rhodococci exhibit a broad capability for degrading diverse pollutants

Several genomic projects involving *Rhodococcus* strains isolated from contaminated environments are now in progress (Table 1). In general, rhodococcal genomes with different sizes (between 5 to 9Mb) are enriched with catabolic genes involved in the degradation of a wide range of pollutants. These genes are usually distributed in the chromosomes as well as in plasmids of different sizes (linear and circular plasmids) (Larkin et al. 2010). The occurrence of this broad repertoire of catabolic genes in rhodococcal genomes supports the ability of these microorganisms for degrading a diversity of organic compounds (Table 2). In general, rhodococci show a high affinity for utilizing *n*-alkanes and alkyl-substituted hydrocarbons as sole carbon and energy sources (Alvarez 2003). In this context, two *Rhodococcus* strains (NRRL B-16531 and Q15), which were isolated from different geographical locations, contained at least four alkane monooxygenase gene homologs (*alk*B1, *alk*B2, *alk*B3, and *alk*B4)

Table 1. Summary of genome projects of pollutants-degrading *Rhodococcus*.

Bacterial strain	Accession	Source	Relevant characteristics
Rhodococcus jostii RHA1	PRJNA58325	University of British Columbia, Canada	Strain RHA1 was isolated from soil contaminated with gamma-hexachlorocyclohexane (lindane: a toxic insecticide) in Japan
Rhodococcus opacus B4	PRJDA34839	National Institute of Technology and Evaluation, Japan	Strain B4 was isolated as an organic solvent-tolerant bacterium from gasoline-contaminated soil
Rhodococcus opacus PD630	PRJNA30413	Broad Institute (MIT), USA	Strain PD630 was isolated from a soil sample at a German gasworks plant after enrichment with phenyldecane
Rhodococcus erythropolis PR4	PRJNA59019	National Institute of Technology and Evaluation, Japan	Strain PR4 is an alkane-degrading bacterium which was isolated from Pacific Ocean seawater. It shows high tolerance to hydrocarbons
Rhodococcus erythropolis XP	PRJNA72225	Shanghai Jiao Tong University	Strain XP is a biodesulfurizing- bacterium
Rhodococcus sp. R04	PRJNA63847	Institute of Biotechnology, Shanxi University, China	Strain R04 is a potent polycholorinated biphenyl-degrading soil actinomycete that catabolizes a wide range of compounds
Rhodococcus sp. F7		University of Patagonia San Juan Bosco, Argentina	Strain F7 is a phenanthrene-degrading bacterium isolated from polluted soil
Rhodococcus sp. JVH1	PRJNA46601	Thompson Rivers University, Canada	Strain JVH1, which uses *bis*-(3-pentafluorophenylpropyl)-sulfide (PFPS) as a sole sulfur source, was isolated from an oil-contaminated environment

in their genome (Whyte et al. 2002b). The presence of multiple alkane monooxygenases in the two rhodococcal strains is reminiscent of other multiple-degradative-enzyme systems reported in *Rhodococcus* (Whyte et al. 2002b). Interestingly, McLeod et al. (2006) reported the occurrence of at least 203 oxygenases probably involved in the degradation pathways of aromatic and steroid compounds in *R. jostii* RHA1. It was postulated that at least 26 different "peripheral aromatic" pathways and eight "central

Table 2. Catabolic spectrum of *Rhodococcus* bacteria.

Type of compounds	Species	References
n-Alkanes	*R. opacus, R. erythropolis, R. ruber, R. fascians, R. baikonurensis, R.* sp.	Whyte et al. 2002, Alvarez 2003, Sameshima et al. 2008, Amouric et al. 2010, Capelletti et al. 2011
Branched alkanes	*R. erythropolis, R.* sp.	Urai et al. 2007, Takei et al. 2008
Phenylalkanes	*R. opacus, R. erythropolis*	Alvarez et al. 1996, Alvarez et al. 2002, Herter et al. 2012
Monoaromatics	*R. aetherivorans*	Hori et al. 2009
Polyaromatics	*R. jostii, R. opacus, R. erythropolis, R. fascians, R.* sp.	Tomás-Gallardo et al. 2006, Patrauchan et al. 2008, Yang et al. 2008, Akhtar et al. 2009, Silva et al. 2010, Araki et al. 2011, Robrock et al. 2011, Yoo et al. 2011
Haloalkanes	*R. rhodochrous, R.* sp.	Newman et al. 1999, Bosma et al. 2003, Lahoda et al. 2011
Polychlorinated-hydrocarbons	*R. jostii, R. aetherivorans, R.* sp.	Taguchi et al. 2007, Puglisi et al. 2010, Ohmori et al. 2011, Yang et al. 2011
Nitriles	*R. ruber, R. rhodochrous*	Yeom et al. 2008, Raj et al. 2010, Kamal et al. 2011
Biodesulfurization of compounds	*R. opacus, R. erythropolis*	Li et al. 2008, Aggarwal et al. 2011, Kawaguchi et al. 2011, Tao et al. 2011
Steroids	*R. rhodochrous, R. jostii, R. ruber*	Fernández de Las Heras et al. 2009, Mathieu et al. 2010, Petrusma et al. 2011
Diesel oil, gasoline	*R. aetherivorans, R. wratislaviensis, R. erythropolis, R.* sp.	Zhang et al. 2007, Etemadifar and Emtiazi 2008, Huang et al. 2008, Auffret et al. 2009

aromatic" pathways are present in the genome of strain RHA1 (McLeod et al. 2006, Larkin et al. 2010). The diversity of degradation pathways and the genes involved in the catabolism of aromatic and steroid compounds are well described specially for *R. jostii* RHA1 (for a review: Yam et al. 2010). Several gene homologues are also present in other members of the genus, such as *R. aetherovorans, R. opacus* or *R. erythropolis* among others, which is a common feature of rhodococci.

Other interesting catabolic genes usually present in rhodococcal genomes are the haloalkane dehalogenases, which catalyze the cleavage of carbon-halogen bonds in halogenated aliphatic compounds. The haloalkane dehalogenase DhaA from *Rhodococcus rhodochrous* NCIMB 13064 can slowly detoxify the industrial pollutant 1,2,3-trichloropropane (TCP) as has been reported by Stsiapanava et al. (2011). On the other hand, a potent dehalogenase activity for conversion of 1-chlorobutane has been reported in a strain of *R. erythropolis* (Erable et al. 2006). Rhodococci are able to degrade recalcitrant pollutants such as polychlorinated compounds. The polychlorinated-biphenyl (PCB) degrader, *Rhodococcus jostii* RHA1, degrades PCBs by cometabolism with biphenyl. A two-component BphS1T1 system encoded by *bph*S1 and *bph*T1 (formerly *bph*S and *bph*T) was responsible for the transcription induction of the five gene clusters, *bph*AaAbAcAdC1B1, *etb*Aa1Ab1CbphD1, *etb*Aa2Ab2AcD2, *etb*AdbphB2, and *etb*D1, which constitute multiple enzyme systems for biphenyl/PCB degradation. In addition to this multiple enzyme systems, RHA1 also employs dual regulatory systems for biphenyl/PCB degradation (Takeda et al. 2010). In another study, Puglisi et al. (2010) analyzed the transcriptional responses of *Rhodococcus aetherivorans* I24 to PCB-contaminated sediments. Although the genome sequences of strain I24 contained many orthologs of the genes in the canonical biphenyl pathway, very few of these genes were up-regulated in response to PCBs or biphenyl. The authors indicated that the transcriptional response of *R. aetherivorans* I24 to PCBs, in both medium and sediment, was primarily directed towards reducing oxidative stress, rather than catabolism (Puglisi et al. 2010).

The genome of rhodococci usually contains a significant number of genes coding nitrilases and nitrile hydratase/amidase enzymes, which are involved in the biodegradation of aliphatic nitriles and benzonitrile herbicides (Martínková et al. 2010). These enzymes possess a significant potential as biocatalysts for the production of bulk and fine chemicals (Martínková et al. 2010). Other interesting feature of rhodococci is their ability to biodesulfurizate fossil fuels, which can have harmful effects on the environment. The biodesulfurization process can remove organically bound sulfur from diesel oil through the bacterial metabolism (Ma 2010).

It is not the aim of this article to describe the rhodococcal degradation routes of the different types of hydrocarbons and their genes involved, but rather to highlight the enormous catabolic potential of rhodococci. The genetics and biochemistry of the biodegradation of diverse pollutant compounds by rhodococci have been well studied and several excellent reviews are available in the literature (Warhurst and Fewson 1994, Larkin et al. 2005, Martínková et al. 2009, Ma 2010, Martínková et al. 2010, Yam et al. 2010, Yam et al. 2011).

Flexibility of the rhodococcal metabolism

All microorganisms living in natural environments must be able to adapt their metabolism to external changes. One interesting feature of rhodococci, which are found in a broad diversity of natural environments, is the flexibility of their metabolism. In this article, flexibility is understood as the ability of rhodococcal cells to respond to potential internal or external changes affecting their capability to thrive and survive in fluctuating environments. In this context, rhodococci seem to have the ability to use alternatively different glycolytic pathways (EMP, ED and PP pathways) according to the conditions (Alvarez 2010). In addition, rhodococcal genomes contain all the necessary genes for a partial or fully functional TCA in both oxidative and reductive directions, which may permit cells to balance the metabolism and to adapt themselves to diverse environments. The switch from oxidative to reductive TCA reactions may facilitate carbon fixation as well as restore the oxidative-reductive balance during environmental fluctuations (Srinivasan and Morowitz 2006). Moreover, analyses of genome databases of diverse *Rhodococcus* members revealed the occurrence of different lithoautotrophic pathways, including genes coding for putative carbon monoxide dehydrogenase (CODH), hydrogenase systems, and a thiocyanate hydrolase enzyme (Alvarez 2010). All these pathways occurring in rhodococci may serve as auxiliary mechanisms for energy metabolism during fluctuating nutritional conditions.

In general, *Rhodococcus* bacteria seem to have a low energy life style showing a relative slow growth even when nutrients are available. These microorganisms seem to posses the ability to conserve metabolic useful energy during catabolism of substrates. Thus, a part of the resulting energy can be used for growth and division and the surplus is channeled into energy storage pathways. In this context, several *Rhodococcus* strains analyzed in different studies were able to accumulate variable amounts of TAG, PHA and glycogen, which were likely produced in a programmed manner (Alvarez et al. 1996, Alvarez et al. 1997, Hernández et al. 2008). Glycogen was produced principally during exponential growth phase, whereas storage lipids biosynthesis predominated during stationary phase (Hernández and Alvarez 2010). All studied strains accumulated TAG as the main storage compounds plus PHA (with 3-hydroxybutyrate and 3-hydroxyvalerate monomers) and glycogen as minor compounds. All key genes for the biosynthesis and mobilization of these storage compounds were identified in the *R. jostii* RHA1 genome database (Hernández et al. 2008). We observed a high redundancy of genes and enzymes involved in storage lipid metabolism. Individual isoforms of enzymes have potentially different substrate specificity, may play distinct functional roles in the pathways of glycerolipid biosynthesis or may be differentially expressed under several environmental conditions (Alvarez et al. 2008, Hernández et al. 2008). Moreover, *R. opacus* PD630, *R. jostii* RHA1 and probably other rhodococci are able to produce polyphosphate, which is deposited in cytoplasm of cells (Alvarez et al. 1996, Hernández et al. 2008). The availability of this high-energy phosphate polymer may enhance the capacity of rhodococci to survive in soil environments by providing phosphate for biosynthesis, maintenance energy or as an osmoprotectant. The metabolic flexibility of rhodococci and their ability to produce diverse storage compounds are traits that enable or enhance the probability of such microorganisms surviving and reproducing in the environment.

Some studies demonstrated that cells of *R. opacus* PD630 and *R. jostii* 602 maintain their ability to be active; although adopting different physiological states, in a broad range of environmental conditions, such as fluctuating temperatures and humidity or in the availability of carbon sources and nutrients (Alvarez et al. 2000, Alvarez et al. 2004, Silva 2009). Starvation experiments demonstrated that *R. opacus* PD630 and *R. jostii* 602 possess specialized mechanisms for turning metabolism down when nutrients are in short supply or when cells are subjected to other stress conditions that normally occur in arid soils (Alvarez et al. 2004, Silva 2009). Metabolic depression may be a relevant physiological mechanism allowing such bacteria to adapt ecologically to poor environments. During nitrogen starvation, cells reduced their metabolic activity and ability to mineralize the carbon source, but significantly increased the biosynthesis and accumulation of TAG. In contrast, under carbon starvation, profound metabolic suppression allowed a slow utilization of stored lipids (Alvarez et al. 2000). The energy obtained by the slow mobilization of stored TAG may support the necessary biochemical and physiological adaptation mechanisms for long periods. According to our studies, we could speculate that *R. opacus* PD630 sustains higher metabolic rates when an external carbon source and other essential nutrients are available in the environment, but during nutrient scarcity cells assume a reduced metabolic state at expenses of the slow mobilization of the energy-rich TAG. Thus, metabolic flux may reflect the "energy status" of these soil bacteria through the pathway of fatty acid biosynthesis, as occur in superior animals (Dowell et al. 2005). Taken together, we may compare the physiological responses of *R. opacus* PD630 to the environment with hibernator organisms to some extent, which deposit large quantities of fat during plenty periods and degrade stored lipids during the hibernation phase using fatty acids as primary fuel (Dark 2005). TAG are excellent reserve materials since their oxidation produce the maximum yields of energy in comparison with other storage compounds such as carbohydrates, since the carbon atoms of the acyl moieties of TAG are in the most reductive form (Alvarez 2006, Alvarez and Steinbüchel 2010). In addition, bacterial TAG may also serve as a reservoir of metabolic water under dry conditions, since fatty acid oxidation releases large amounts of water.

Since rhodococci usually undergo simultaneously diverse stresses in natural environments, such as nutrient scarcity, fluctuating temperatures and desiccation, among others; they developed metabolic strategies to cope with such environments. Some of these mechanisms may be, (a) the accumulation of storage compounds that can be utilized by cells as endogenous carbon sources and electron donors during periods of nutritional scarcity, (b) the occurrence of metabolic gene and enzyme redundancy in genomes, (c) the reduction of energy requirements in response to starvation and other stress conditions and (d) the formation of cell aggregates, which promotes a relative isolation from the surrounding environment by the presence of an EPS. The EPS probably provides protection and prevent the population from dispersing in the environment. This "multicellular biological system" with the availability of a variety of storage compounds, compatible solutes, pigments and other oxidative protection systems may be advantageous in fluctuating environments. These processes may provide cells of energetic autonomy and a temporal independence from the environment and contribute to cell survival when they do not have access to energy resources in soil.

Rhodococci are able to degrade pollutants under growth-restricting conditions

Rhodococci are able to degrade a diversity of hydrocarbons still under unbalanced growth conditions, as occur frequently in the environment. Rhodococcal cells continue degrading diverse hydrocarbons when an essential nutrient like nitrogen is lacking. A similar situation can be found in natural environments when an oil spill occurs in soil containing a low amount of nitrogen. Under such conditions, rhodococci and other related actinobacteria usually couple hydrocarbon catabolism with lipid biosynthesis, principally neutral lipids such as TAG and wax esters (WS) (Fig. 1). In contrast, many bacteria inhibit the lipid metabolism during limitation of essential nutrients (Huisman et al. 1993). The oxidation of hydrocarbons by rhodococci seem not to be complete under nitrogen-limiting conditions, since different intermediates such as fatty acids or fatty alcohols derived from hydrocarbon oxidation are used for the biosynthesis of TAG and/or WS with a chemical structure related to that of the hydrocarbon used as carbon source (Table 3). The degradation and conversion of diverse hydrocarbons into lipids, as well as the metabolic pathways involved, are discussed in more detail below.

Aliphatic hydrocarbons

Although crude oil composition varies widely depending on the reservoir, saturated or unsaturated aliphatic hydrocarbons are the usual constituents of oil. Rhodococci exhibit a high capacity to degrade aliphatic hydrocarbons, and are able to transform them in intracellular TAG/WS during growth under nitrogen starvation. *R. opacus* PD630, which is a known oleaginous bacterium, was able to accumulate up to 40% cellular dry weight of TAG

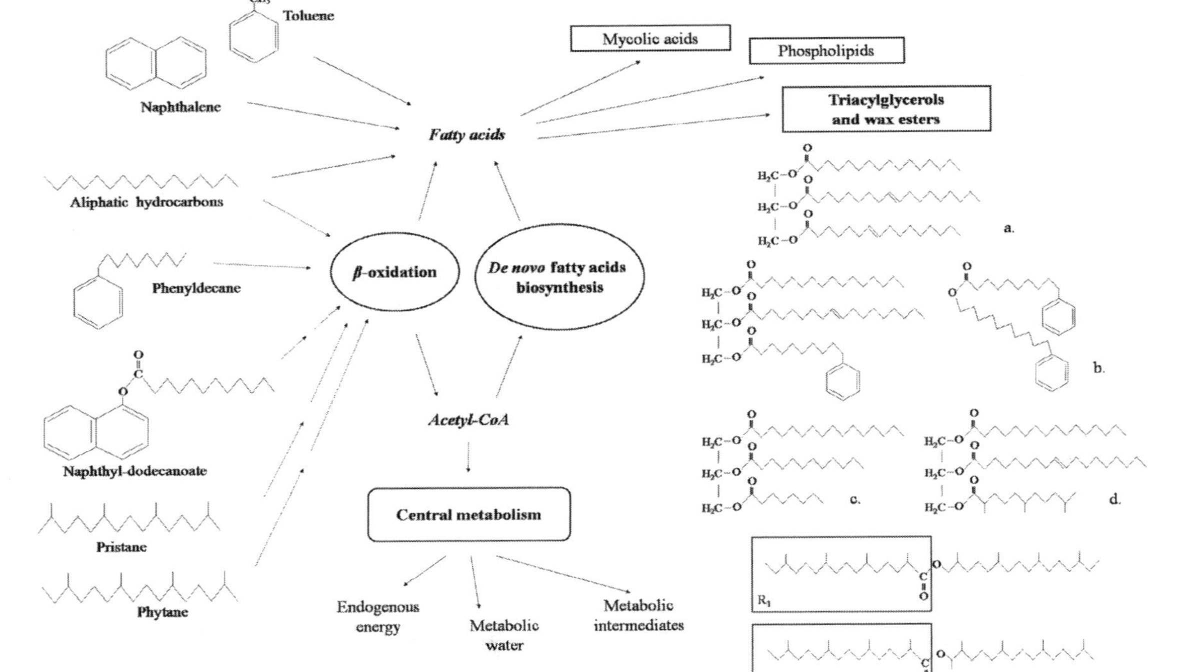

Figure 1. Synthesis of TAG and wax ester from hydrocarbons metabolism. R_1: 4,8-dimethyldecanoic, 2,6,10-trimethyldodecanoic, 4,8,12-trimethyltetradecanoic, 4,8,12-trimethyltetradecenoic, 2,6,10,14-tetramethylhexadecanoic or an unsaturated C20 isoprenoid (different from phytenic acid); R_2: 4,8-dimethyldecanoic, 2,6,10-trimethyldodecanoic, 2,6,10,14-tetramethylhexadecanoic, 3,7,11,15-tetramethylhexadecanoic or an unsaturated C20 isoprenoid (different from phytenic acid).

Table 3. Occurrence of neutral lipids in hydrocarbon-degrading actinobacteria.

Bacteria and carbon sources	TAG	WS	Fatty acids detected into reserve lipid compounds	References
R. aetherivorans IAR1				
Toluene	+	–	$C_{14:0}$, $C_{16:0}$, $C_{16:1}$, $C_{17:0}$, $C_{18:0}$ and $C_{18:1}$	Hori et al. (2009)
R. erythropolis 17				
n-Pentadecane	+	nr	$C_{13:0}$, $C_{15:0}$ and $C_{15:1}$	Alvarez et al. (2003)
n-Hexadecane	+	nr	$C_{14:0}$, $C_{16:0}$, $C_{16:1}$, $C_{18:0}$ and $C_{18:1}$	Alvarez et al. (2003)
Gas-oil	+	nr	$C_{14:0}$, $C_{15:0}$, $C_{16:0}$, $C_{16:1}$, $C_{17:0}$, $C_{17:1}$, $C_{18:0}$ and $C_{18:1}$	Alvarez et al. (2003)
R. fascians 123				
n-Pentadecane	+	nr	$C_{13:0}$ and $C_{15:0}$	Alvarez et al. (2003)
n-Hexadecane	+	nr	$C_{14:0}$, $C_{16:0}$, $C_{16:1}$ and $C_{18:1}$	Alvarez et al. (2003)
Gas-oil	+	nr	$C_{14:0}$, $C_{15:0}$, $C_{16:0}$, $C_{17:0}$, $C_{18:0}$ and $C_{18:1}$	Alvarez et al. (2003)
R. opacus PD630				
n-Pentadecane	+	nr	$C_{13:0}$, $C_{15:0}$ and $C_{15:1}$	Alvarez et al. (1996)
n-Hexadecane	+	nr	$C_{12:0}$, $C_{14:0}$, $C_{16:0}$ and $C_{16:1}$	Alvarez et al. (1996)
n-Heptadecane	+	nr	$C_{13:0}$, $C_{15:0}$, $C_{15:1}$, $C_{17:0}$, and $C_{17:1}$	Alvarez et al. (1996)
n-Octadecane	+	nr	$C_{12:0}$, $C_{14:0}$, $C_{16:0}$, $C_{16:1}$, $C_{18:0}$ and $C_{18:1}$	Alvarez et al. (1996)
Phenyldecane	+	+	Phenyl-$C_{10:0}$, phenyl-$C_{9:0}$, phenyl-$C_{8:0}$, phenyl-$C_{6:0}$ and phenyl-$C_{2:0}$ $C_{16:0}$, $C_{18:0}$ and $C_{19:0}$	Alvarez et al. (1996, 2002)
Rhodococcus sp. 602				
n-Hexadecane	+	–	$C_{14:0}$, $C_{16:0}$ and $C_{16:1}$	Silva et al. (2010)
Naphthalene	+	–	$C_{14:0}$, $C_{15:0}$, $C_{16:0}$, $C_{16:1}$, $C_{17:0}$, $C_{17:1}$, $C_{18:0}$ and $C_{18:1}$	Silva et al. (2010)

Naphthyl-1-dodecanoate	+	–	$C_{8:0}$, $C_{10:0}$, $C_{12:0}$ and $C_{14:1}$	Silva et al. (2010)
Rhodococcus sp. 20				
n-Hexadecane	+	nr	$C_{14:0}$, $C_{16:0}$ and $C_{16:1}$	Alvarez et al. (2003)
D. maris 53				
n-Hexadecane	+	nr	$C_{14:0}$, $C_{16:0}$ and $C_{16:1}$	Alvarez et al. (2003)
G. amarae 106				
n-Hexadecane	+	nr	$C_{14:0}$, $C_{16:0}$ and $C_{16:1}$	Alvarez et al. (2003)
N. asteroides 419				
n-Pentadecane	+	nr	$C_{13:0}$, $C_{15:0}$ and $C_{15:1}$	Alvarez et al. (2003)
n-Hexadecane	+	nr	$C_{14:0}$, $C_{16:0}$ and $C_{16:1}$	Alvarez et al. (2003)
Gas-oil	+	nr	$C_{14:0}$, $C_{15:0}$, $C_{16:0}$ and $C_{18:1}$	Alvarez et al. (2003)
N. globerula 432				
n-Hexadecane	+	nr	$C_{14:0}$, $C_{16:0}$, $C_{16:1}$ and $C_{18:1}$	Alvarez et al. (2003)
2,6,10,14-tetramethyl pentadecane (pristane)	+	–	$C_{16:0}$, $C_{17:0}$, $C_{18:0}$, $C_{18:1}$, 4,8,12-trimethyl-C_{13}	Alvarez et al. (2001)
N. restricta 560				
n-Hexadecane	+	nr	$C_{14:0}$, $C_{16:0}$ and $C_{16:1}$	Alvarez et al. (2003)
M. ratisbonense SD4				
2,6,10,14-tetramethyl hexadecane (phytane)	–	+	4,8-dimethyl-C_{10}, 2,6,10-trimethyl-C_{12}, 4,8,12-trimethyl-C_{14}, 2,6,10,14-tetramethyl-C_{16}, 3,7,11,15-tetramethyl-C_{16}	Silva et al. (2007)

Abbreviations: *D.*, *Dietzia*; *G.*, *Gordonia*; *M.*, *Mycobacterium*; *N.*, *Nocardia*; nr, not reported; *R.*, *Rhodococcus*; TAG, triglycerides; WS, Wax esters.

during cultivation on *n*-hexadecane as sole carbon source (Alvarez et al. 1996). In general, during cultivation of rhodococcal cells on *n*-alkanes, the main fatty acids occurring in cells are related to the chain length of the substrate, in addition to β-oxidation-derived fatty acids. Even fatty acids ($C_{12:0}$, $C_{14:0}$, $C_{16:0}$ and $C_{16:1}$) occurred in *n*-hexadecane-grown cells of diverse rhodococci, whereas odd-numbered fatty acids were the main components of TAG during the growth of cells on *n*-pentadecane or *n*-heptadecane as sole carbon sources (Alvarez et al. 1996, Alvarez 2003) (Table 3). These results suggested that the *n*-alkanes were incorporated into cellular lipids by rhodococcal cells, after monoterminal oxidation and without complete degradation to acetyl-CoA level. The β-oxidation pathway, solely or tightly coupled to the *de novo* fatty acid biosynthesis route, serve as the main source of acyl-compounds for lipid biosynthesis during growth of cells on *n*-alkanes. Cultivation experiments performed with acrylic acid, which is an inhibitor of β-oxidation pathway, demonstrated that this catabolic route is the main source of fatty acid for the synthesis of TAG in cells cultivated on *n*-alkanes (Alvarez et al. 1997).

Aliphatic substituted hydrocarbons

Rhodococcus members are able to degrade phenylalkane derivatives not only during optimal growth conditions, but also during unbalanced growth conditions (Alvarez et al. 2002, Herter et al. 2012). *R. opacus* PD630, which was isolated from a soil sample after enrichment with phenyldecane (Alvarez et al. 1996), was able to grow on this substituted hydrocarbon and produce TAG and WS with a diversity of chemical structures. After cultivation of PD630 on phenyldecane under nitrogen-limiting conditions, cells accumulated three different types of neutral lipids: (a) a mixture of TAG containing usual odd- and even-straight chain length-fatty acids (from $C_{12:0}$ to $C_{19:0}$), (b) a mixture of TAG in which one fatty acid was replaced by a phenylalkanoic acid residue and (c) a single WS formed by the esterification of phenyldecanol and phenyldecanoic acid residues (Figs. 1 and 2) (Table 3) (Alvarez et al. 2002). In addition, lipid inclusions accumulated by this strain contained the unmodified hydrocarbon and minor amounts of α- or β-oxidation intermediates, diacylglycerols, phospholipids and proteins (Alvarez et al. 1996). These results indicated that strain PD630 degraded phenyldecane mainly by monoterminal oxidation of the alkyl side-chain to phenyldecanoic acid (Fig. 2). The degradation of the phenylalkane by PD630 was induced by phenyldecane as well as by *n*-hexadecane, suggesting the involvement of the same enzymatic system for the catabolism of both hydrocarbons (Alvarez et al. 2002). The presence of phenylnonanoic acid and phenylpropionic acid in lipid inclusions suggested the occurrence of α-oxidation in phenyldecane-grown cells, to some extent. The resulting

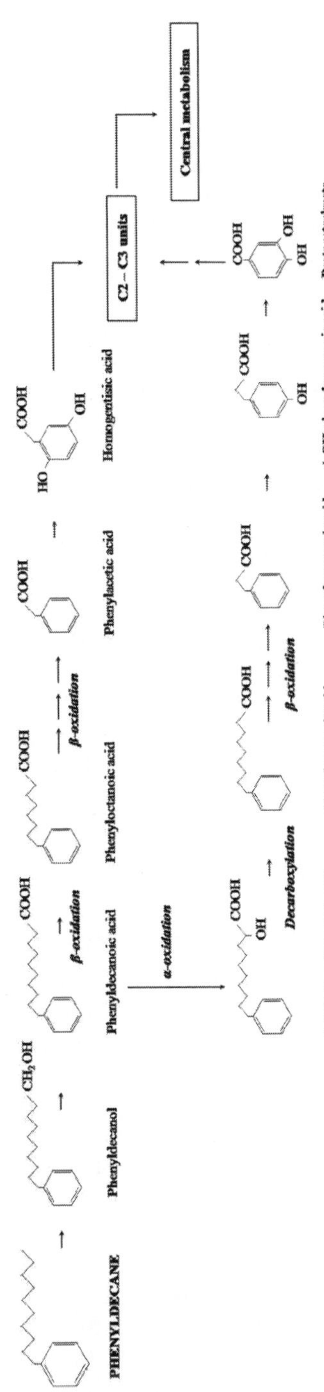

Figure 2. Phenyldecane catabolic pathway in *Rhodococcus opacus* PD630.

phenyl-fatty acids were subsequently catabolized by β-oxidation to phenylacetic or phenylpropionic acid, respectively. These intermediates were probably hydroxylated prior cleavage of the aromatic ring producing central metabolites (Fig. 2) (Alvarez et al. 1996). Although strain PD630 possesses the necessary enzymes for the aromatic ring fission of phenylacetic acid or phenylpropionic acid intermediates, these enzymes were not utilized for the attack of the aromatic ring of phenyldecane, which was rather degraded firstly at the aliphatic side chain. This suggested that the presence of the alkyl side-chain in the aromatic molecule may hinder the oxidative attack by PD630 cells. Similar results have been reported for *R. erythropolis* during cultivation on phenylalkanes (Herter et al. 2012).

On the other hand, the presence of TAG containing fatty acid profiles, similar to that of gluconate-grown cells indicated that *de novo* fatty acid biosynthesis continues active during growth on phenyldecane under nitrogen-limiting conditions; thus, PD630 used acyl-CoA residues produced from the oxidation of phenyldecanoic acid for *de novo* synthesizing fatty acids, which were then incorporated into TAG. Moreover, PD630 cells directly utilized the monoterminal-oxidation products of phenyldecane (phenyldecanoic acid and phenyldecanol) for the biosynthesis of aromatic WS (Alvarez et al. 2002). These novel TAG and WS molecules containing phenyl-residues have been only reported for rhodococci (Fig. 1b).

Monoaromatic hydrocarbons

Benzene, toluene, xylene isomers and ethylbenzenes are usual components of crude oil and derived hydrocarbon mixtures. These hydrocarbons are commonly known as BTEX. Biodegradation pathways for BTEX are well-known in bacteria. The first step is the oxidative attack of the aromatic ring to form catechol which is further metabolized *via* an *ortho-* or *meta-*cleavage (between both hydroxyl groups or next to them, respectively) to produce intermediates from central metabolism (Martínková et al. 2009, Yam et al. 2010). *Rhodococcus* sp. DK17, which is able to grow on toluene, possesses both degradation routes (Kim et al. 2002). However, transformation of monoaromatic hydrocarbons in neutral lipids under nitrogen starvation was poorly studied. Hori et al. (2009) investigated TAG and poly (3-hydroxybutyrate-co-3-hydroxyvalerate) (PHBV) biosynthesis from toluene in *R. aetherivorans* IAR1. This strain was able to degrade toluene under growth-restricting conditions and synthesize TAG after approximately 80 hours of incubation. The accumulated TAG were composed by saturated and unsaturated odd- and even-fatty acids, with carbon chain length between C_{14} and C_{18} (Fig. 1a) (Table 3). This fatty acid composition was similar to that obtained from culturing strain IAR1 in acetate as the sole carbon source (Hori et al. 2009). These results suggested that toluene-grown cells produce metabolic precursors for the *de novo* fatty acid biosynthesis, which generates acyl-residues for TAG biosynthesis.

Polyaromatic compounds

PAH are compounds widely distributed in polluted environments which can possess between two to seven aromatic rings. Many of them are considered as priority pollutants to the Environmental Protection Agency (EPA) due to their toxicity, mutagenicity and carcinogenicity. Naphthalene is the simplest PAH (two aromatic rings) and it is generally used as a model compound for PAH degradation studies. A rhodococcal strain isolated from a chronically hydrocarbon-polluted soil in Patagonia (Argentina) (strain 602) was able to degrade naphthyl-hydrocarbons (naphthalene and naphthyl-1-dodecanoate) still in the total absence of the nitrogen source (Silva et al. 2010). Under such conditions, naphthalene and naphthyl-1-dodecanoate were used by cells not for growth, but for the production of TAG, which were intracellularly accumulated. Naphthalene-grown cells produced TAG containing odd- and even-numbered linear fatty acids with carbon chain length ranging from C_{14} to C_{18} carbon atoms, similar to fatty acids synthesized by this strain from gluconate as sole carbon source (Fig. 1a) (Table 3). On the other hand, strain 602, which was taxonomically close related to *R. jostii,* was able to accumulate TAG containing C_8 to C_{16} fatty acids (Fig. 1c). These results suggested that naphthyl-1-dodecanoate molecule was firstly attacked by an esterase enzyme yielding 1-naphthol and dodecanoic acid residues as oxidation products. Short fatty acids (C_8, C_{10} and C_{12}) occurring in TAG from naphthyl-1-dodecanoate-grown cells were derived from β-oxidation of dodecanoic acid residues (Silva et al. 2010). In addition to the lipids, some oxidation products of naphthalene and naphthyl-1-dodecanoate were detected in cells after chemical analyses, with 1-naphthol, 1,2,3,4-tetrahydro-1-hydroxynaphthalene and 4-hydroxy-1-tetralone as main compounds (Fig. 3) (Silva et al. 2010). These results allowed us to propose new oxidation reactions for naphthyl-hydrocarbons in rhodococci. Naphthalene catabolism has been extensively studied with reference to bacteria. Generally, bacteria use dioxygenase enzymes to catalyze a ring cleavage producing 1,2- or 2,3-dihydroxynaphthalene firstly, and then salicylic or benzoic acid, respectively (Bosch et al. 1999, Annweiler et al. 2000, Kulakov et al. 2005). Salycilate metabolism can further continue *via*-catechol or *via*-gentisate, as has been reported for *Pseudomonas, Ralstonia* and *Rhodococcus*, respectively (Grund et al. 1992, Bosch et al. 2000, Di Gennaro et al. 2001, Zhou et al. 2001, Kulakov et al. 2005). On the other hand, catabolism of benzoic acid usually proceeds *via*-catechol in naphthalene-degrading bacteria (Annweiler et al. 2000, Martínková et al. 2009). In *R. jostii* RHA1, the previous intermediate to benzoic acid, phthalate, can be incorporated in

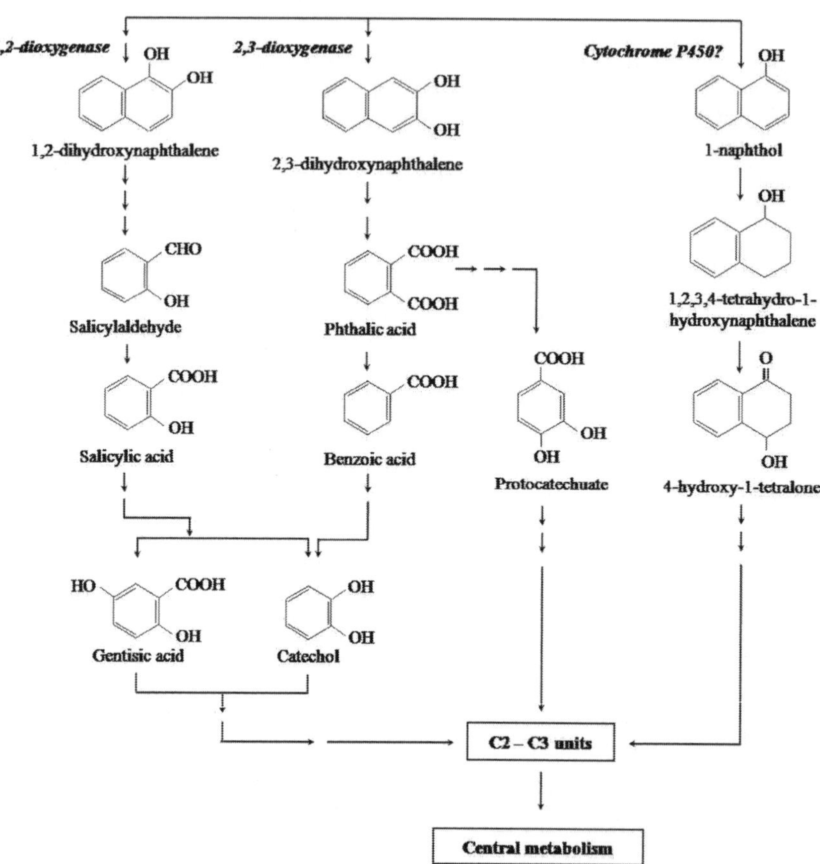

Figure 3. Naphthalene catabolic pathways in rhodococci.

central metabolism after conversion to protocatechuate (Fig. 3) (Patrauchan et al. 2008). *R. jostii* strain 602 utilized a different catabolic pathway for naphthalene degradation, in which a monooxygenase converts naphthalene in 1-naphthol as the first oxidative step (Fig. 3). The same reaction was also reported for *B. cereus* and *Streptomyces griseus* during growth on this aromatic hydrocarbon (Cerniglia et al. 1984, Gopishetty et al. 2007). Naphthalene may be transformed in 1,2-naphthalene oxide intermediate which can spontaneously isomerize to 1-naphthol by a reaction known as NIH shift (Cerniglia et al. 1984). Afterwards, strain 602 may degrade 1-naphthol by a series of hydroxylations and reductions of the aromatic ring yielding 1,2,3,4-

tetrahydro-1-hydroxynaphthtalene and 4-hydroxy-1-tetralone (Fig. 3) (Silva et al. 2010). These compounds are completely metabolized by the TCA cycle. These oxidation intermediates were also reported in *Pseudomonas* and *Streptomyces* strains during degradation of phenanthrene and naphthalene, respectively (Samanta et al. 1999, Gopishetty et al. 2007).

Isoprenoid hydrocarbons

Pristane and phytane, the most known branched isoprenoid hydrocarbons, are ubiquitous and normally occur in crude oil. In addition, they are derived from the biodegradation of the phytol-side chain of chlorophyll. In general, they are used as biomarkers of different habitats, including crude oil source rock depositional environments (Hughes et al. 1995) and as controls for the progress of biodegradation processes because they are considered recalcitrant hydrocarbons (Ferguson et al. 2003, Quek et al. 2006).

Pristane (2,6,10,14-tetramethylpentadecane) biodegradation has been well studied in actinobacteria (McKenna and Kallio 1971, Cox et al. 1972, Nakajima and Sato 1983, Warhurst and Fewson 1994, Alvarez et al. 2001, Nhi-Cong et al. 2009), including in rhodococci (Whyte et al. 1998, Sharma and Pant 2000, Rapp and Gabriel-Jürgens 2003, Kunihiro et al. 2005). Pristane is degraded by at least three catabolic pathways *via* mono-, di- (monoterminal oxidation followed by ω-oxidation) and subterminal oxidation (Fig. 4) (Nhi-Cong et al. 2009). These pathways yield 2,6,10,14-tetramethylpentadecanol, 2,6,10,14-tetramethylpentadecanediol and 2,6,10,14-tetramethylpentadecan-3-ol as first oxidation products, respectively (Nhi-Cong et al. 2009). These intermediates can be subsequently oxidized by different reactions up to 2-methylpropanoic, 2-methylpropanedioic and 2-methylbutanedioic acid, respectively (Fig. 4). 2,6,10,14-tetramethylpentadecan-3-one originated from subterminal oxidation is firstly transformed by a Baeyer-Villiger reaction to an intermediate ester which is spontaneously split by an esterase to yield 3,7,11-trimethyldodecanol or 4,8,12-trimethyltridecanoic acid. Then, the isoprenoid alcohol is transformed to the corresponding fatty acid and ω-oxidized to a dioic acid. Both branched fatty acids (2,6,10-trimethyldodecanedioic and 4,8,12-trimethyltridecanoic acid) are further β-oxidized up to methyl-C_4-dioic or methyl-C_3 fatty acids (Fig. 4) (Nhi-Cong et al. 2009). 4,8,12-trimethyltridecanoic acid can also be derived from β-oxidation of 2,6,10,14-tetramethylpentadecanoic acid. On the other hand, 2,6,10,14-tetramethylpentadecanedioic acid from di-terminal oxidation is degraded by β-oxidation to methyl-C_3-dioic acid (Nhi-Cong et al. 2009). These C_3–C_4 fatty acids are then metabolized *via* central metabolism.

Similar to rhodococci, related actinobacteria belonging to *Nocardia* and *Mycobacterium* genera were able to degrade branched *n*-alkanes such as pristane and phytane under nitrogen-limiting conditions and under absence

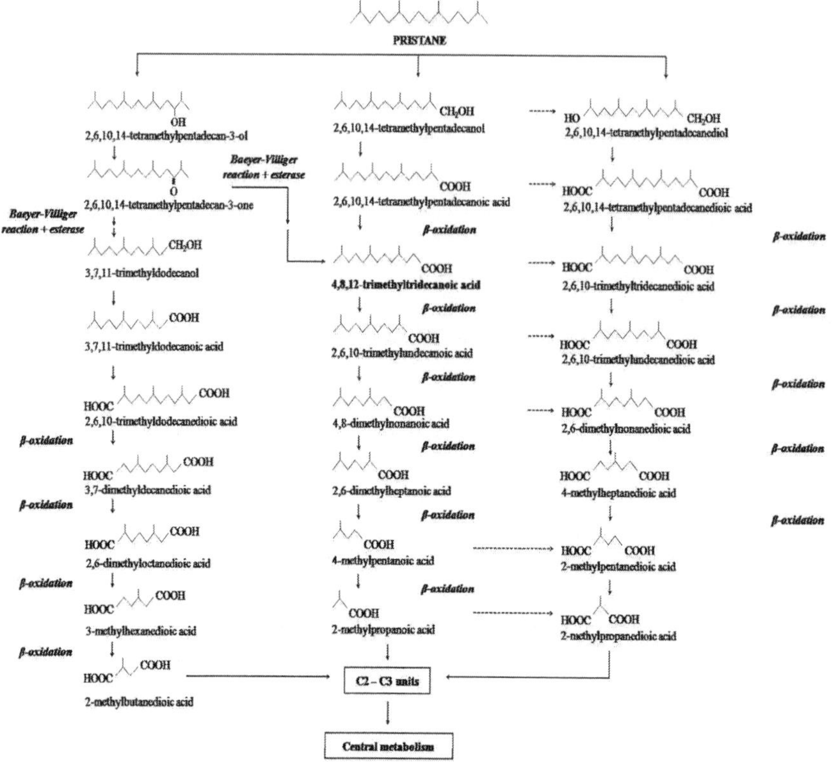

Figure 4. Pristane catabolic pathways in members of *Rhodococcus* and *Mycobacterium* genera. Compounds in bold were incorporated in triacylglycerols.

of the nitrogen source. In this context, *Nocardia globerula* 432 was able to transform pristane into a mixture of TAG containing isoprenyl-residues (Alvarez et al. 2001). Pristane-grown cells produced TAG containing saturated and unsaturated odd- and even-fatty acids (C_{14} to C_{18}) plus 4,8,12-trimethyltridecanoic acid residues (Fig. 1d). Although 2,6,10,14-tetramethylpentadecanoic acid and 4,8,12-trimethyltridecan-4-olide were also detected in cells of strain 432, they were not present in the TAG fraction (Alvarez et al. 2001).

In a similar way, *Mycobacterium ratisbonense* SD4, which is a well known branched alkane-degrader (Berekaa and Steinbüchel 2000), was able to degrade phytane (2,6,10,14-tetramethylhexadecane) still under growth restricting conditions, and produce a complex mixture of WS as storage lipids from this isoprenoid hydrocarbon (Silva et al. 2007). Strain SD4 synthesized nine isoprenoid wax esters from phytane, three of which were detected in high proportions. The most abundant WS (43.8 and 34.8% of the total WS) were formed by 2,6,10,14-tetramethylhexadecanol and

2,6,10,14-tetramethylhexadecanoic or 4,8,12-trimethyltetradecanoic acid (Silva et al. 2007). The chemical structure of the third major WS (16.4%) could not be totally clarified, although may be constituted by 2,6,10,14-tetramethylhexadecanol or 3,7,11,15-tetramethylhexadecan-2-ol with one of its corresponding fatty acids (Silva et al. 2007). Other additional fatty acids detected in WS to a lesser extent were 3,7,11-trimethyldodecanoic, 2,6,10-trimethyldodecanoic or 4,8-dimethyldecanoic acids, and an unsaturated C_{20} isoprenoid acid (different from phytenic acid) (Silva et al. 2007). No TAG were detected in phytane-grown cells of strain SD4, despite the accumulation of TAG during cell cultivation on the non-branched analogous of phytane, *n*-hexadecane (Silva et al. 2007). Apparently, strain SD4 was not able to produce the intermediate key for TAG biosynthesis, glycerol-3-phosphate, in an efficient way from phytane. In contrast, isoprenoid alcohols and acids derived from the oxidation of phytane were available in the metabolism for the biosynthesis of WS, which are synthesized by the same bacterial acyltransferase enzyme like TAG, as discussed below.

According to the results of different studies, phytane can be degraded at least by four different pathways in actinobacteria (Fig. 5) (Nakajima et al. 1985, Silva et al. 2007). The main route is the monoterminal oxidation at the C1 of the isopropyl side yielding 2,6,10,14-tetramethylhexadecan-1-ol (Fig. 5). This compound is subsequently degraded by β-oxidation up to C_2-C_3 carbon units. A secondary pathway involves oxidation at the subterminal carbon of the ethyl-terminus (Cω-1), yielding 3,7,11,15-tetramethylhexadecan-2-ol (Fig. 5). This alcohol is oxidized to the corresponding ketone which is further transformed by a series of reactions (including a Baeyer-Villiger oxidation with peracids) to 3,7,11-trimethyldodecanoic acid; subsequent decarboxylation and β-oxidation of this last compound produce C_2–C_3 carbon units (Fig. 5) (Silva et al. 2007). Oxidation at the terminal carbon of the ethyl-terminus (Cω) and at the methyl group situated α to this ethyl-terminus also occur to a minor extent, yielding 3,7,11-trimethyldodecanoic acid and 2-ethyl-6,10,14-trimethylpentadecanoic acid, respectively (Fig. 5) (Silva et al. 2007). Sakai et al. (2004) proposed that a specific α-methylacyl-CoA racemase is involved in β-oxidation of isoprenoid alkanes in *Mycobacterium*. Bioinformatic analyses of rhodococcal genomes confirmed the existence of two to four paralogous of this enzyme in *R. jostii* RHA1, *R. erythropolis* PR4, *R. opacus* B4, *R. equi* 103S and *R. fascians* F7. The enzyme reported by Sakai et al. (2004) is steoreospecific for the (*S*) configuration of the methyl-branched alkanes, meanwhile their (*R*) isomers inhibit this desaturation step and constitute themselves end-products. An alternative route to detoxify these (*R*) isomers may be their incorporation into TAG or WS.

All these studies highlight the flexibility of actinobacterial metabolism for degrading diverse hydrocarbons still under growth restricting conditions and for producing TAG and/or WS from their oxidation products. The low

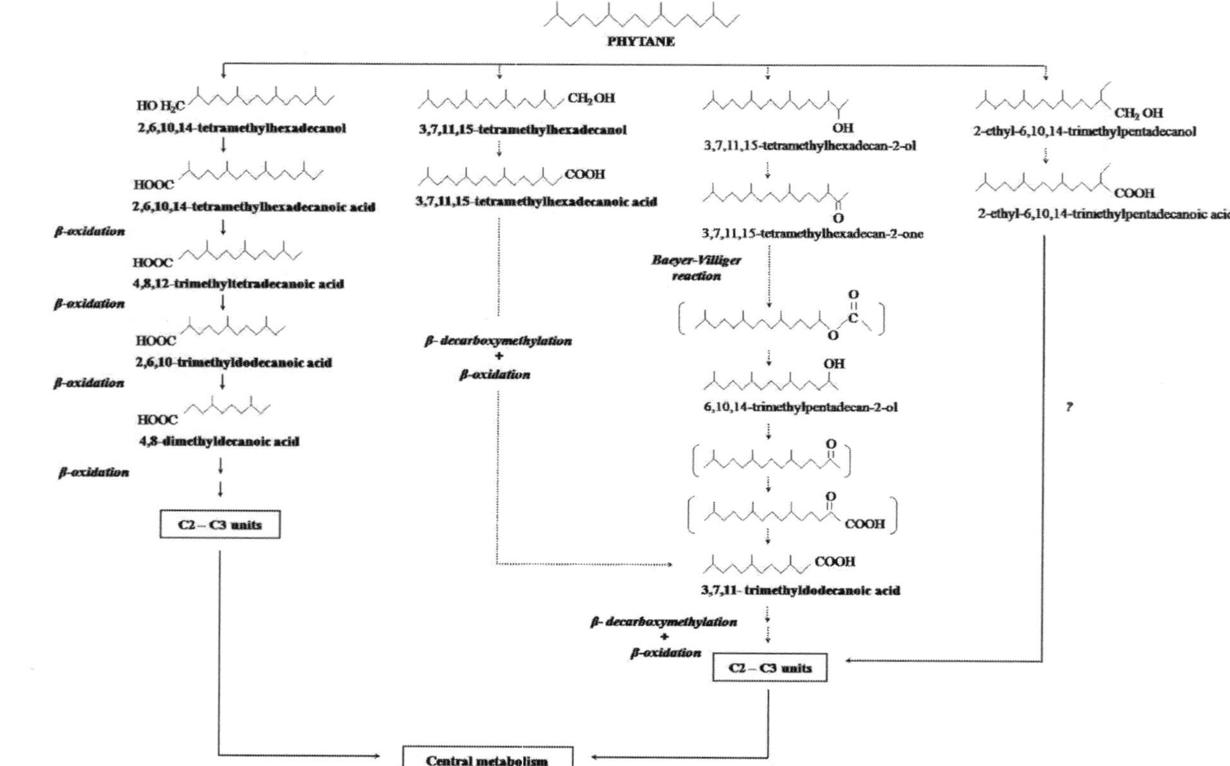

Figure 5. Phytane catabolic pathways in *Mycobacterium ratisbonense* SD4. Compounds in bold were incorporated in wax esters.

specificity of WS/DGAT enzymes, which catalyzes the last reaction for the biosynthesis of WS and/or TAG in bacteria (Fig. 6) (Kalscheuer and Steinbüchel. 2003), seems to be responsible for the ability of actinobacteria to produce WS and TAG from a high diversity of carbon sources with different chemical structures. WS/DGAT enzymes, which simultaneously exhibit both diacylglycerol acyltransferase and acyl-CoA:fatty alcohol acyltransferase (wax ester synthase, WS) activities are usually very promiscuous enzymes with the ability to accept a high diversity of acyl- and alcohol-residues as substrates (Wältermann et al. 2007). Whereas only one or a few WS/DGAT occur in Gram-negative bacteria able to produce WS and TAG, a high redundancy of these enzymes occurs in most TAG-accumulating actinomycetes bacteria, such as the genera *Mycobacterium*, *Nocardia* and *Rhodococcus*. Daniel et al. (2004) identified 15 genes as putative WS/DGAT in *M. tuberculosis* strain H37Rv, which exhibited acyltransferase activity when expressed in *E. coli*. Moreover, a genome-wide bioinformatic analysis of key genes encoding metabolism of diverse storage lipids by *R. jostii* RHA1 identified 14 genes encoding putative WS/DGAT enzymes likely involved in TAG and wax esters biosynthesis (Hernández et al. 2008), whereas 10 WS/DGAT genes were identified in *R. opacus* PD630 (Alvarez et al. 2008). The high redundancy of WS/DGAT genes in actinobacteria may permit cells to maintain active lipid accumulation in different environments and/or to incorporate alcohol- and acyl-residues with different chemical structures into storage lipids. The role of TAG/WS as acceptor of unusual fatty acids, which may be generated by the catabolism of hydrocarbon-degrading rhodococci and related actinobacteria, could be considered as a mechanism for protecting the integrity and functionality of cellular membranes (Alvarez and Steinbüchel 2002, Alvarez 2006). These kinds of

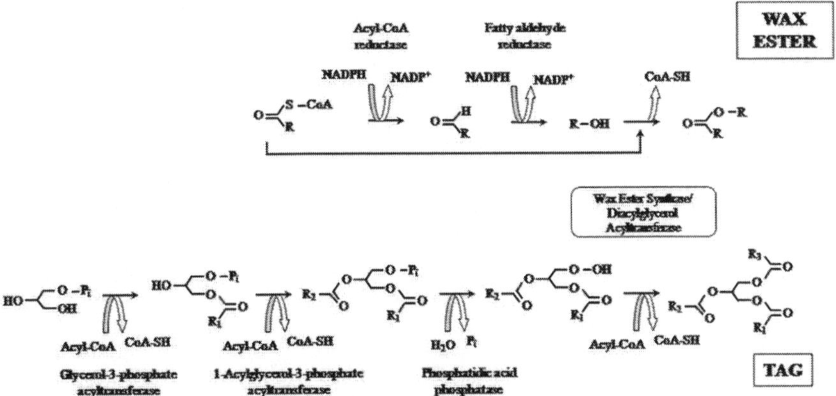

Figure 6. Biosynthesis of TAG and wax ester in bacteria. R, acyl residue; Pi, inorganic phosphate, CoA, coenzyme A. Figure adapted from Wältermann et al. (2007).

fatty acids may otherwise disturb membrane fluidity during degradation of hydrocarbons under conditions that normally occur in the environment (Alvarez and Steinbüchel 2002, Alvarez 2006). Thus, rhodococci and related actinomycetes seem to possess metabolic mechanisms that permit cells to maintain the physiological conditions of cytoplasmic membranes during degradation of hydrocarbons under growth restricting conditions as usually occur in natural environments.

Concluding Remarks

The degradation of diverse hydrocarbons is a common feature among rhodococci. These microorganisms are endowed with multiple catabolic enzymes and routes for biodegradation of a broad range of hydrocarbons. In addition, their physiology and metabolism seem to be flexible enough to continue hydrocarbon degradation in a wide range of environmental conditions. In this context, rhodococci and related actinobacteria are still able to maintain catabolic activity under conditions that restrict growth, utilizing the intermediates generated by the incomplete oxidation of the hydrocarbon for the biosynthesis and accumulation of storage lipids. This is an interesting property for considering such microorganisms as candidates for bioremediation of polluted environments. In addition, the metabolism of these non-growing cells seems to be adaptable to a high diversity of hydrocarbons, such as *n*-alkanes, branched alkanes, phenylalkanes, monoaromatic hydrocarbons and polyaromatic compounds, among other possible pollutants. Thus, rhodococcal cells are able to balance their metabolism according to the changes of environmental conditions. Metabolic diversity and flexibility seems to be essential components of the adaptation of rhodococci to soil environment, and probably permit cells to survive under a wide range of environmental conditions.

Acknowledgements

H.M. Alvarez and R.A. Silva are career investigators of the Consejo Nacional de Investigaciones Científicas y Técnicas (CONICET), Argentina.

References

Aggarwal, S., I.A. Karimi and D.Y. Lee. 2011. Reconstruction of a genome-scale metabolic network of *Rhodococcus erythropolis* for desulfurization studies. Mol. Biosyst. 7: 3122–3131.

Akhtar, N., M.A. Ghauri, M.A. Anwar and K. Akhtar. 2009. Analysis of the dibenzothiophene metabolic pathway in a newly isolated *Rhodococcus* spp. FEMS Microbiol. Lett. 301: 95–102.

Alonso-Gutiérrez, J., A. Figueras, J. Albaigés, N. Jiménez, M. Viñas, A.M. Solanas and B. Novoa. 2009. Bacterial communities from shoreline environments (costa da morte, northwestern Spain) affected by the prestige oil spill. Appl. Environ. Microbiol. 75: 3407–3418.

Alvarez, H.M. 2003. Relationship between β-oxidation pathway and the hydrocarbon-degrading profile in actinomycetes bacteria. Inter. Biodeter. Biodeg. 52: 35–42.

Alvarez, H.M. Bacterial Triacylglycerols. pp. 159–176. *In*: L.T. Welson [ed.]. 2006. Triglycerides and Cholesterol Research. Nova Science Publishers, Inc., Hauppauge, New York, USA.

Alvarez, H.M. Central metabolism of species of the genus *Rhodococcus*. pp. 91–108. *In*: H.M. Alvarez [ed.]. 2010. Biology of *Rhodococcus*. Microbiology Monographs Series. Springer Verlag, Heidelberg, Germany.

Alvarez, H.M. and A. Steinbüchel. 2002. Triacylglycerols in prokaryotic microorganisms. Appl. Microbiol. Biotechnol. 60: 367–376.

Alvarez, H.M. and A. Steinbüchel. Physiology, biochemistry and molecular biology of triacylglycerol accumulation by *Rhodococcus*. pp. 263–290. *In*: H.M. Alvarez [ed.]. 2010. Biology of *Rhodococcus*. Microbiology Monographs Series. Springer Verlag, Heidelberg, Germany.

Alvarez, H.M., F. Mayer, D. Fabritius and A. Steinbüchel. 1996. Formation of intracytoplasmic lipid inclusions by *Rhodococcus opacus* strain PD630. Arch. Microbiol. 165: 377–386.

Alvarez, H.M., R. Kalscheuer and A. Steinbüchel. 1997. Accumulation of storage lipids in species of *Rhodococcus* and *Nocardia* and effect of inhibitors and polyethylene glycol. Fett/Lipid. 99: 239–246.

Alvarez, H.M., R. Kalscheuer and A. Steinbüchel. 2000. Accumulation and mobilization of storage lipids by *Rhodococcus opacus* PD630 and *Rhodococcus ruber* NCIMB 40126. Appl. Microbiol. Biotechnol. 54: 218–223.

Alvarez, H.M., M.F. Souto, A. Viale and O.H. Pucci. 2001. Biosynthesis of fatty acids and triacylglycerols by 2,6,10,14-tetramethyl pentadecane-grown cells of *Nocardia globerula* 432. FEMS Microbiol. Lett. 200: 195–200.

Alvarez, H.M., H. Luftmann, R.A. Silva, A.C. Cesari, A. Viale, M. Wältermann and A. Steinbüchel. 2002. Identification of phenyldecanoic acid as constituent of triacylglycerols and wax ester produced by *Rhodococcus opacus* PD630. Microbiology. 148: 1407–1412.

Alvarez, H.M., R.A. Silva, A.C. Cesari, A.L. Zamit, S.R. Peressutti, R. Reichelt, U. Keller, U. Malkus, C. Rasch, T. Maskow, F. Mayer and A. Steinbüchel. 2004. Physiological and morphological responses of the soil bacterium *Rhodococcus opacus* strain PD630 to water stress. FEMS Microbiol. Ecol. 50: 75–86.

Alvarez, A.F., H.M. Alvarez, R. Kalscheuer, M. Wältermann and A. Steinbüchel. 2008. Cloning and characterization of a gene involved in triacylglycerol biosynthesis and identification of additional homologous genes in the oleaginous bacterium *Rhodococcus opacus* PD630. Microbiology. 154: 2327–2335.

Amouric, A., M. Quéméneur, V. Grossi, P.P. Liebgott, R. Auria and L. Casalot. 2010. Identification of different alkane hydroxylase systems in *Rhodococcus ruber* strain SP2B, an hexane-degrading actinomycete. J. Appl. Microbiol. 108: 1903–1916.

Andreoni, V., L. Cavalca, M.A. Rao, G. Nocerino, S. Bernasconi, E. Dell'Amico, M. Colombo and L. Gianfreda. 2004. Bacterial communities and enzyme activities of PAHs polluted soils. Chemosphere. 57: 401–412.

Annweiler, E., H.H. Richnow, G. Antranikian, S. Hebenbrock, C. Garms, S. Franke, W. Francke and W. Michaelis. 2000. Naphthalene degradation and incorporation of naphthalene-derived carbon into biomass by the thermophile *Bacillus thermoleovorans*. Appl. Environ. Microbiol. 66: 518–523.

Araki, N., Y. Niikura, K. Miyauchi, D. Kasai, E. Masai and M. Fukuda. 2011. Glucose-mediated transcriptional repression of PCB/biphenyl catabolic genes in *Rhodococcus jostii* RHA1. J. Mol. Microbiol. Biotechnol. 20: 53–62.

Auffret, M., D. Labbé, G. Thouand, C.W. Greer and F. Fayolle-Guichard. 2009. Degradation of a mixture of hydrocarbons, gasoline, and diesel oil additives by *Rhodococcus aetherivorans* and *Rhodococcus wratislaviensis*. Appl. Environ. Microbiol. 75: 7774–7782.

Berekaa, M.M. and A. Steinbüchel. 2000. Microbial degradation of the multiply branched alkane 2,6,10,15,19,23-hexamethyltetracosane (squalane) by *Mycobacterium fortuitum* and *Mycobacterium ratisbonense*. Appl. Environ. Microbiol. 66: 4462–4467.

Bosch, R., E. García-Valdés and E.R.B. Moore. 1999. Genetic characterization and evolutionary implications of a chromosomally encoded naphthalene-degradation upper pathway from *Pseudomonas stutzeri* AN10. Gene. 236: 149–157.

Bosch, R., E. García-Valdés and E.R. Moore. 2000. Complete nucleotide sequence and evolutionary significance of a chromosomally encoded naphthalene-degradation lower pathway from *Pseudomonas stutzeri* AN10. Gene. 245: 65–74.

Bosma, T., M.G. Pikkemaat, J. Kingma, J. Dijk and D.B. Janssen. 2003. Steady-state and pre-steady-state kinetic analysis of halopropane conversion by a *Rhodococcus* haloalkane dehalogenase. Biochemistry. 42: 8047–8053.

Cappelletti, M., S. Fedi, D. Frascari, H. Ohtake, R.J. Turner and D. Zannoni. 2011. Analyses of both the *alk*B gene transcriptional start site and *alk*B promoter-inducing properties of *Rhodococcus* sp. strain BCP1 grown on *n*-alkanes. Appl. Environ. Microbiol. 77: 1619–1627.

Cerniglia, C.E., J.P. Freeman and F.E. Evans. 1984. Evidence for an arene oxide-NIH shift pathway in the transformation of naphthalene to 1-naphthol by *Bacillus cereus*. Arch. Microbiol. 138: 283–286.

Chaîneau, C.H., J. Morel, J. Dupont, E. Bury and J. Oudot. 1999. Comparison of the fuel oil biodegradation potential of hydrocarbon-assimilating microorganisms isolated from a temperate agricultural soil. Sci. Total Environ. 227: 237–247.

Cox, R.E., J.R. Maxwell, R.G. Ackman and S.N. Hooper. 1972. Stereochemical studies of acyclic isoprenoid compounds. III. The stereochemistry of naturally occurring (Marine) 2,6,10,14-tetramethylpentadecane. Can. J. Biochem. 50: 1238–1241.

Daniel, J., C. Deb, V.S. Dubey, T.D. Sirakova, B. Abomoelak, H.R. Morbidoni and P.E. Kolattukudy. 2004. Induction of a novel class of diacylglycerol acyltransferases and triacylglycerol accumulation in *Mycobacterium tuberculosis* as it goes into a dormancy-like state in culture. J. Bacteriol. 186: 5017–5030.

Dark, J. 2005. Annual lipid cycles in hibernators: integration of physiology and behaviour. Annu. Rev. Nutr. 25: 469–497.

Di Gennaro, P., E. Rescalli, E. Galli, G. Sello and G. Bestetti. 2001. Characterization of *Rhodococcus opacus* R7, a strain able to degrade naphthalene and *o*-xylene isolated from a polycyclic aromatic hydrocarbon-contaminated soil. Res. Microbiol. 152: 641–651.

Dowell, P., Z. Hu and M.D. Lane. 2005. Monitoring energy balance: metabolites of fatty acid synthesis as hypothalamic sensors. Annu. Rev. Biochem. 74: 515–534.

Erable, B., I. Goubet, S. Lamare, M.D. Legoy and T. Maugard. 2006. Bioremediation of halogenated compounds: comparison of dehalogenating bacteria and improvement of catalyst stability. Chemosphere. 65: 1146–1152.

Etemadifar, Z. and G. Emtiazi. 2008. Microtitre plate assay for biofilm formation, production and utilization of hydroxybiphenyl by *Rhodococcus* sp. isolated from gasoline-contaminated soil. Z. Naturforsch. C. 63: 599–604.

Ferguson, S.H., P.D. Franzmann, A.T. Revill, I. Snape and J.L. Rayner. 2003. The effects of nitrogen and water on mineralisation of hydrocarbons in diesel-contaminated terrestrial Antarctic soils. Cold Reg. Sci. Technol. 37: 197–212.

Fernández de Las Heras, L., E. García Fernández, J.M. Navarro Llorens, J. Perera and O. Drzyzga. 2009. Morphological, physiological, and molecular characterization of a newly isolated steroid-degrading actinomycete, identified as *Rhodococcus ruber* strain Chol-4. Curr. Microbiol. 59: 548–553.

Gopishetty, S.R., J. Heinemann, M. Deshpande and J.P.N. Rosazza. 2007. Aromatic oxidations by *Streptomyces griseus*: biotransformations of naphthalene to 4-hydroxy-1-tetralone. Enzyme Microb. Technol. 40: 1622–1626.

Grund, E., B. Denecke, and R. Eichenlaub. 1992. Naphthalene degradation via salicylate and gentisate by *Rhodococcus* sp. strain B4. Appl. Environ. Microbiol. 58: 1874–1877.

Heald, S.C., P.F. Brandão, R. Hardicre and A.T. Bull. 2001. Physiology, biochemistry and taxonomy of deep-sea nitrile metabolising *Rhodococcus* strains. Antonie Van Leeuwenhoek. 80: 169–183.

Hernández, M.A., W.W. Mohn, E. Martínez, E. Rost, A.F. Alvarez and H.M. Alvarez. 2008. Biosynthesis of storage compounds by *Rhodococcus jostii* RHA1 and global identification of genes involved in their metabolism. BMC Genomics. 12; 9(1): 600.

Hernández, M.A. and H.M. Alvarez. 2010. Glycogen formation by *Rhodococcus* species and effect of inhibition of lipid biosynthesis on glycogen accumulation in *R. opacus* PD630. FEMS Microbiol. Lett. 312: 93–99.

Hernandez-Raquet, G., H. Budzinski, P. Caumette, P. Dabert, K. Le Ménach, G. Muyzer and R. Duran. 2006. Molecular diversity studies of bacterial communities of oil polluted microbial mats from the Etang de Berre (France). FEMS Microbiol. Ecol. 58: 550–562.

Herter, S., A. Mikolasch and F. Schauer. 2012. Identification of phenylalkane derivatives when *Mycobacterium neoaurum* and *Rhodococcus erythropolis* were cultured in the presence of various phenylalkanes. Appl. Microbiol. Biotechnol. 93: 343–355.

Hori, K., M. Abe and H. Unno. 2009. Production of triacylglycerol and poly(3-hydroxybutyrate-co-3-hydroxyvalerate) by the toluene-degrading bacterium *Rhodococcus aetherivorans* IAR1. J. Biosci. Bioeng. 108: 319–324.

Huang, L., T. Ma, D. Li, F.L. Liang, R.L. Liu and G.Q. Li. 2008. Optimization of nutrient component for diesel oil degradation by *Rhodococcus erythropolis*. Mar. Pollut. Bull. 56: 1714–1718.

Hughes, W.B., A.G. Holba and L.I.P. Dzou. 1995. The ratios of dibenzothiophene to phenanthrene and pristane to phytane as indicators of depositional environment and lithology of petroleum source rocks. Geochim. Cosmochim. 59: 3581–3598.

Huisman, G.W., D.A. Siegele, M.M. Zambrano and R. Kolter. Morphological and physiological changes during stationary phase. pp. 1672–1682. *In*: F.C. Neidhardt, J.L. Ingraham, K.B. Low, B. Magsanik, M. Schaechter, and H.E. Umbarger [eds.]. 1993. *Escherichia coli* and *Salmonella typhimurium*: Cellular and Molecular Biology. American Society for Microbiology, Washington, DC, USA.

Ivshina, I.B., T.N. Kamenskikh and Y.E. Liapunov. 1994. IEGM catalogue of strains of regional specialized collection of alkanotrophic microorganisms. Nauka, Moscow.

Ivshina, I.B., M.V. Berdichevskaya, L.V. Zvereva, L.V. Rybalka and E.A. Elovikova. 1995. Phenotypic characterization of alkanotrophic rhodococci from various ecosystems. Microbiology. 64: 507–513.

Kalscheuer, R. and A. Steinbüchel. 2003. A novel bifunctional wax ester synthase/acyl-CoA:diacylglycerol acyltransferase mediates wax ester and triacylglycerol biosynthesis in *Acinetobacter calcoaceticus* ADP1. J. Biol. Chem. 278: 8075–8082.

Kamal, A., M.S. Kumar, C.G. Kumar and T. Shaik. 2011. Bioconversion of acrylonitrile to acrylic acid by *Rhodococcus ruber* strain AKSH-84. J. Microbiol. Biotechnol. 21: 37–42.

Kawaguchi, H., H. Kobayashi and K. Sato. 2011. Metabolic engineering of hydrophobic *Rhodococcus opacus* for biodesulfurization in oil-water biphasic reaction mixtures. J. Biosci. Bioeng. (in press).

Kim, D., Y.S. Kim, S.K. Kim, S.W. Kim, G.J. Zylstra, Y.M. Kim and E. Kim. 2002. Monocyclic Aromatic Hydrocarbon Degradation by *Rhodococcus* sp. Strain DK17. Appl. Environ. Microbiol. 68: 3270–3278.

Koronelli, T.V., V.V. Il'inskiĭ, V.A. Ianushka and T.I. Krasnikova. 1987. Hydrocarbon-oxidizing microflora from the water of the Baltic sea and Kurshsky bay polluted after a fuel oil spill. Mikrobiologiia. 56: 472–478.

Kulakov, L.A., S. Chen, C.C.R. Allen and M.J. Larkin. 2005. Web-Type Evolution of *Rhodococcus* Gene Clusters Associated with Utilization of Naphthalene. Appl. Environ. Microbiol. 71: 1754–1764.

Kunihiro, N., M. Haruki, K. Takano, M. Morikawa and S. Kanaya. 2005. Isolation and characterization of *Rhodococcus* sp. strains TMP2 and T12 that degrade 2,6,10,14-

tetramethylpentadecane (pristane) at moderately low temperatures. J. Biotechnol. 115: 129–136.

Kuyukina, M.S. and I.B. Ivshina. Application of *Rhodococcus* in bioremediation of contaminated environments. pp. 231–262. *In*: H.M. Alvarez [ed.]. 2010. Biology of *Rhodococcus*. Microbiology Monographs Series. Springer Verlag, Heidelberg, Germany.

Lahoda, M., R. Chaloupkova, A. Stsiapanava, J. Damborsky and I.K. Smatanova. 2011. Crystallization and crystallographic analysis of the *Rhodococcus rhodochrous* NCIMB 13064 DhaA mutant DhaA31 and its complex with 1,2,3-trichloropropane. Acta Crystallogr. Sect. F. Struct. Biol. Cryst. Commun. 67: 397–400.

Larkin, M.J., L.A. Kulakov and C.C.R. Allen. 2005. Biodegradation and *Rhodococcus*-masters of catabolic versatility. Curr. Opin. Biotechnol. 16: 282–290.

Larkin, M.J., L.A. Kulakov and C.C.R. Allen. Genomes and Plasmids in *Rhodococcus*. pp. 73–90. *In*: H.M. Alvarez [ed.]. 2010. Biology of *Rhodococcus*. Microbiology Monographs Series. Springer Verlag, Heidelberg, Germany.

Li, G.Q., S.S. Li, M.L. Zhang, J. Wang, L. Zhu, F.L. Liang, R.L. Liu and T. Ma. 2008. Genetic rearrangement strategy for optimizing the dibenzothiophene biodesulfurization pathway in *Rhodococcus erythropolis*. Appl. Environ. Microbiol. 74: 971–976.

Lin, T.C., C.C. Young, M.J. Ho, M.S. Yeh, C.L. Chou, Y.H. Wei and J.S. Chang. 2005. Characterization of floating activity of indigenous diesel-assimilating bacterial isolates. J. Biosci. Bioeng. 99: 466–472.

Luz, A.P., V.H. Pellizari, L.G. Whyte and C.W. Greer. 2004. A survey of indigenous microbial hydrocarbon degradation genes in soils from Antarctica and Brazil. Can. J. Microbiol. 50: 323–333.

Ma, T. The desulfurization pathway in *Rhodococcus*. pp. 207–230. *In*: H.M. Alvarez [ed.]. 2010. Biology of *Rhodococcus*. Microbiology Monographs Series. Springer Verlag, Heidelberg, Germany.

Mahmoud, H.M., P. Suleman, N.A. Sorkhoh, S. Salamah and S.S. Radwan. 2011. The potential of established turf cover for cleaning oily desert soil using rhizosphere technology. Int. J. Phytoremediation. 13: 156–167.

Marcos, M.S., M. Lozada and H.M. Dionisi. 2009. Aromatic hydrocarbon degradation genes from chronically polluted Subantarctic marine sediments. Lett. Appl. Microbiol. 49: 602–608.

Margesin, R., D. Labbé, F. Schinner, C.W. Greer and L.G. Whyte. 2003. Characterization of hydrocarbon-degrading microbial populations in contaminated and pristine Alpine soils. Appl. Environ. Microbiol. 69: 3085–3092.

Martínková, L., B. Uhnáková, M. Pátek, J. Nešvera and V. Křen. 2009. Biodegradation potential of the genus *Rhodococcus*. Environ. Int. 35: 162–177.

Martínková, L., M. Pátek, A.B. Schlosserová, O. Kaplan, B. Uhnáková and J. Nesvera. Catabolism of nitriles in *Rhodococcus*. pp. 171–206. *In*: H.M. Alvarez [ed.]. 2010. Biology of *Rhodococcus*. Microbiology Monographs Series. Springer Verlag, Heidelberg, Germany.

Mathieu, J.M., W.W. Mohn, L.D. Eltis, J.C. LeBlanc, G.R. Stewart, C. Dresen, K. Okamoto and P.J. Alvarez. 2010. 7-Ketocholesterol catabolism by *Rhodococcus jostii* RHA1. Appl. Environ. Microbiol. 76: 352–355.

McKenna, E.J. and R.E. Kallio. 1971. Microbial metabolism of the isoprenoid alkane pristane. Proc. Natl. Acad. Sci. USA. 68: 1552–1554.

McLeod, M.P., R.L. Warren, W.W. Hsiao, N. Araki, M. Myhre, C. Fernandes, D. Miyazawa, W. Wong, A.L. Lillquist, D. Wang, M. Dosanjh, H. Hara, A. Petrescu, R.D. Morin, G. Yang, J.M. Stott, J.E. Schein, H. Shin, D. Smailus, A.S. Siddiqui, M.A. Marra, S.J. Jones, R. Holt, F.S. Brinkman, K. Miyauchi, M. Fukuda, J.E. Davies, W.W. Mohn and L.D. Eltis. 2006. The complete genome of *Rhodococcus* sp. RHA1 provides insights into a catabolic powerhouse. Proc. Natl. Acad. Sci. USA. 103: 15582–15587.

Nakajima, K. and A. Sato. 1983. Microbial metabolism of the isoprenoid alkane pristane. Nippon Nogeikagaku Kaishi. 57: 299–305.

Nakajima, K., A. Sato, Y. Takahara and T. Iida. 1985. Microbial oxidation of isoprenoid alkanes, phytane, norpristane and farnesane. Agric. Biol. Chem. 49: 1993–2002.

Newman, J., T.S. Peat, R. Richard, L. Kan, P.E. Swanson, J.A. Affholter, I.H. Holmes, J.F. Schindler, C.J. Unkefer and T.C. Terwilliger. 1999. Haloalkane dehalogenases: structure of a *Rhodococcus* enzyme. Biochemistry. 38: 16105–1614.

Nhi-Cong, L.T., A. Mikolasch, H.P. Klenk and F. Schauer. 2009. Degradation of the multiple branched alkane 2,6,10,14-tetramethyl-pentadecane (pristane) in *Rhodococcus ruber* and *Mycobacterium neoaurum*. Inter. Biodeter. Biodeg. 63: 201–207.

Ohmori, T., H. Morita, M. Tanaka, K. Miyauchi, D. Kasai, K. Furukawa, K. Miyashita, N. Ogawa, E. Masai and M. Fukuda. 2011. Development of a strain for efficient degradation of polychlorinated biphenyls by patchwork assembly of degradation pathways. J. Biosci. Bioeng. 111: 437–442.

Patrauchan, M.A., C. Florizone, S. Eapen, L. Gómez-Gil, B. Sethuraman, M. Fukuda, J. Davies, W.W. Mohn and L.D. Eltis. 2008. Roles of ring-hydroxylating dioxygenases in styrene and benzene catabolism in *Rhodococcus jostii* RHA1. J. Bacteriol. 190: 37–47.

Peng, F., Y. Wang, F. Sun, Z. Liu, Q. Lai and Z. Shao. 2008. A novel lipopeptide produced by a Pacific Ocean deep-sea bacterium, *Rhodococcus* sp. TW53. J. Appl. Microbiol. 105: 698–705.

Peressutti, S.R., H.M. Alvarez and O.H. Pucci. 2003. Dynamic of hydrocarbon-degrading bacteriocenosis of an experimental oil pollution on Patagonic soil. Inter. Biodeter. Biodeg. 52: 21– 30.

Petrusma, M., G. Hessels, L. Dijkhuizen and R. van der Geize. 2011. Multiplicity of 3-Ketosteroid-9α-Hydroxylase enzymes in *Rhodococcus rhodochrous* DSM43269 for specific degradation of different classes of steroids. J. Bacteriol. 193: 3931–3940.

Puglisi, E., M.J. Cahill, P.A. Lessard, E. Capri, A.J. Sinskey, J.A. Archer and P. Boccazzi. 2010. Transcriptional response of *Rhodococcus aetherivorans* I24 to polychlorinated biphenyl-contaminated sediments. Microb. Ecol. 60: 505–515.

Quek, E., Y.P. Ting and H.M. Tan. 2006. *Rhodococcus* sp. F92 immobilized on polyurethane foam shows ability to degrade various petroleum products. Biores. Technol. 97: 32–38.

Radwan, S.S., N.A. Sorkhoh, F. Fardoun and R.H. Al-Hasan. 1995. Soil management enhancing hydrocarbon biodegradation in the polluted Kuwaiti desert. Appl. Microbiol. Biotechnol. 44: 265–270.

Raj, J., S. Prasad, N.N. Sharma, and T.C. Bhalla. 2010. Bioconversion of acrylonitrile to acrylamide using polyacrylamide entrapped cells of *Rhodococcus rhodochrous* PA-34. Folia Microbiol (Praha). 55: 442–446.

Rapp, P. and L.H. Gabriel-Jürgens. 2003. Degradation of alkanes and highly chlorinated benzenes, and production of biosurfactants, by a psychrophilic *Rhodococcus* sp. and genetic characterization of its chlorobenzene dioxygenase. Microbiology. 149: 2879–2890.

Robrock, K.R., W.W. Mohn, L.D. Eltis and L. Alvarez-Cohen. 2011. Biphenyl and ethylbenzene dioxygenases of *Rhodococcus jostii* RHA1 transform PBDEs. Biotechnol. Bioeng. 108: 313–321.

Ruberto, L.A.M., S. Vazquez, A. Lobalbo and W.P. Mac Cormack. 2005. Psychrotolerant hydrocarbon-degrading *Rhodococcus* strains isolated from polluted Antarctic soils. Antarctic Science. 17: 47–56.

Sakai, Y., H. Takahashi, Y. Wakasa, T. Kotani, H. Yurimoto, N. Miyachi, P.P. Van Veldhoven and N. Kato. 2004. Role of alpha-methylacyl coenzyme A racemase in the degradation of methyl-branched alkanes by *Mycobacterium* sp. strain P101. J. Bacteriol. 186: 7214–7220.

Samanta, S.K., A.K. Chakraborti and R.K. Jain. 1999. Degradation of phenanthrene by different bacteria: evidence for novel transformation sequences involving the formation of 1-naphthol. Appl. Microbiol. Biotechnol. 53: 98–107.

Sameshima, Y., K. Honda, J. Kato, T. Omasa and H. Ohtake. 2008. Expression of *Rhodococcus opacus alk*B genes in anhydrous organic solvents. J. Biosci. Bioeng. 106: 199–203.

Sharma, S.L. and A. Pant. 2000. Biodegradation and conversion of alkanes and crude oil by a marine *Rhodococcus* sp. Biodegradation. 11: 289–294.

Silva, R.A. 2009. Relación entre el metabolismo de lípidos y la adaptación a la presencia de contaminantes y a condiciones ambientales en bacterias autóctonas. PhD Thesis. University of Patagonia San Juan Bosco, Comodoro Rivadavia, Argentina.

Silva, R.A., V. Grossi and H.M. Alvarez. 2007. Biodegradation of phytane (2,6,10,14-tetramethylhexadecane) and accumulation of related isoprenoid wax esters by Mycobacterium ratisbonense strain SD4 under nitrogen-starved conditions. FEMS Microbiol. Lett. 272: 220–8.

Silva, R.A., V. Grossi, N. Olivera and H.M. Alvarez. 2010. Characterization of the indigenous *Rhodococcus* sp. 602, a strain able to accumulate triacylglycerides from naphthyl-compounds under nitrogen-starved conditions. Res. Microbiol. 161: 198–207.

Sorkhoh, N.A., R.H. Al-Hasan, M. Khanafer and S.S. Radwan. 1995. Establishment of oil-degrading bacteria associated with cyanobacteria in oil-polluted soil. J. Appl. Bacteriol. 78: 194–199.

Srinivasan, V. and H.J. Morowitz. 2006. Ancient genes in contemporary persistent microbial pathogens. Biol. Bull. 210: 1–9.

Stsiapanava, A., R. Chaloupkova, A. Fortova, J. Brynda, M.S. Weiss, J. Damborsky and I.K. Smatanova. 2011. Crystallization and preliminary X-ray diffraction analysis of the wild-type haloalkane dehalogenase DhaA and its variant DhaA13 complexed with different ligands. Acta Crystallogr. Sect. F. Struct. Biol. Cryst. Commun. 67: 253–257.

Taguchi, K., M. Motoyama, T. Iida and T. Kudo. 2007. Polychlorinated biphenyl/biphenyl degrading gene clusters in *Rhodococcus* sp. K37, HA99, and TA431 are different from well-known *bph* gene clusters of Rhodococci. Biosci. Biotechnol. Biochem. 71: 1136–1144.

Takeda, H., J. Shimodaira, K. Yukawa, N. Hara, D. Kasai, K. Miyauchi, E. Masai and M. Fukuda. 2010. Dual two-component regulatory systems are involved in aromatic compound degradation in a polychlorinated-biphenyl degrader, *Rhodococcus jostii* RHA1. J. Bacteriol. 192: 4741–4751.

Takei, D., K. Washio and M. Morikawa. 2008. Identification of alkane hydroxylase genes in *Rhodococcus* sp. strain TMP2 that degrades a branched alkane. Biotechnol. Lett. 30: 1447–1452.

Tao, F., P. Zhao, Q. Li, F. Su, B. Yu, C. Ma, H. Tang, C. Tai, G. Wu and P. Xu. 2011. Genome sequence of *Rhodococcus erythropolis* XP, a biodesulfurizing bacterium with industrial potential. J. Bacteriol. 193: 6422–6423.

Tomás-Gallardo, L., I. Canosa, E. Santero, E. Camafeita, E. Calvo, J.A. López and B. Floriano. 2006. Proteomic and transcriptional characterization of aromatic degradation pathways in *Rhodococcus* sp. strain TFB. Proteomics. 6: 119–132.

Urai, M., H. Yoshizaki, H. Anzai, J. Ogihara, N. Iwabuchi, S. Harayama, M. Sunairi and M. Nakajima. 2007. Structural analysis of an acidic, fatty acid ester-bonded extracellular polysaccharide produced by a pristane-assimilating marine bacterium, *Rhodococcus erythropolis* PR4. Carbohydr. Res. 342: 933–942.

Wagner-Döbler, I., A. Bennasar, M. Vancanneyt, C. Strömpl, I. Brümmer, C. Eichner, I. Grammel and E.R. Moore. 1998. Microcosm enrichment of biphenyl-degrading microbial communities from soils and sediments. Appl. Environ. Microbiol. 64: 3014–3022.

Wältermann, M., T. Stöveken and A. Steinbüchel. 2007. Key enzymes for biosynthesis of neutral lipid storage compounds in prokaryotes: properties, function and occurrence of wax ester synthases/acyl-CoA: diacylglycerol acyltransferases. Biochimie. 89: 230–242.

Warhurst, A.M. and C.A. Fewson. 1994. Biotransformations catalyzed by the genus *Rhodococcus*. Crit. Rev. Biotechnol. 14: 29–73.

Warton, B., J.N. Matthiessen and M.M. Roper. 2001. The soil organisms responsible for the enhanced biodegradation of metham sodium. Biol. Fertil. Soils. 34: 264–269.

Whyte, L.G., C.W. Greer and W.E. Inniss. 1996. Assessment of the biodegradation potential of psychrotrophic microorganisms. Can. J. Microbiol. 42: 99–106.

Whyte, L.G., J. Hawari, E. Zhou, L. Bourbonnière, W.E. Inniss and C.W. Greer. 1998. Biodegradation of variable-chain-length alkanes at low temperatures by a psychrotrophic *Rhodococcus* sp. Appl. Environ. Microbiol. 64: 2578–2584.

Whyte, L.G., S.J. Slagman, F. Pietrantonio, L. Bourbonnière, S.F. Koval, J.R. Lawrence, W.E. Inniss and C.W. Greer. 1999. Physiological adaptations involved in alkane assimilation at a low temperature by *Rhodococcus* sp. strain Q15. Appl. Environ. Microbiol. 65: 2961–2968.

Whyte, L.G., A. Schultz, J.B. Beilen, A.P. Luz, V. Pellizari, D. Labbé and C.W. Greer. 2002a. Prevalence of alkane monooxygenase genes in Arctic and Antarctic hydrocarbon-contaminated and pristine soils. FEMS Microbiol. Ecol. 41: 141–150.

Whyte, L.G., T.H. Smits, D. Labbé, B. Witholt, C.W. Greer and J.B. van Beilen. 2002b. Gene cloning and characterization of multiple alkane hydroxylase systems in *Rhodococcus* strains Q15 and NRRL B-16531. Appl. Environ. Microbiol. 68: 5933–5942.

Yam, K.C., R. van der Geize and L.D. Eltis. Catabolism of aromatic compounds and steroids by *Rhodococcus*. pp. 133–169. *In*: H.M. Alvarez [ed.]. 2010. Biology of *Rhodococcus*. Microbiology Monographs Series. Springer Verlag, Heidelberg, Germany.

Yam, K.C., S. Okamoto, J.N. Roberts and L.D. Eltis. 2011. Adventures in *Rhodococcus*—from steroids to explosives. Can. J. Microbiol. 57: 155–168.

Yang, X., F. Xie, G. Zhang, Y. Shi and S. Qian. 2008. Purification, characterization, and substrate specificity of two 2,3-dihydroxybiphenyl 1,2-dioxygenase from *Rhodococcus* sp. R04, showing their distinct stability at various temperature. Biochimie. 90: 1530–1538.

Yang, X., R. Xue, C. Shen, S. Li, C. Gao, Q. Wang and X. Zhao. 2011. Genome sequence of *Rhodococcus* sp. strain R04, a polychlorinated-biphenyl biodegrader. J. Bacteriol. 193: 5032–5033.

Yeom, S.J., H.J. Kim, J.K. Lee, D.E. Kim and D.K. Oh. 2008. An amino acid at position 142 in nitrilase from *Rhodococcus rhodochrous* ATCC 33278 determines the substrate specificity for aliphatic and aromatic nitriles. Biochem. J. 415: 401–407.

Yoo, M., D. Kim, G.J. Zylstra, B.S. Kang and E. Kim. 2011. Biphenyl hydroxylation enhanced by an engineered o-xylene dioxygenase from *Rhodococcus* sp. strain DK17. Res. Microbiol. 162: 724–728.

Zhang, Q., M.Y. Tong, Y.S. Li, H.J. Gao and X.C. Fang. 2007. Extensive desulfurization of diesel by *Rhodococcus erythropolis*. Biotechnol. Lett. 29: 123–127.

Zhou, N.Y., S.L. Fuenmayor and P.A. Williams. 2001. *nag* Genes of *Ralstonia* (formerly *Pseudomonas*) sp. strain U2 encoding enzymes for gentisate catabolism. J. Bacteriol. 183: 700–708.

CHAPTER 15

Cold-active Enzymes Bioprospecting from Actinobacteria Isolated from Beagle Channel, in South Extreme of Argentina

Adriana E. Alvarenga,[1,*] Claudia E. Pereira,[1] Héctor A. Cristóbal[2] and Carlos M. Abate[1,3,4]

Introduction

About 70% of the surface of the earth is water in the oceans. Across it, the temperature at depths of 1000 m or more is a constant 4°C, constituting a vast environment populated by a diverse group of psychrophilic ("cold-loving") microorganisms.

The coastal areas are attractive to study because they are easily accessible and temperature of the water ranges between 4°C and 10°C, which are optimal conditions for microorganisms that have adapted them to cold.

[1]Planta Piloto de Procesos Industriales y Microbiológicos (PROIMI-CONICET), Av. Belgrano y Pje. Caseros.
[2]Instituto de Investigaciones para la Industria Química – INIQUI-CONICET, Avenida Bolivia 5150, (4400). Salta, Argentina.
[3]Facultad de Bioquímica, Química y Farmacia, UNT Batalla de Ayacucho 471 - CPA T4000INI, Tucumán–Argentina.
[4]Facultad de Ciencias Naturales e IML. UNT. Miguel Lillo 205, T4001MVB, Tucumán, Argentina.
*Corresponding author: aalvarenga@proimi.org.ar

The Beagle Channel located in the southern South America, Tierra del Fuego, Argentina, has an environment with a constant moderately cold climate. The temperature of the surface seawater near its capital city, Ushuaia, ranged from 9.7ºC in January to 4ºC in July (http://www.hidro.gov.ar/ceado/ceado.asp). Microbiological studies undertaken in this sub-Antarctic area are limited in comparison to the extensive literature about microorganisms that colonize extreme cold environments, such as those mentioned above. As these coasts are exposed to oil pollution resulting from petroleum exploitation and transport (Esteves and Amin 2004), microbiological research was mainly focused on hydrocarbon-degrading communities.

The marine environment is extremely diverse and microorganisms are exposed to extreme pressure, temperature, salinity and nutrient availability. These distinct marine environmental niches are likely to have highly diverse bacterial communities.

The old impression that diversity of actinobacteria in the oceans was small and restricted has been completely dispelled by 16S rRNA phylogenetic diversity inventories and estimates and cultivation approaches. Thus, deep-sea sediments were found to contain >1300 different actinobacterial operational taxonomic units, a great proportion of which are predicted to represent novel species and genera (Stach and Bull 2005). Complementing this strategy are intelligent approaches to sample handling and growth conditions, which have led to the recovery in culture of many new taxa (Bull and Stach 2007).

Enzymes isolated from bacteria from such environments are likely to have a range of quite diverse biochemical and physiological characteristics that have allowed microbial communities to adapt and ultimately thrive in these conditions (Kennedy et al. 2008).

For examples, during baking, enzymes such as xylanases, proteases, amylases, lipases and glucose oxidases, can modify the hemicellulose, gluten, starch and free sulfhydryl groups, respectively, during dough preparation and processing which generally take place at temperatures below 35°C. The combined actions of these enzymes can result in improved elasticity and machinability of the dough, resulting in a larger volume and improved crumb structure. A cold-adapted family 8 xylanase was recently shown to be more efficient in baking and yielded a larger loaf when compared with a widely-used commercial mesophilic enzyme preparation, indicating another benefit cold-adapted enzymes hold for the baking industry (Dutron et al. 2005, Collins et al. 2006).

Several industrial processes request for alternatives to improve efficiency, reduce costs and increase the availability of some enzymes of biotechnological interest; these are compelling justifications to conduct research. Therefore, the search of new programs of isolation and selection of microorganisms from marine environments with high production of termostable enzyme is important to assess their optimal growing conditions, production and enzyme activities (Sánchez et al. 2004).

Isolation and molecular identification

Samples were collected aseptically in July, 2001 and February, 2002 in Tierra del Fuego, Argentina. Subsurface seawater samples (20 m depth) were retrieved from various coastal areas in the Beagle Channel (55°S; 67°W): Ushuaia Bay (54°50′01″S, 68°15′48″W), Ensenada Bay (54°51′11″S, 68°29′59″W) and Punta Segunda (54°51′27″S, 68°27′41″W).

Isolation of microorganism were carried out from enriched cultures in liquid R2A medium (Suzuki et al. 1997) incubated at 4°C and 20°C in an orbital shaker (200 rpm) for 5 and 3 days, respectively. Then, 100 µl of each enriched culture were spread on Luria-Bertani (LB) and R2A plates and incubated at 4°C and 20°C. All the culture media were formulated using seawater. Twenty-four isolates were selected based on their morphology in different culture media.

Genomic DNA (gDNA) extraction was based on a protocol described by Hoffman and Winston (1987). The 16S rDNA were amplified using universal primers 27f and 1492r (Weisburg et al. 1991). Sequences belonging to the same genus or validly published close related species, available in public databases (GenBank and RDP II) were aligned and phylogenetic and molecular evolutionary analyses based on 16S rDNA were inferred by the Maximum Likelihood method based on Tamura-Nei model (Mega5, Tamura et al. 2011). Phylogenetic trees were constructed with Mega5 and the bootstrap consensus tree was inferred from 1000 replicates (Felsenstein 1985) to represent the evolutionary history of the taxa analyzed.

Phylogenetic analyses based on 16S rDNA gene sequence confirmed that 24 marine bacteria studied belong to the robust well-defined monophyletic taxon of *Actinobacteria.* Three clusters were established on the basis of nearly full-length sequences of 16S rDNA from six strains (Fig. 1), where phylogenetic relationships are established for the following genera: *Microbacterium* (cluster I), *Salinibacterium* (cluster II) and *Rhodococcus* (cluster III). In cluster I, *Microbacterium* sp. Ci8, *Microbacterium* sp. Ci12 and *Microbacterium* sp. Ci14 were strongly related to *M. liquefaciens* DSM 20638^{T}, *M. maritypicum* DSM 12512^{T} and *M. oxydans* DSM 20578^{T} (99% of identities according to the BLAST analysis). *Salinibacterium* sp. Ci20 from cluster II was associated to *S. amurskyense* KMM 3670 and *S. amurskyense* KMM 3673 (99%) In cluster III, *Rhodococcus* sp. Ci6 and *Rhodococcus* sp. Ci13 were closely related to *Nocardia coeliaca* DSM44595^{T}, *R. erythropolis* DSM43188^{T}, *R. erythropolis* DSM 43200^{T} (99%), but also associated to *R. globerulus* DSM 4954^{T} and *R. marinonascens* DSM 43752^{T} (98%).

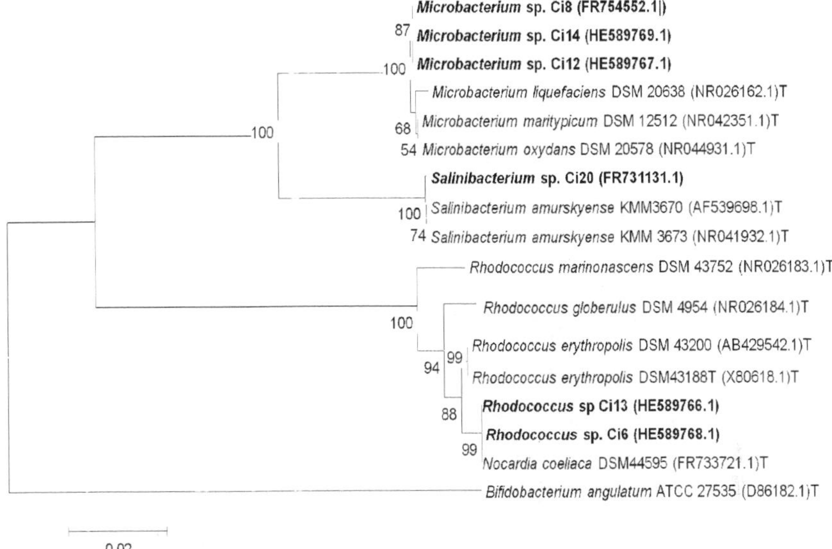

Figure 1. The evolutionary history was inferred using maximum likelihood method based on Tamura-Nei model. The percentage of replicate trees in which the associated taxa clustered together in the bootstrap test (1000 replicates) is shown next to the branches. Evolutionary analyses were conducted in MEGA5 (Tamura et al. 2011). *Bifidobacterium angulatum* ATCC 27535 was used as the outgroup microorganism.

Evidence supporting the existence of marine actinomycetes came from the description of *Rhodococcus marinonascens*, the first marine actinobacteria specie to be characterized (Helmke and Weyland 1984). Further support has come from the discovery that some strains display specific marine adaptations (Jensen et al. 1991), whereas others appear to be metabolically active in marine sediments (Moran et al. 1995).

Recent data from culture-dependent studies have shown that indigenous marine actinobacteria indeed exist in the oceans. These include members of the genera *Dietzia, Rhodococcus, Streptomyces, Salinispora, Marinophilus, Solwaraspora, Salinibacterium, Aeromicrobium marinum, Williamsia maris* and *Verrucosispora* (Bull et al. 2005, Jensen et al. 2005a,b, Magarvey et al. 2004, Stach et al. 2004).

Morphological and Physiological Characterization

Colony description and formation of melanoid pigment

For the description of the isolated strains four parameters were used: growth, reverse colour, aerial mycelium and soluble pigment, according to Wink (2002).

Rhodococcus sp. Ci6 and *Rhodococcus* sp. Ci13 showed good growth in medium ISP, *Microbacterium* sp. Ci8, *Microbacterium* sp. Ci12 and *Microbacterium* sp. Ci14 and *Salinibacterium* sp. Ci20 presented sparse growth (Fig. 2).

Reverse colour (Table 1) was described by the colours of the RAL-code (edition of 1990). All studied isolates produced no melanoid pigments and do not form aerial mycelium.

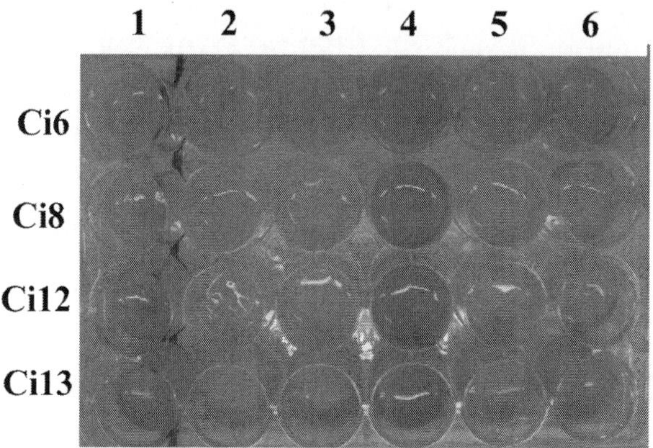

Figure 2. Photography of selected strains cultured in different media in a twelve wells plate. The filling pattern is: column 1: ISP 4; column 2: ISP 5; column 3: Synthetic medium Suter without tyrosine; column 4: ISP 6; column 5: ISP 7 and column 6: Synthetic medium Suter with tyrosine (Wink 2002).

Color image of this figure appears in the color plate section at the end of the book.

Table 1. Reverse colours of selected strains according to RAL-code.

ISOLATES	RAL-CODE
Ci6	1001
Ci8	1000
Ci12	1000
Ci13	1001
Ci14	1000
Ci20	1002

Physiological tests

Resistance to sodium chloride

Resistance toward NaCl (Kutzner 1981) is a helpful tool for differentiating between marine and halophilic actinobacteria, as well as for distinguishing different actinobacteria species.

On a six well plate the growth on 0, 2.5, 5, 7.5 and 10% of sodium chloride could be checked.

Microbacterium sp. Ci12 and *Microbacterium* sp. Ci14 showed to be tolerant to all tested NaCl concentrations.

An obligate requirement for Na ions and either obligate requirements or tolerance of oligotrophic substrate concentrations, low temperatures and elevated pressures for growth would provide evidence of indigenicity from marine environment (Bull and Stach 2007).

Table 2. Test results of tolerance to different NaCl concentrations. The + and – notation is used to represent presence or absence of bacterial growth.

ISOLATES	NaCl CONCENTRATIONS			
	2.5 %	5.0 %	7.5 %	10.0 %
Ci6	+	–	–	–
Ci8	+	–	–	–
Ci12	+	+	+	+
Ci13	+	–	–	–
Ci14	+	+	+	+
Ci20	–	–	–	–

Resistance to lysozyme

Gram positive cells are in principle sensitive to lysozyme. However, because of additional layers like teiconic acids or modifications of the peptidoglycan molecule, some Gram-positive bacteria became resistant to lysozyme. Therefore *Mycobacterium*, *Nocardia* and *Streptoverticillium* are lysozyme resistant, while most of the *Streptomyces* species are lysozyme sensitive (Kutzner 1981).

The final lysozyme concentrations tested were 0, 10, 25, 50, 75 and 100 μg mL^{-1}.

Table 3. Test results of resistance to different concentrations of lysozyme. The + and – notation is used to represent presence or absence of bacterial growth.

ISOLATES	LYSOZYME CONCENTRATIONS					
	0	10	25	50	75	100
Ci6	+	+	+	+	+	+
Ci8	+	+	–	–	–	–
Ci12	+	+	–	–	–	–
Ci13	+	–	–	–	–	–
Ci14	+	–	–	–	–	–
Ci20	+	–	–	–	–	–

Lysozyme resistance is characteristic of the genus *Nocardia,* giving evidence that Ci6 strain might belong to this genus. As shown, it is possible to achieve a reliable taxonomic identification by combining different methods (poliphasic approach).

Utilization of carbohydrates

The utilization of carbon sources is an important characteristic to identify different actinobacteria species (Bennedict et. al 1955).

The ability of strains to use ten compounds was tested in a twelve well plate according to the method of Shirling and Gottlieb (1966), by using the basal agar 5338. The following carbon sources were used: Glucose, Arabinose, Sucrose, Xylose, Inositol, Mannitol, Fructose, Rhamnose, Raffinose and Cellulose. Different profiles of studied strains are showed in Table 4.

Table 4. Utilization of carbon sources.

	Ci6	Ci8	Ci12	Ci13	Ci14	Ci20
Glucose	+	+	+	+	+	+
Arabinose	+	–	–	+	–	+
Sucrose	–	–	+	–	–	+
Xylose	+	–	+	+	–	–
Inositol	+	–	–	+	–	–
Mannitol	+	+	+	+	+	+
Fructose	+	+	+	+	–	+
Rhamnose	+	–	+	–	+	+
Raffinose	–	–	+	–	–	–
Cellulose	+	–	+	+	+	–

Enzymes screening

β-glucosidase, cellulase, hemicelullase and xylanase

Productions of different glycosyl hydrolases enzymes (β-glucosidases, celullases, hemicelullases and xylanases) by marine bacteria were evaluated on BMM (Brunner Minimal Medium) plates containing cellobiose, cellulose, hemicellulose (sugarcane bagasse) and xylan respectively. A qualitative system in agar plate was used to test enzymatic activities by assay of Congo red staining (Teather and Wood 1982).

All strains were able to produce xylanase and β-glucosidase activites. *Rhodococcus* sp. Ci6 and *Rhodococcus* sp. Ci13 showed cellulase activity. Hemicellulase activity was detected in *Rhodococcus* sp. Ci6 and *Rhodococcus* sp. Ci13 and *Microbacterium* sp. Ci14 (Table 5, Fig. 3).

Table 5. Qualitative determination of enzymatic activities at 20°C. The + and – notation is used to represent presence or absence of bacterial growth.

	Ci6	Ci8	Ci12	Ci13	Ci14	Ci20
α-L-Rhamnosidase	+	+	+	–	+	+
Amylase	+	+	+	+	+	+
Lipase	–	–	–	–	–	–
Protease	–	+	–	+	+	–
β-Xylanase	+	+	+	+	+	+
Cellulase	+	–	–	+	–	–
β-Glucosidase	+	+	+	+	+	+
Hemicellulase	+	–	–	+	+	–

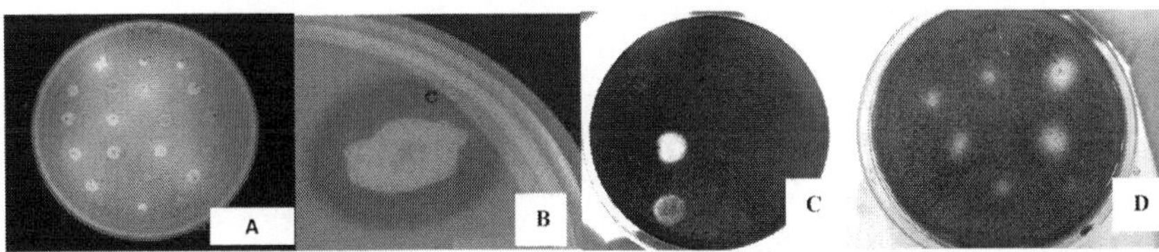

Figure 3. Photographs of the dishes. (A) α-L-rhamnosidase activity. (B) Protease Activity. (C) (D) Congo red stain was used to detect xylanase and hemicellulase activities, respectively.

α-L-rhamnosidase

Selection of α-L-rhamnosidase producer microorganisms was performed in plates with Brunner Minimal Medium (MMB). Qualitative determination of α -L-rhamnosidase activity was evaluated in Petri dishes supplemented with of 4-Methylumbelliferyl α-L-Rhamnopyranoside (MUR, 10 mM) at 4 and 20°C for 120 and 72 hr, respectively. Positive colonies showed fluorogenic characteristics clearly distinguishable by the cleavage of the free 4-methylumbelliferyl moiety when exposed to long-wave UV radiation. Positives strains were selected for quantitative assay (Table 5, Fig. 3).

For α-L-rhamnosidase activity quantification, each reaction contained 15 μL of substrate (100 mmol L^{-1} of 4-nitrophenyl α-L- Rhamnopyranoside in dimethylformamide), 900 μL of buffer (0.1 mol L^{-1} Tris–HCl pH 5 or pH 7) and 100 μL of cell free extract or supernatant (enzyme). The reaction was performed for 1 hr at 20°C and stopped by adding 100 μL of sodium hydroxide (0.1 mol L^{-1}). One enzyme unit (U) was defined as the amount of enzyme that released 1 μmol of 4-nitrophenol in 1 h at the indicated temperature (Orrillo et al. 2007). Each reaction was carried out in triplicate. Protein concentration was determined by Bradford assay (Bradford 1976).

Enzymatic activities were not detected in the supernatants. *Salinibacterium* sp. Ci20 showed the highest specific activity at tested pH values, 7.74 U mg protein^{-1} at pH 5 and 10.41 U mg protein^{-1} at pH 7 (Fig. 4).

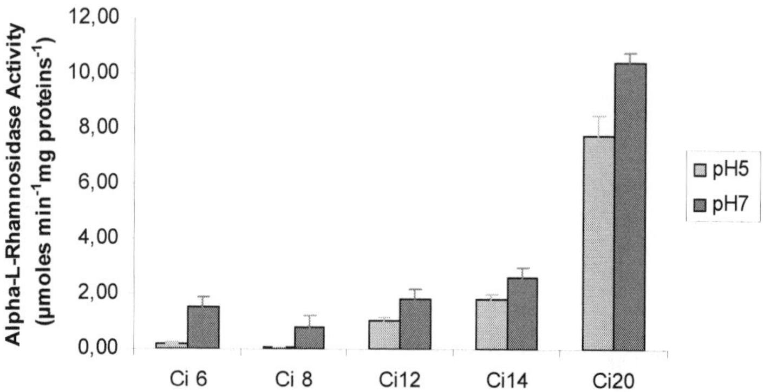

Figure 4. α-L-rhamnosidase activity determined in cell free extract of the different studied strains. Error bars represent the standard deviation calculated from three independent experiments ($p<0.05$).

These specific activity values detected are higher than those reported in *Pseudomonas paucimobilis* FP2001 (9.7 U mg protein^{-1}, Miake et al. 2000), *Pichia angusta* X349 (0.7 U mg protein^{-1}, Yanai and Sato 2000), *Aspergillus nidulans* (2.8 U mg protein^{-1}, Manzanares et al. 2000), *Fusarium solani* and *F. sambucinum* (2.10 and 0.19 U mg protein^{-1}, respectively, Scaroni et al. 2002).

This is the first report from cold active α-L-rhamnosidase activity in Actinobacteria class.

Protease, amylase and lipase

Qualitative proteolytic, amilolytic and lipolytic assays were carried out on Petri dishes white milk, starch or tributyrin as carbon source, respectively, where the hydrolysis halo around the colonies was considered as positive reaction to estimate the degree of enzymatic activity. Colonies were incubated at 4 and 20°C for 120 and 72 hr, respectively.

All studied strains were positive to amylolytic activity. *Microbacterium* sp. Ci8, *Rhodococcus* sp. Ci13 and *Microbacterium* sp. Ci14 showed proteolytic activity at mentioned conditions (Table 5, Fig. 3). No lipolytic activity was detected after 120 hr (Table 5).

Conclusions

A variety of cold-active enzymes (cellulases, proteases, xylanases, hemicellulases, α-L-rhamnosidases, glucosidases and amylases) was detected in the majority of isolated strains. It is expected that isolation and characterization of marine actinobacteria of the Beagle Channel will be useful for the identification of new enzymes. Actually, cold-active enzymes present a great potential biotechnological established for numerous industrial applications and process.

References Cited

Bennedict, R.G., T.G. Pridham, L.A. Lindenfelster, H.H. Hall and T.W. Jackson. 1955. Further studies in the evaluation of carbohydrate utilization tests as aides in the differentiation of species of *Streptomyces*. J. Appl. Microbiol. 3: 1–6.

Bradford, M. 1976. A rapid and sensitive method for the quantization of microgram quantities of protein utilizing the principle of protein dye binding. Anal. Biochem. 72: 248–252.

Bull, A.T., J.E.M. Stach, A.C. Ward and M. Goodfellow. 2005. Marine actinobacteria: perspectives, challenges, future directions. Antonie Van Leeuwenhoek. 87: 65–79.

Bull, A.T and J.E.M. Stach. 2007. Review. Marine actinobacteria: new opportunities for natural product search and discovery. Trends Microbiol. 15: 491–499.

Collins, T., A. Hoyoux, A. Dutron, J. Georis, B. Genot, T. Dauvrin, F. Arnaut, C. Gerday and G. Feller. 2006. Use of glycoside family 8 xylanases in baking. J. Cereal Sci. 43: 79–84.

Dutron, A., J. Georis, T. Dauvrin, T. Collins, A. Hoyoux and G. Feller. 2005. Use of family 8 enzymes with xylanolytic activity in baking. Patent No. MXPA05002751.

Esteves, J.L. and O. Amin. 2004. Evaluación de la Contaminación Urbana de las Bahías de Ushuaia, Encerrada y Golondrina (Provincia de Tierra del Fuego, Antártica e Islas del Atlántico Sud). Consolidación e Implementación del Plan de la Zona Costera Patagónica (PMZCP)-ARG/02/G31-GEF/PNUD, CD-ROM, pp. 64.

Felsenstein, J. 1985. Confidence Limits on Phylogenies: An Approach Using the Bootstrap. Evolution. 39: 783–791.

Helmke, E. and H. Weyland. 1984. *Rhodococcus marinonascens* sp. nov., an actinomycete from the sea. Int. J. Syst. Bacteriol. 34: 127–138.

Hoffman, C.S. and F. Winston. 1987. A ten-minute DNA preparation from yeast efficiently releases autonomous plasmids for transformation of *Escherichia coli*. Gene. 57: 267–272.

Jensen, P.R., R. Dwight and W. Fenical. 1991. Distribution of actinomycetes in near-shore tropical marine sediments. Appl. Environ. Microbiol. 57: 1102–1108.

Jensen, P.R., E. Gontang, C. Mafnas, T.J. Mincer and W. Fenical. 2005a. Culturable marine actinomycete diversity from tropical Pacific Ocean sediments. Environ. Microbiol. 7: 1039–1048.

Jensen, P.R., T.J. Mincer, P.G. Williams and W. Fenical. 2005b. Marine actinomycete diversity and natural product discovery. Antonie Van Leeuwenhoek. 87: 43–48.

Kennedy, J., J.R. Marchesi and A.D. Dobson. 2008. Marine metagenomics: strategies for the discovery of novel enzymes with biotechnological applications from marine environments. Microb. Cell Fact. 7: 27.

Kutzner, H.J. The family *Streptomycetaceae*. pp. 2028–2090. *In*: M.P. Starr, H. Stolp, H.G. Trüper, A. Balons and H.G. Schlegel [eds.]. 1981. The Prokaryotes. A handbook on habitats, isolation and identification of bacteria. Springer Verlag. Berlin.

Magarvey, N.A., J.M. Keller, V. Bernan, M. Dworkin and D.H. Sherman. 2004. Isolation and characterization of novel marine-derived actinomycete taxa rich in bioactive metabolites. Appl. Environ. Microbiol. 70: 7520–7529.

Manzanares, P., M. Orejas, E. Ibañez, S. Vallés and D. Ramón. 2000. Purification and characterization of an α-L-rhamnosidase from *Aspergillus nidulans*. Lett. Appl. Microbiol. 31: 198–202.

Miake, F., T. Satho, H. Takesue, F. Yanagida, N. Kashige, K. Watanabe. 2000. Purification and characterization of intracellular α-L-rhamnosidase from *Pseudomonas paucimobilis* FP2001. Arch. Microbiol. 173: 65–70.

Moran, M.A., L.T. Rutherford and R.E. Hodson. 1995. Evidence for indigenous *Streptomyces* populations in a marine environment determined with a 16S rRNA probe. Appl. Environ. Microbiol. 61: 3695–3700.

Orrillo, G., P. Ledesma, O. Delgado, G. Spagna and J. Breccia. 2007. Cold-active α -L-rhamnosidase from psychrotolerant bacteria isolated from a sub-Antarctic ecosystem. Enzyme Microb. Tech. 40: 236–241.

Sánchez, T., J. León, J. Woolcott and K. Arauco. 2004. Extracellular proteases produced by marine bacteria isolated from sea water contaminated with fishing effluents. Rev. Peru Biol. 11: 179–186.

Scaroni, E., C. Cuevas, L. Carrillo and G. Ellenrieder. 2002. Hydrolytic properties of crude α-L-rhamnosidases produced by several wild strains of mesophilic fungi. Appl. Microbiol. 34: 461–465.

Shirling, E.B. and D. Gottlieb. 1966. Methods for characterization of *Streptomyces* species. Int. J. Syst. Bacteriol. 16: 313–340.

Stach, J.E.M., Maldonado, L.A., Ward, A.C., Bull, A.T. and M. Goodfellow. 2004. *Williamsia maris* sp. nov., a novel actinomycete isolated from the Sea of Japan. Int. J. Syst. Evol. Micr. 54: 191–194.

Stach, J.E.M. and A.T. Bull. 2005. Estimating and comparing the diversity of marine actinobacteria. Antonie Van Leeuwenhoek. 87: 3–9.

Suzuki, M., M. Rappé, Z. Haimberger, H. Winfield, N. Adair, J. Strobel and S. Giovannoni. 1997. Bacterial diversity among small-subunit rRNA gene clones and cellular isolates from the same seawater sample. Appl. Environ. Microbiol. 63: 983–989.

Tamura, K., D. Peterson, N. Peterson, G. Stecher, M. Nei and S. Kumar. 2011. MEGA5: Molecular Evolutionary Genetics Analysis Using Maximum Likelihood, Evolutionary Distance, and Maximum Parsimony Methods. Mol. Biol. Evol. 28: 2731–2739.

Teather, R.M. and P.J. Wood. 1982. Use of Congo red-polysaccharide interactions in enumeration and characterization of cellulolytic bacteria from bovine rumen. Appl. Environ. Microbiol. 43: 777–780.

Wink, J. 2002. CD: An order in the class of actinobacteria important to the pharmaceutical industry.

Electronic manual. The Actinomycetales. Copyright Aventis Pharma Dutschand Gmbh.

Yanai, T. and S. Sato. 2000. Purification and characterization of a novel α-L-arabinofuranosidase from *Pichia capsulata* X91. Biosci. Biotechnol. Biochem. 64:1181–1188.

Weisburg, W., M. Barns, D. Pelleteried and D. Lane. 1991. 16S Ribosomal DNA amplification for the phylogenetic study. J. Bacteriol. 173: 697–703.

Index

Color Plate Section

Chapter 10

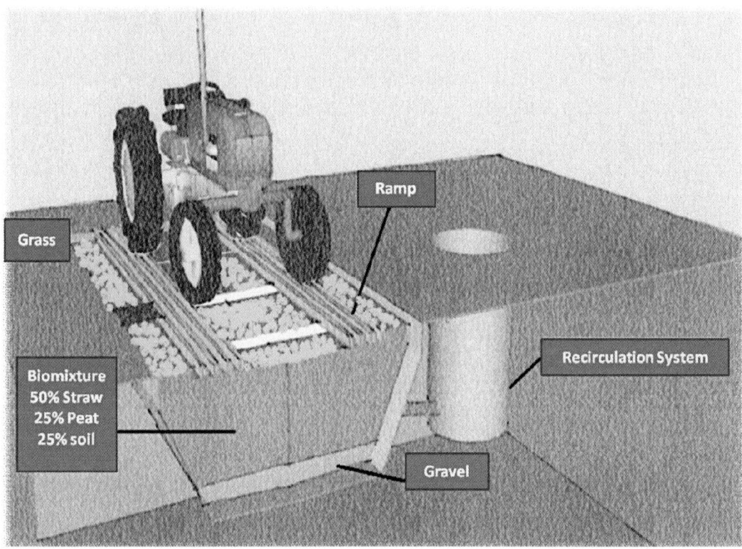

Figure 1. Diagram of a biobed

Chapter 15

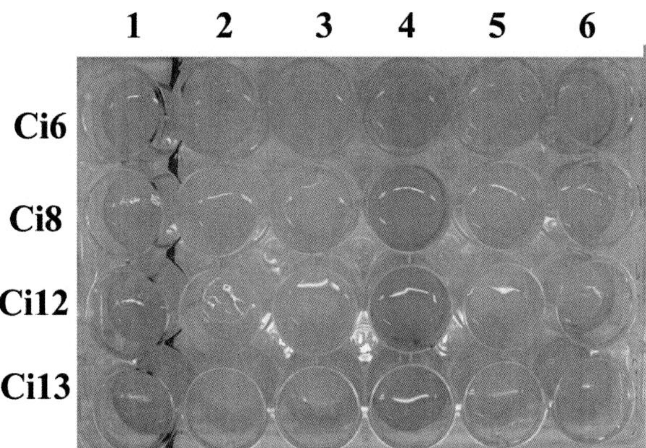

Figure 2. Photography of selected strains cultured in different media in a twelve wells plate. The filling pattern is: column 1: ISP 4; column 2: ISP 5; column 3: Synthetic medium Suter without tyrosine; column 4: ISP 6; column 5: ISP 7 and column 6: Synthetic medium Suter with tyrosine (Wink 2002).